Transgenic Animals

BIOTECHNOLOGY

JULIAN E. DAVIES, *Editor*
Pasteur Institute
Paris, France

Editorial Board

L. Bogorad	Harvard University, Cambridge, USA
J. Brenchley	Pennsylvania State University, University Park, USA
P. Broda	University of Manchester Institute of Science and Technology, Manchester, United Kingdom
A.L. Demain	Massachusetts Institute of Technology, Cambridge, USA
D.E. Eveleigh	Rutgers University, New Brunswick, USA
D.H. Gelfand	Cetus Corporation, Emeryville, California, USA
D.A. Hopwood	John Innes Institute, Norwich, United Kingdom
S.-D. Kung	University of Maryland, College Park, USA
J.-F. Martin	University of Leon, Leon, Spain
C. Nash	Schering-Plough Corporation, Bloomfield, New Jersey, USA
T. Noguchi	Suntory, Ltd., Tokyo, Japan
W. Reznikoff	University of Wisconsin, Madison, USA
R.L. Rodriguez	University of California, Davis, USA
A.H. Rose	University of Bath, Bath, United Kingdom
P. Valenzuela	Chiron, Inc., Emeryville, California, USA
D. Wang	Massachusetts Institute of Technology, Cambridge, USA

BIOTECHNOLOGY SERIES

1. R. Saliwanchik — *Legal Protection for Microbiological and Genetic Engineering Inventions*

2. L. Vining (editor) — *Biochemistry and Genetic Regulation of Commercially Important Antibiotics*

3. K. Herrmann and R. Somerville (editors) — *Amino Acids: Biosynthesis and Genetic Regulation*

4. D. Wise (editor) — *Organic Chemicals from Biomass*

5. A. Laskin (editor) — *Enzymes and Immobilized Cells in Biotechnology*

6. A. Demain and N. Solomon (editors) — *Biology of Industrial Microorganisms*

7. Z. Vaněk and Z. Hošťálek (editors) — *Overproduction of Microbial Metabolites: Strain Improvement and Process Control Strategies*

8. W. Reznikoff and L. Gold (editors) — *Maximizing Gene Expression*

9. W. Thilly (editor) — *Mammalian Cell Technology*

10. R. Rodriguez and D. Denhardt (editors) — *Vectors: A Survey of Molecular Cloning Vectors and Their Uses*

11. S.-D. Kung and C. Arntzen (editors) — *Plant Biotechnology*

12. D. Wise (editor) — *Applied Biosensors*

13. P. Barr, A. Brake, and P. Valenzuela (editors) — *Yeast Genetic Engineering*

14. S. Narang (editor) *Protein Engineering: Approaches to the Manipulation of Protein Folding*

15. L. Ginzburg (editor) *Assessing Ecological Risks of Biotechnology*

16. N. First and F. Haseltine (editors) *Transgenic Animals*

Transgenic Animals

Proceedings of the Symposium on
Transgenic Technology in Medicine and Agriculture

Sponsored by the
Center for Population Research,
National Institute of Child Health and Human Development
held at the
National Institutes of Health
Bethesda, Maryland
December 12–15, 1988

Edited by

Neal L. First
Department of Meat and Animal Science
University of Wisconsin
Madison, Wisconsin

Florence P. Haseltine
Center for Population Research
National Institute of Child Health and Human Development
National Institutes of Health
Bethesda, Maryland

Butterworth–Heinemann
Boston   London   Singapore   Sydney   Toronto   Wellington

Copyright © 1991 by Butterworth–Heinemann, a division of Reed Publishing (USA) Inc.
All rights reserved.

No part of this publication may be reproduced, stored in a retrieval system, or transmitted, in any form or by any means, electronic, mechanical, photocopying, recording, or otherwise, without the prior written permission of the publisher.

Recognizing the importance of preserving what has been written, it is the policy of Butterworth–Heinemann to have the books it publishes printed on acid-free paper, and we exert our best efforts to that end.

Editorial and production supervision by Science Tech Publishers, Madison, WI 53705.

Library of Congress Cataloging-in-Publication Data
Symposium on Transgenic Technology in Medicine and Agriculture (1988 : National Institutes of Health)
  Transgenic animals : proceedings of the Symposium on Transgenic Technology in Medicine and Agriculture ; edited by Neal First, Florence P. Haseltine.
    p. cm. — (Biotechnology series ; #16)
  Includes bibliographical references.
  ISBN 0-409-90189-X
  1. Transgenic animals—Congresses. 2. Gene expression—Congresses. 3. Medicine—Research—Methodology—Congresses.
  4. Domestic animals—Genetic engineering—Congresses. I. First, Neal L. II. Haseltine, Florence. III. Title. IV. Series. V. Series: Biotechnology ; 16.
    [DNLM: 1. Animals, Transgenic—congresses. 2. Gene Expression Regulation—congresses. 3. Genetic Engineering—congresses.
  4. Transcription, Genetic—congresses. 5. Transformation, Genetic—congresses. W1 BI918M no. 16 / QY 50 S9913t 1988]
  QH442.6.S96 1988
  660'.65—dc20
  DNLM/DLC   90-1730
  for Library of Congress   CIP

British Library Cataloguing in Publication Data
Symposium on Transgenic Technology in Medicine and Agriculture *(1988: Bethesda, MD)*
  Transgenic animals.
  1. Livestock. Genetic engineering
  I. Title II. First, Neal III. Haseltine, Florence P.
  IV. Series
  636.0824

  ISBN 0-409-90189-X

Butterworth–Heinemann
80 Montvale Avenue
Stoneham, MA 02180

10 9 8 7 6 5 4 3 2 1

Printed in the United States of America

To Dr. Frank Ruddle

In 1976, he outlined, with a pencil on a yellow pad, many experiments that would become possible if animals could be transformed by changing their complement of genetic information. The possibilities were awesome then and their realization has opened vistas beyond our wildest imagination.

CONTRIBUTORS

Jan W. Abramczuk
Laboratory of Oral Medicine
National Institute of Dental
 Research
National Institutes of Health
Bethesda, Maryland

A.G. Amador
Department of Physiology
School of Medicine
Southern Illinois University
Carbondale, Illinois

P. Bain
Department of Physiology
School of Medicine
Southern Illinois University
Carbondale, Illinois

A. Bartke
Department of Physiology
School of Medicine
Southern Illinois University
Carbondale, Illinois

Karen A. Biery
Granada Genetics, Inc.
Texas A&M University Research
 Park
College Station, Texas

Leonard Bogarad
Department of Biology
Yale University
New Haven, Connecticut

Kenneth R. Bondioli
Granada Genetics, Inc.
Texas A&M University Research
 Park
College Station, Texas

Jan L. Breslow
Laboratory of Biochemical
 Genetics and Metabolism
The Rockefeller University
New York, New York

Guerard W. Byrne
Department of Biology
Yale University
New Haven, Connecticut

Sally A. Camper
Human Genetics
University of Michigan Medical
 School
Ann Arbor, Michigan

V. Chandrashekar
Department of Physiology
School of Medicine
Southern Illinois University
Carbondale, Illinois

Howard Chen
Merck Sharpe & Dohme Research
 Laboratories
Rahway, New Jersey

Thomas T. Chen
Center of Marine Biotechnology
University of Maryland
Baltimore, Maryland

Frank Costantini
Department of Genetics and
 Development
College of Physicians and
 Surgeons
Columbia University
New York, New York

Lyman B. Crittenden
USDA-ARS Regional Poultry
 Research Laboratory
East Lansing, Michigan

Chella S. David
Department of Immunology
Mayo Clinic
Rochester, Minnesota

Franko J. De Mayo
Department of Cell Biology
Baylor College of Medicine
Houston, Texas

Eva Derman
Department of Developmental
 and Structural Biology
The Public Research Institute of
 the City of New York, Inc.
New York, New York

Thomas C. Doetschman
College of Medicine
University of Cincinnati
Cincinnati, Ohio

Joseph P. Dougherty
Department of Molecular
 Genetics and Microbiology
University of Medicine and
 Dentistry of New Jersey–
 Robert Wood Johnson Medical
 School
Piscataway, New Jersey

Rex A. Dunham
Department of Fisheries and
 Allied Aquacultures
Auburn University
Auburn, Alabama

Timothy J. Dyer
Department of Zoological and
 Biomedical Sciences
Ohio University
Athens, Ohio

Rachel Ehrlich
Experimental Immunology
 Branch
National Cancer Institute,
 National Institutes of Health
Bethesda, Maryland

Anthony J. Faras
Department of Microbiology
Institute for Human Genetics
University of Minnesota
St. Paul, Minnesota

Neal L. First
Department of Meat and Animal
 Science
University of Wisconsin
Madison, Wisconsin

William I. Frels
Agricultural Research Center
USDA
Beltsville, Maryland

Hanan Galski
The Institute of Life Sciences
Department of Zoology
The Hebrew University of
 Jerusalem
Jerusalem, Israel

Lucia Irene Gonzalez-Villasenor
Department of Biology
Johns Hopkins University
Baltimore, Maryland

Jon W. Gordon
Department of Geriatrics and Adult Development, and Department of Obstetrics, Gynecology and Reproductive Science
Mt. Sinai School of Medicine
New York, New York

Michael M. Gottesman
Laboratory of Molecular Biology
National Cancer Institute,
National Institutes of Health
Bethesda, Maryland

Kevin S. Guise
Department of Animal Science
University of Minnesota
St. Paul, Minnesota

Perry B. Hackett, Jr.
Department of Genetics and Cell Biology
Institute for Human Genetics
University of Minnesota
St. Paul, Minnesota

Florence P. Haseltine
Center for Population Research
National Institute of Child Health and Human Development
National Institutes of Health
Bethesda, Maryland

Jan Heideman
University of Wisconsin
Biotechnology Center
Madison, Wisconsin

Lothar Hennighausen
Laboratory of Biochemistry and Metabolism
NIDDK, National Institutes of Health
Bethesda, Maryland

Keith G. Hill
Granada Genetics, Inc.
Texas A&M University Research Park
College Station, Texas

Rosemarie Hunziker
Molecular Biology Section, Laboratory of Immunology
NIAID, National Institutes of Health
Bethesda, Maryland

Luis M. Isola
Polly Annenberg Levy Division of Hematology, Department of Medicine
Mt. Sinai School of Medicine
New York, New York

Yasushi Ito
Laboratory of Biochemical Genetics and Metabolism
The Rockefeller University
New York, New York

Karen B. Jones
Granada Genetics, Inc.
Texas A&M University Research Park
College Station, Texas

Claudia Kappen
Department of Biology
Yale University
New Haven, Connecticut

Anne A.R. Kapuscinski
Department of Fisheries and Wildlife
University of Minnesota
St. Paul, Minnesota

Contributors

Evelyn M. Karson
Laboratory of Molecular
 Hematology
NHLBI, National Institutes of
 Health
Bethesda, Maryland

David L. Kooyman
Department of Zoological and
 Biomedical Sciences
Ohio University
Athens, Ohio

John J. Kopchick
Department of Zoology
Edison Animal Biotechnology
 Center
Ohio University
Athens, Ohio

John M. Leonard
Laboratory of Molecular
 Microbiology
NIAID, National Institutes of
 Health
Bethesda, Maryland

F. Leung
Merck Sharpe & Dohme Research
 Laboratories
Rahway, New Jersey

Roger Little
Department of Immunology
Mayo Clinic
Rochester, Minnesota

F. Macken
Hubbard Farms
Walpole, New Hampshire

David H. Margulies
Molecular Biology Section,
 Laboratory of Immunology
NIAID, National Institutes of
 Health
Bethesda, Maryland

Javier Martin
Department of Immunology
Mayo Clinic
Rochester, Minnesota

Malcolm A. Martin
Laboratory of Molecular
 Microbiology
NIAID, National Institutes of
 Health
Bethesda, Maryland

A. Mayerhofer
Department of Physiology
School of Medicine
Southern Illinois University
Carbondale, Illinois

Glenn T. Merlino
Laboratory of Molecular Biology
National Cancer Institute,
 National Institutes of Health
Bethesda, Maryland

Ed Mills
Hubbard Farms
Walpole, New Hampshire

Bert W. O'Malley
Department of Cell Biology
Baylor College of Medicine
Houston, Texas

Ira Pastan
Laboratory of Molecular Biology
National Cancer Institute,
 National Institutes of Health
Bethesda, Maryland

David S. Pezen
Laboratory of Molecular
 Microbiology
NIAID, National Institutes of
 Health, and
Howard Hughes Medical Institute
Bethesda, Maryland

Carl A. Pinkert
DNX, Inc.
Athens, Ohio

Christoph W. Pittius
Laboratory of Biochemistry and
  Metabolism
NIDDK, National Institutes of
  Health
Bethesda, Maryland

Dennis A. Powers
Hopkins Marine Station
Stanford University
Pacific Grove, California

Lisa J. Raines
Industrial Biotechnology
  Association
Washington, D.C.

Caird E. Rexroad, Jr.
Reproduction Laboratory
Livestock and Poultry Sciences
  Institute
Beltsville Agricultural Research
  Center
Beltsville, Maryland

Charles Rosenblum
Merck Sharpe & Dohme Research
  Laboratories
Rahway, New Jersey

Frank H. Ruddle
Department of Biology
Yale University
New Haven, Connecticut

Donald W. Salter
Department of Microbiology and
  Public Health
Michigan State University
East Lansing, Michigan

Lakshmanan Sankaran
Laboratory of Biochemistry and
  Metabolism
NIDDK, National Institutes of
  Health
Bethesda, Maryland

Suresh Savarirayan
Department of Immunology
Mayo Clinic
Rochester, Minnesota

Klaus Schughart
Department of Biology
Yale University
New Haven, Connecticut

Kate Shahan
Department of Developmental
  and Structural Biology
The Public Health Research
  Institute of the City of New
  York, Inc.
New York, New York

Moshe Shani
Institute of Animal Science
Agricultural Research
  Organization
The Volcani Center
Bet Dagan, Israel

J.G.M. Shire
Department of Biology
University of Essex
Colchester, England

Dinah S. Singer
Experimental Immunology
  Branch
National Cancer Institute,
  National Institutes of Health
Bethesda, Maryland

Jim Smith
Hubbard Farms
Walpole, New Hampshire

H. Jin Son
Molecular Neurobiology
Burke Rehabilitation Center
White Plains, New York

R.W. Steger
Department of Physiology
School of Medicine
Southern Illinois University
Carbondale, Illinois

K. Tang
Department of Physiology
School of Medicine
Southern Illinois University
Carbondale, Illinois

Joyce Taylor
Merck Sharpe & Dohme Research
 Laboratories
Rahway, New Jersey

Howard M. Temin
Department of Oncology
McArdle Laboratory
University of Wisconsin
Madison, Wisconsin

Kirk Thomas
Biology Department
University of Utah
Salt Lake City, Utah

Shirley M. Tilghman
Department of Biology
Princeton University
Princeton, New Jersey

Manuel Utset
Department of Biology
Yale University
New Haven, Connecticut

T.E. Wagner
Edison Animal Biotechnology
 Center
Ohio University
Athens, Ohio

Annemarie Walsh
Laboratory of Biochemical
 Genetics and Metabolism
The Rockefeller University
New York, New York

Christoph Westphal
Laboratory of Biochemistry and
 Metabolism
NIDDK, National Institutes of
 Health
Bethesda, Maryland

Bing-Yuan Wei
Department of Immunology
Mayo Clinic
Rochester, Minnesota

J.S. Yun
Edison Animal Biotechnology
 Center
Ohio University
Athens, Ohio

Peijun Zhang
Hopkins Marine Station
Stanford University
Pacific Grove, California

CONTENTS

Preface xxiii

PART I. GENE TRANSFER: APPROACHES, METHODS, AND TOOLS 1

1. **Transgenic Animals: A New Era in Developmental Biology and Medicine** 3
   Luis M. Isola and Jon W. Gordon
   1.1 Integration of Microinjected Sequences into the Embryo 5
   1.2 Expression of Foreign Genes in Transgenic Mice 7
   1.3 Specific Applications of Transgenic Mice 10
   1.4 Transgenic Mice in Developmental Genetics 16
   1.5 Summary and Conclusions 17
   References 18

2. **Determination of Retroviral Mutation Rates Using Spleen Necrosis Virus-Based Vectors and Helper Cells** — 21
   Joseph P. Dougherty and Howard M. Temin
   2.1 Materials and Methods — 22
   2.2 Results — 23
   2.3 Discussion — 30
   References — 32

3. **Steroid Hormone Receptors as Transactivators of Gene Expression** — 35
   Bert W. O'Malley
   References — 41

PART II. GENE EXPRESSION — 43

4. **Targeted Mutagenesis in Embryo-Derived Stem Cells** — 45
   Kirk Thomas
   4.1 Homologous Recombination in Cultured Mouse Fibroblasts — 46
   4.2 Targeted Mutagenesis of the *HPRT* Locus in Mouse ES Cells — 47
   4.3 Targeted Mutagenesis of Nonselectable Genes — 49
   4.4 Conclusion — 53
   References — 53

5. **Application of Germline Transformation to the Study of Myogenesis** — 55
   Moshe Shani
   5.1 Expression of Rat Muscle-Specific Genes in Transgenic Mice — 56
   5.2 Expression of Muscle-Specific Genes Introduced into ES Cells — 59
   5.3 Gene Transfer Mediated by Retroviral Vectors: Expression of an Internal Muscle-Specific Promoter — 61
   References — 63

6. **Regulation of Expression of Genes for Milk Proteins** — 65
   Lothar Hennighausen, Christoph Westphal, Lakshmanan Sankaran, and Christoph W. Pittius
   6.1 Expression of the WAP and β-Casein Genes During Mammary Development — 66
   6.2 Tissue-Specific Expression of the Genes for WAP and β-Casein — 68
   6.3 Hormone-Induced Accumulation of WAP and β-Casein RNA in Mammary Glands of Pregnant Mice — 71

	6.4	Summary	71
		References	73

7. Expression of a Silent MUP Gene in Transgenic Mice — 75
H. Jin Son, Kate Shahan, Eva Derman, and Frank Costantini

    7.1 Results   76
    7.2 Discussion   77
         References   80

8. The Activation and Silencing of Gene Transcription in the Liver — 81
Sally A. Camper and Shirley M. Tilghman

    8.1 Function of Multiple AFP Enhancer Elements in Tissue Specificity and Gene Activation   83
    8.2 Role of the AFP Promoter-Proximal Region   84
    8.3 Regulatory Elements of the Albumin Gene   84
    8.4 Developmental Regulation of Expression of AFP and Albumin Genes   85
         References   86

9. Gene Targeting in Embryonic Stem Cells — 89
Thomas C. Doetschman

    9.1 ES Cells In Vitro   90
    9.2 ES Cells In Vivo   93
    9.3 Gene Targeting   95
    9.4 Future Prospects for Animal Modeling via Gene Targeting   100
         References   100

10. Expression of a Human Multidrug-Resistance cDNA (*MDR*1) under the Control of a β-Actin Promoter in Transgenic Mice — 103
Hanan Galski, Glenn T. Merlino, Michael M. Gottesman, and Ira Pastan

    10.1 Construction and Transfection of β-Actin Promoter-*MDR*1 Plasmids   104
    10.2 Production of Mice that Carry the βAP-*MDR*1 Transgene   108
    10.3 Transmission of *MDR*1 to Progeny   109
    10.4 Expression Studies in *MDR*1 Transgenic Mice   110
    10.5 Expression of P-Glycoprotein Detected by Immunofluorescence   115
    10.6 Studies In Vivo   115
    10.7 Expression of the Transgene: No Obvious Phenotypic Effects   119
    10.8 Discussion   119

xviii Contents

	References	123
11.	**Insertion of a Disease Resistance Gene into the Chicken Germline**	**125**
	Donald W. Salter and Lyman B. Crittenden	
11.1	Materials and Methods	126
11.2	Results and Discussion	126
11.3	Conclusions	130
	References	131

PART III.   USE OF TRANSGENICS IN UNDERSTANDING BIOLOGY — **133**

12.	**Analysis of Regulatory Genes Using the Transgenic Mouse System**	**135**
	Guerard W. Byrne, Claudia Kappen, Klaus Schughart, Manuel Utset, Leonard Bogarad, and Frank H. Ruddle	
12.1	Discovery of the Homeobox	136
12.2	Evolution of Homeobox Genes	137
12.3	Expression of Homeobox Genes	139
12.4	Regulation of Homeobox Genes	143
12.5	Transgenic Mice and the Function of Homeobox Genes	144
12.6	Multiplex Gene Regulation	145
12.7	Future Prospects	150
	References	151

13.	**Regulation of Expression of a Class I MHC Transgene**	**153**
	Dinah S. Singer, William I. Frels, and Rachel Ehrlich	
	References	158

14.	**Generation of Transgenic Mice with Major Histocompatibility Class II Genes**	**161**
	Javier Martin, Bing-Yuan Wei, Roger Little, Suresh Savarirayan, and Chella S. David	
14.1	Materials and Methods	163
14.2	Results	165
14.3	Discussion	171
	References	173

15.	**Mice Transgenic for a Gene That Encodes a Soluble, Polymorphic Class I MHC Antigen**	**175**
	Rosemarie Hunziker and David H. Margulies	
15.1	Experimental Procedures	177
15.2	Results and Discussion	178
15.3	Conclusions	182

		Contents	xix
		References	184

PART IV. USE OF TRANSGENICS IN MEDICINE — 187

16. Principles of Gene Transfer and the Treatment of Disease — 189
Evelyn M. Karson
- 16.1 Which Diseases are Candidates for Treatment? — 190
- 16.2 Bone Marrow-Mediated Gene Transfer in the Mouse — 198
- 16.3 Gene Transfer via Autologous Bone Marrow Transplantation in the Adult Primate — 198
- 16.4 Protocol for Gene Transfer into Fetal Lambs In Utero — 201
- 16.5 Gene Transfer into Fetal Human and Nonhuman Primate Hematopoietic Cells — 206
- 16.6 Gene Transfer into Tumor-Infiltrating Lymphocytes (TIL) — 209
- References — 210

17. Transgenic Mice Carrying HIV Proviral DNA — 213
David S. Pezen, John M. Leonard, Jan W. Abramczuk, and Malcolm A. Martin
- References — 226

18. Apolipoprotein A-I Gene Expression in Transgenic Mice — 227
Annemarie Walsh, Yasushi Ito, and Jan L. Breslow
- 18.1 Materials and Methods — 228
- 18.2 Results — 229
- 18.3 Discussion — 234
- References — 235

19. Effects of Human Growth Hormone on Reproductive and Neuroendocrine Functions in Transgenic Mice — 237
A. Bartke, J.G.M. Shire, V. Chandrashekar, R.W. Steger, A. Mayerhofer, A.G. Amador, P. Bain, K. Tang, J.S. Yun, and T.E. Wagner
- 19.1 Production of Transgenic Mice and Their Characteristics — 238
- 19.2 Reproductive and Neuroendocrine Functions in Transgenic Females — 240
- 19.3 Reproductive and Neuroendocrine Functions in Transgenic Males — 241
- 19.4 Conclusions — 246
- References — 247

xx   Contents

PART V.   USE OF TRANSGENICS IN ANIMAL AGRICULTURE AND OTHER ANIMALS USED AS RESEARCH MODELS   249

20.   Enhanced Growth Performance in Transgenic Swine   251
Carl A. Pinkert, David L. Kooyman, and Timothy J. Dyer
    20.1   Discussion   253
    20.2   Prospects   255
    20.3   Conclusion   256
           References   256

21.   Production of Sheep Transgenic for Growth Hormone Genes   259
Caird E. Rexroad, Jr.
    21.1   Insertion of Genes   259
    21.2   Growth-Related Genes for Insertion into Ova   261
    21.3   Expression of Transgenes   261
    21.4   Physiology of Integrated Genes   262
    21.5   Transmission of Transgenes   262
    21.6   Conclusion   263
           References   263

22.   Production of Transgenic Cattle by Pronuclear Injection   265
Kenneth R. Bondioli, Karen A. Biery, Keith G. Hill, Karen B. Jones, and Franko J. De Mayo
    22.1   Materials and Methods   266
    22.2   Results and Discussion   269
           References   272

23.   Methods for the Introduction of Recombinant DNA into Chicken Embryos   275
John J. Kopchick, Ed Mills, Charles Rosenblum, Joyce Taylor, F. Macken, F. Leung, Jim Smith, and Howard Chen
    23.1   Materials and Methods   276
    23.2   Results   280
    23.3   Discussion   289
           References   293

24.   Gene Transfer in Fish   295
Kevin S. Guise, Anne A.R. Kapuscinski, Perry B. Hackett, Jr., and Anthony J. Faras
    24.1   Gene Transfer by Microinjection   298
    24.2   Success of Transfer and Expression of Transgenes   301
    24.3   Production of Sterile Transgenic Fish   302
    24.4   Future of Gene Transfer Technology in Fish   303
           References   305

25.	**Studies on Transgenic Fish: Gene Transfer, Expression, and Inheritance**	**307**
	Dennis A. Powers, Lucia Irene Gonzalez-Villasenor,	
	Peijun Zhang, Thomas T. Chen, and Rex A. Dunham	
	25.1 Materials and Methods	308
	25.2 Construction of Transgenic Fish	312
	25.3 Transgenic Fish Containing Genes for Growth Hormones	320
	References	323
26.	**Transgenic Rats: A Discussion**	**325**
	Jan Heideman	
	26.1 Need for Transgenic Rats	326
	26.2 Rats Versus Mice	327
	26.3 Collection and Transfer of Embryos	328
	26.4 Transfer of Embryos to Rat Oviducts	328
	26.5 Technique for Microinjection of Rat Embryos	329
	26.6 Increasing the Yield of Embryos	331
	26.7 Increasing the Size of Litters	331
	References	332
	PART VI. TRANSGENICS: SOCIETY, COMMERCE, AND THE ENVIRONMENT	**333**
27.	**The Mouse That Roared**	**335**
	Lisa J. Raines	
	27.1 An Abundance of Uses	336
	27.2 The Rules Change, Slowly	337
	27.3 The Nature of Patents	338
	27.4 Animals as Tennis Balls?	340
	27.5 Down on the Farm	342
	27.6 Time to Decide	344
	Recommended Reading	345
Index		*347*

PREFACE

A workshop on transgenic technology was held at the Center for Population Research, National Institutes of Health, Bethesda, MD, in December 1988. The Center for Population Research of the National Institute of Child Health and Human Development fosters programs aimed at promoting research in reproductive biology and medicine that will generate technological advances and lead to the improved reproductive health of the human population. The purpose of the workshop was to provide a forum at which scientists could discuss the many newly recognized applications of transgenic technology and the use of transgenics to enhance our understanding of the control of physiological functions, which are so important to our everyday health.

Transgenic animals are those with a foreign gene introduced into their genome. Such animals were given this designation by Gordon and his colleagues in 1980 (*Proc. Natl. Acad. Sci. USA*, 77, 7380–7384) when they reported the birth of the first mouse with a new gene that had been introduced into its genome by microinjection of that gene into the pronuclei of the early mouse embryo. This technology has subsequently been utilized to produce more than 400 strains of transgenic mice, as well as transgenic amphibians, rats, rabbits, sheep, goats, swine, cattle, poultry, and fish.

The production of transgenic mice was the culmination of previous advances in the areas of recombinant DNA and micromanipulation of mammalian cells. The techniques and their development are described by Dr. Gordon in Chapter 1. Much of the workshop was then devoted to the vast number of possible applications of such techniques.

As discussed in subsequent chapters, transgenic mice have provided a powerful model for the examination of the regulation of gene expression and for investigations into the genetic regulation of cellular and physiological functions. Transgenic mice have provided models for genetic diseases and have helped in the design of strategies for therapy. Other models have been generated for resistance to animal disease and novel or improved products or productivity of domestic animals. Gene transfer into fish, domestic animals, and birds suggests that it may be possible to increase the rate and efficiency of animal growth as well as to enhance disease resistance and to produce new or improved products.

The microinjection method commonly used to produce transgenics is still very inefficient. More efficient or versatile methods are being developed. The most promising techniques involve the use of replication-defective viral vectors for the efficient transfer of genes into cells or embryos and the use of the embryonic stem-cell transfer technique, which consists of the transfection, microinjection, or infection of genes into a nontransformed replicating line of embryonic stem cells. Stem cells enriched after selection for expression of a reporter gene are then incorporated into the germline of a blastocyst-stage embryo as chimeras, with the subsequent birth of mice derived from the stem-cell line. Such gene transfer, mediated by stem cells, when combined with recombination of homologous DNA sequences with native gene sequences, as described in Chapters 4 and 9, has opened new frontiers for site-specific gene transfer, gene deletion, and targeted gene therapy.

Great advances are occurring in our understanding both of promoter-enhancer sequences involved in the intrinsic control of gene expression and in the control of gene expression by external transcription-regulatory proteins, as discussed in Chapters 1 through 4. The introduction of foreign genes into animals has allowed scientists to test theories of genetic regulation and expression in vivo. Major questions about gene regulation, locations of chromosomal material, and the importance of *trans*-acting signals have been studied using this system, and other questions of basic developmental relevance have been posed. Fooling the animal into turning on stage-specific genes whose products have reporter molecules attached has permitted scientists to look more closely at the mechanisms that regulate physiological processes (see Chapters 3, 12, and 26). Three major factors influence the expression of genes in developing eukaryotes: *cis*-acting elements in or around the gene; *trans*-acting factors, which interact with genes in open chromosomal domains and stimulate transcription; and the location of the specific gene within the host genome. Because these same factors must operate in the regulation of transferred genes, analysis of the expression of foreign genes in transgenic mice has revealed the relative importance of these factors in determining developmental timing, efficiency, and tissue distribution of gene expression.

Ultimately, the generation of transgenics that are useful in medicine, agriculture, or biology requires the ability to target gene expression to a specific tissue and to control the timing and level of expression of specific genes. It is becoming possible to manipulate such factors in several tissues of the body, such as the mammary gland, skeletal muscle, eye lens, liver, hemopoietic tissue, etc., as discussed in Chapters 4 through 11.

The possibility of somatic-cell correction of genetic defects and new medical therapies is clearly of salient importance (see Chapters 16 through 18). The targeting of specific tissues for gene expression is also important in agriculture: for example, the production of new or improved products in milk, improvement of the tenderness and reduction in the fat content of meat, or changing the growth patterns of animals, as discussed in Chapters 12, 19, 20, 22, 24, and 25.

Resistance to disease is important to humans and animals. The importance of the major histocompatibility genes in such resistance and strategies for enhancing resistance, by the addition of genes of the MHC complex or by interfering with the recognition of pathogens in birds and animals, are discussed in Chapters 11, 13, 14, 15, and 23.

No matter how useful transgenic animals are to scientists in their quest for a complete understanding of biology and nature, the potential release of transgenics into the natural environment for agricultural, animal conservation, or medical purposes is of public concern. The environmental, public safety, and patent aspects of transgenics are presented in Chapters 16 and 27.

Our understanding of cellular and molecular biology and its application to the whole animal through "transgenics" has grown rapidly throughout the 1980s. Applications of the results described in this volume will proliferate rapidly at an ever increasing rate during the 1990s, and the word "transgenic" will no longer require definition.

We would like to thank Dr. Ann Korner and Ms. Marge Perikles for their help in editing, typing, and assembling the manuscripts.

Neal L. First
Florence P. Haseltine

PART I

Gene Transfer: Approaches, Methods, and Tools

CHAPTER 1

Transgenic Animals: A New Era in Developmental Biology and Medicine

Luis M. Isola
Jon W. Gordon

In recent years a wealth of knowledge has accumulated as a result of the ability of researchers to introduce genetic material into intact mammalian organisms. The production of transgenic mice by pronuclear microinjection (Gordon et al. 1980) has provided an especially powerful new tool for the study of the regulation of gene expression in mammalian development. Animals carrying new genes were first called "transgenic" by Gordon and Ruddle (1983), a term that is now used to describe particular variants of species into which a new gene(s) has been inserted. Transgenic technology has allowed the rapid elucidation of many fundamental principles of the tissue-specific expression of certain genes and has also been particularly useful in immunology, oncology, and studies of the genetic basis of disease. In addition to its importance for studies of gene expression, the introduction

---

Work by the authors described in this manuscript was supported by grants from the National Institutes of Health (HD20484 and CA42103) and from the March of Dimes (1-1026) to JWG. L. Isola is the recipient of the American Society of Hematology and the Henry M. and Lillian Stratton Foundation Research Awards.

of DNA into the host genome has the potential for disruption of the integrity of host genes. The identification of insertional mutations and the cloning of the affected host genes, made possible by the insertion of defined sequences of DNA into the affected loci, is another important application of transgenic technology.

The production of transgenic mice was the culmination of previous advances in the areas of recombinant DNA and the micromanipulation of mammalian cells. Although the production of allophenic mice could be considered to be the first step toward the introduction of genetic material into intact mammalian organisms, this approach only allowed the introduction of whole genomes into chimeric mice by aggregation of genetically different embryos at the cleavage stage (Tarkowski 1961; Mintz 1962). Another approach to the transfer of genes into the germline resulted in the insertion of discrete sequences of DNA. This technique is based on the introduction of teratocarcinoma cells into blastocyst cavities (Papaiannou et al. 1975; Mintz and Illmensee 1975). These cells can be subjected to DNA-mediated transfer of genes in vitro and can also become phenotypically normalized and contribute to tissues of chimeric mice (Pellicer et al. 1980). The major drawback of this approach was the fact that the transformed cells rarely contributed to the germline of chimeric mice (Mintz and Cronmiller 1981). A more efficient method is based on the use of embryonal stem (ES) cells (Evans and Kaufman 1981; Martin 1981). These cells can be transformed or selected for a particular phenotype and they also contribute frequently to the germ cell population of chimeric animals. This new and powerful approach is discussed in detail elsewhere in this volume.

Retroviruses have been successfully used to transform embryos by infection (Jaenisch 1976). Evidence for expression of retroviruses introduced in this manner stems from the fact that Moloney murine leukemia virus (MoMuLV) is frequently activated later in development and causes leukemia (Jaenisch 1979). Expression of heterologous genes in recombinant retroviruses has been unpredictable and generally less efficient than such expression after microinjection of gene-specific DNA (Jahner et al. 1985). However, this technology is extremely valuable for tracing cell lineages and the evaluation of insertional mutants; it is discussed in detail by other contributors to this volume.

Because the ES cell system is relatively new and more demanding to manage, and because the retroviral system has not yielded consistently high levels of expression of genes, microinjection has thus far been the most widely exploited method for the transfer of genes to the germ line. In this chapter, we shall review the most important achievements that have resulted from microinjection of genes, citing examples from our own laboratory. No effort will be made to review the entire literature, but reports illustrative of the major accomplishments of microinjection technology will be cited. In the course of this discussion, we will emphasize the many questions that remain to be answered about the control of gene expression and insertional mutagenesis. The implications of this technology for the future of science, agriculture, and medicine will also be discussed briefly at the end of the chapter. A much more detailed review and discussion of all these problems can be found in another publication by one of the authors of this chapter (Gordon 1989).

1.1 INTEGRATION OF MICROINJECTED SEQUENCES INTO THE EMBRYO

A detailed description of the method for the production of transgenic mice is beyond the scope of this chapter and can be found elsewhere (Gordon et al. 1980; Gordon and Ruddle 1983; Hogan et al. 1986). The procedure is illustrated in Figure 1–1. Briefly, immature female mice are made to superovulate by sequential administration of FSH/LH and hCG and mated to fertile males. One-celled embryos are flushed from the oviducts and placed in a drop of medium and viewed by phase-contrast or interference microscopy. While the embryo at the pronuclear stage is held in place by suction, a microneedle loaded with a suspension of plasmid DNA is introduced through the zona pellucida and plasma membrane into the most accessible pronucleus (usually the male) and several hundred molecules of the recombinant DNA are injected in a volume of approximately 1 picoliter (pl). Embryos surviving the injection are transferred to foster mothers and

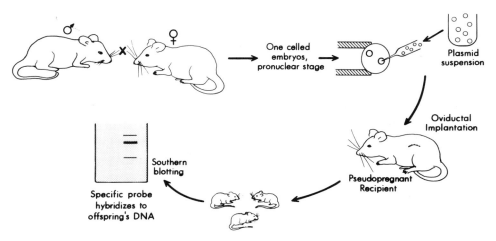

FIGURE 1–1 Production of transgenic mice by pronuclear microinjection (Gordon et al. 1980). One-celled fertilized eggs are recovered and the most accessible pronucleus, usually the male, is microinjected. Embryos are then tranferred to the oviducts of pseudopregnant females and evaluated after birth by Southern blot hybridization, or a related technique, for retention of foreign DNA.

allowed to develop to term. After birth, the presence of foreign material is studied by DNA hybridization with appropriate probes.

Since integration of microinjected DNA is an event unique to each embryo, and since most of the attention of researchers in laboratories that produce transgenic mice is focused on the expression of transferred genes, little is known about the mechanisms that operate in the fertilized oocyte and allow it to integrate the microinjected DNA. However, several features have been recognized as characteristic of the integration process: the microinjected molecules tend to form long, head-to-tail concatamers (Costantini and Lacy 1981) and, thereafter, to integrate at one or, rarely, two sites within the host genome; integration may or may not be delayed for several cleavage divisions after microinjection; and insertion of microinjected DNA is associated with rearrangements both of the donor material and the recipient genome.

It has been proposed that the rate-limiting step for the integration of microinjected sequences is the disruption of the host genome, which is necessary to provide a target site for insertion, after which additional copies integrate by homologous recombination (Brinster et al. 1985). This model is not entirely consistent with the observation that restriction fragments indicative of rearrangements within the donor DNA are often reduplicated in the concatemers. In addition, Burki and Ulrich (1981) noted integration of concatamers into different sites within placental and somatic DNA of single embryos. Because integration was rare in their experiments, these results must be interpreted to indicate that formation of concatamers is linked to integration. It is, therefore, logical to assume that formation of concatamers can precede integration and, as in the case of DNA-mediated transfer of genes into cells in tissue culture (reviewed by Scangos and Ruddle 1981), is an intermediate step in the integration process. Nonetheless, since all concatamers thus far reported contain at least one entire copy of the donor fragment of DNA, homologous recombination between donor molecules, which occurs in cultured cells prior to integration (Small et al. 1985), most probably also takes place. However, results of formal experiments designed to clarify the mechanism of integration have never been reported.

Integration of microinjected DNA can ultimately result in deletions and duplications within host DNA as well as interspersion of "islands" of genomic DNA within the donor material (Covarrubias et al. 1986). At the chromosomal level, breakage and translocations at the insertion site have also been documented (Overbeek et al. 1986). These phenomena are particularly vexing when attempts are made to analyze mutations caused by insertion of microinjected material.

Even after insertion, transgene sequences may be subjected to structural changes, which can cause loss of copies of the tandem array, perhaps by "looping out" of identical subunits that recombine with each other (Palmiter et al. 1982). Other reported alterations include transient amplification (Shani 1986) of transgenes, dispersion of donor material throughout the genome

(Rubinstein and Gordon 1987), and maintenance of foreign material in the absence of integration (Rassoulzadegan et al. 1986). Because of the very large number of transgenic pedigrees in existence and the relatively rare incidence of these "abnormal" patterns of integration, we can assume that the pathway of integration of foreign genes is generally characterized by concatamerization, opening of the host genome, and integration of the injected DNA.

It is unknown at present whether microinjected genes integrate randomly within the host genome, even though such has generally been assumed to be the case. However, recent work with retrovirus-mediated transfer of genes strongly suggests that retroviruses, previously assumed to integrate at random, do not do so (Vijaya et al. 1986). Thus, analysis of microinjected fragments of flanking DNA could determine whether integration is random, and if not, whether the favored sites of integration are also those predisposed to attack by retroviruses.

1.2 EXPRESSION OF FOREIGN GENES IN TRANSGENIC MICE

Three major factors influence the expression of genes in developing eukaryotes: *cis*-acting elements in or around genes; *trans*-acting factors, which interact with genes in open chromosomal domains and stimulate transcription; and the location of a gene within the host genome. Because these same factors must operate to determine the regulation of expression of transferred genes, analysis of expression of foreign genes in transgenic mice produced by microinjection has provided important and basic information about the relative importance of these factors in the determination of the timing with respect to development, the efficiency, and the tissue distribution of gene expression.

1.2.1 *cis*-Acting Elements

Early experiments by Brinster et al. (1981) clearly demonstrated that discrete, *cis*-acting elements, often located near the 5' (promoter) regions of structural genes, dictate the tissue distribution of expression. In these experiments, the 5' region of the mouse metallothionein–1 (MT-1) gene was linked to the coding sequence for thymidine kinase (TK) from Herpes virus. Expression of the gene for TK in these animals followed a pattern typical of MT-1, while TK, like MT-1, was inducible by the heavy metal cadmium. This result was observed despite the fact that the integration site was not within the endogenous MT locus (Palmiter et al. 1982). These experiments, and many similar ones that followed, illustrate an important principle of mammalian gene regulation: It is *not* the chromosomal environment that determines whether a gene will be accessible to *trans*-activators of expression at a given place or time, it is *cis*-acting elements of genes that determine the state of the chromosomal environment and, consequently, the tissue distribution and timing of expression. Of course, chromosomal position is not entirely irrelevant, as demonstrated by the fact that a given gene may fail to express in one or more lines of animals, while in others the same construct is active. The extent of *cis*-acting influences related to the chromosomal site of integration appears to vary depending upon the strength of the individual *cis*-acting element. When genomic clones encompassing the human β-globin gene are introduced into mice, their expression is highly variable between lines, suggesting that the site of integration exerts a major influence (Chada et al. 1985; Townes et al. 1985). Immunoglobulin light-chain genes, by contrast, appear to be expressed far more consistently (Brinster et al. 1983). Genes such as pancreatic elastase-1 are expressed at a high level, which is roughly correlated with the number of copies of the transgene present in the animals (Swift et al. 1984). Another measure of the relative strength of promoter/enhancer elements is their sensitivity to inhibition by certain of the plasmid or bacteriophage sequences with which they may be introduced. Genes for β-globin are strongly inhibited by sequences derived from cloning vectors (Chada et al. 1985; Townes et al. 1985), while immunoglobulin genes are not (Brinster et al. 1983; Storb 1987). An important issue to be pursued relates to these apparent differences in strengths of promoter/enhancer elements. Recent findings by Grosveld et al. (1987) suggest that all genes may, in fact, have strong enhancers of expression, but that in some cases these elements may be relatively distant from the coding sequence. This possibility was suggested by the finding that the level of expression of the human β-globin gene is high, independent of chromosomal position, and directly correlated with copy number, once elements as far away as 50 kb are approximated to the gene in new recombinant constructs.

Genes with complex patterns of expression may be regulated by multiple, *cis*-acting, regulatory elements, one for each tissue or subcomponent of each tissue. Thy-1, a cell-surface protein of the immunoglobulin family of genes, is expressed throughout the brain in mice and is also produced in the thymus, peripheral T cells and fibroblasts. When we introduced a construct of mouse genomic Thy-1 into transgenic animals, a reproducible failure of expression was observed in discrete subregions of the brain, and expression was also undetectable in peripheral T cells (Gordon et al. 1987). These results indicate that the introduced gene was separated from specific enhancers when cloned from the genome. Multiple *cis*-acting elements may also be differentially sensitive to inhibition by cloning vector DNA. When introduced with plasmid material, the α-fetoprotein gene specifically loses its ability to express in the yolk sac, while expression in the fetal liver is preserved (Krumlauf et al. 1985). Similar observations have been made with class II genes of the major histocompatibility complex (MHC) in transgenic mice (Widera et al. 1987).

1.2.2 *Trans*-Acting Factors

These elusive, *trans*-acting factors interact with DNA with resultant activation of gene expression. Because *trans*-acting elements are themselves gene products, it is logically impossible that each gene is activated by a unique *trans*-acting factor. Thus, because each of these factors must control expression of several genes, they are expected to be highly conserved over the course of evolution. Data obtained in the transgenic mouse system are highly illustrative of this principle. Transgenic mice carrying the human Thy-1 gene display a pattern of transgene expression identical to that of the human, even though several sites at which the gene is expressed are not characteristic of mouse Thy-1 (Gordon et al. 1987). Expression of human Thy-1 in mouse organs that do not express Thy-1 indicates that the *trans*-activating factors needed for expression are present in those tissues. These experiments also indicate that changes in the *cis*-acting sequences are most likely responsible for the different patterns of expression seen between the two species.

Transgenic mice have been used to illustrate the possibility that aberrant developmental regulation of transgene expression may be due to evolutionary changes in the timing of production of *trans*-acting factors. The human β-globin gene complex contains three different genes, expressed sequentially in the embryonic yolk sac, fetal liver, and adult bone marrow. This orchestrated regulation follows a different pathway in the mouse, where a distinct embryonic gene does not exist. Transgenic mice carrying the human fetal globin gene express it in the yolk sac, and this expression corresponds to an embryonic pattern of globin expression (Chada et al. 1986). This result indicates that, in order to produce embryonic globin, the mouse produces a *trans*-activator of expression of fetal gene in the yolk sac, and that this factor interacts with the transferred, human, fetally active gene.

Clearly, the important area for future research is the isolation and characterization of these *trans*-activators of gene expression. Isolation of such factors appears more likely to be accomplished from experiments in tissue culture (Davis et al. 1987). However, once genes encoding such factors have been cloned, their introduction into transgenic animals under various modes of regulation should usher in a new era in the use of transgenic animals for exploration of problems in gene regulation and for genetic engineering.

1.2.3 Site of Integration

As noted above, the site of integration can influence transgene expression, with the degree of modification dependent upon the strength of promoter/enhancer elements in the donor construct. It also appears that, regardless of the potency of enhancer elements on the transgene, some sites of integration are totally nonpermissive with respect to expression.

Some integration sites actually favor low levels of expression. In cases where transgenes integrate within regions that are later opened for activation of an endogenous gene, "leaky" expression may occur at that time. One example of this phenomenon was a line of mice in which a rabbit β-globin gene was expressed in the skeletal muscle (Lacy et al. 1983).

1.2.4 Aberrant Expression of Genes

Given the dynamic interaction between promoter/enhancer elements, the *trans*-acting factors present in cells for activation of a variety of endogenous genes, and the potential sites of integration of foreign genes, it is not surprising that abnormal patterns of expression can be observed. When artificial recombinant constructs containing a promoter from one gene linked to the coding sequence of another are introduced, and when both portions of the donor molecule are derived from other species or even classes of organisms, it is not surprising that activation of transgenes in unexpected sites may occur. For example, the MT-1 promoter linked to the Herpes *tk* gene is active primarily in liver and kidney (Brinster et al. 1981), but MT-1-hypoxanthine phosphoribosyltransferase (HPRT) constructs are expressed in the central nervous system (Stout et al. 1985), and MT-1 ornithine transcarbamylase chimeric genes are active in testis (Kelley et al. 1988). In the case of genes for immunoglobulin heavy chains, rearrangement prior to microinjection results in expression in T cells. Moreover, the μ enhancer is associated with reproducible and aberrant expression in sites such as the heart (Grosschedl et al. 1984) and the brain (Nussenzweig et al. 1987). While, at our present state of sophistication, these phenomena may be disconcerting, they also suggest that a greater knowledge of interaction between *cis*- and *trans*-acting factors may lead to design of recombinant constructs with unique and exquisitely well-defined patterns of expression, an eventuality that would significantly improve genetic engineering strategies.

1.3 SPECIFIC APPLICATIONS OF TRANSGENIC MICE

In the following section we shall describe some of the experiments that have contributed invaluable knowledge in several fields of developmental biology. Because of the ever-increasing number of publications about transgenic mice, we cite only a limited number of studies, which serve to illustrate the power of this technology.

1.3.1 Gene Expression

1.3.1.1 Transgenic Mice in Immunology The ability of the immune system to distinguish and mount a response to myriad foreign antigens while recognizing "self" as nonantigenic has always been one of the most fascinating areas of biology. Information about the molecular genetics of the genes for immunoglobulins (Ig) and the major histocompatibility complex

(MHC) has increased dramatically, and a major contribution to such information has come from studies with transgenic mice. This field has recently been elegantly reviewed by Storb (1987).

Introduction of rearranged Ig genes into transgenic mice has helped enormously to clarify mechanisms of rearrangements of Ig genes and the importance of this phenomenon to the regulation of their expression. For example, insertion of rearranged κ light-chain genes resulted in blockage of rearrangement and allelic exclusion of the endogenous counterparts when secretion of a mature Ig molecule containing the product of the transgene occurred (Ritchie et al. 1984). However, rearrangement was able to proceed when no Ig molecule was assembled and expression of endogenous genes was inhibited only by the presence of transgenic light chains in the cytoplasm.

Microinjection of rearranged heavy-chain genes led to the conclusion that, in this case, allelic exclusion depends on the ability of the product of the transgene to bind to the cell membrane. When the sequences for membrane-binding components of heavy chains are removed and transferred, allelic exclusion does not occur (Storb et al. 1986); correspondingly, deletion of signals for secretion is associated with allelic exclusion (Nussenzweig et al. 1987). Furthermore, expression of rearranged heavy-chain genes in T cells indicates that the inability of the T cell to express heavy-chain genes is related to the absence of gene rearrangement in these lineages (Grosschedl et al. 1984). This result is all the more intriguing since T cells are able to rearrange the T-cell receptor gene, which shares several features with the Ig gene.

One of the classical concepts in immunology, namely, that rearrangement of κ genes inhibits rearrangement of λ genes (isotypic exclusion) was challenged by the finding that the presence of transgenic rearranged κ genes did not, in all cases, result in isotypic exclusion. This observation led to the hypothesis that two separate B cell lineages exist, one that is capable of rearranging both κ and λ genes and one that can only rearrange κ genes (Storb 1987).

One of the mechanisms that allows for the development of the immune repertoire is somatic hypermutation. This phenomenon was not clearly understood in terms of the rearrangement of Ig genes until it was demonstrated in transgenic mice. In these mice, rearranged κ Ig genes, which were microinjected in the rearranged configuration, were mutated (O'Brien et al. 1987).

1.3.1.2 Transgenic Mice in Oncology The study of transferred oncogenes has always been hampered by the fact that cell lines in culture have already been transformed to an abnormal phenotype. Thus, the transforming potential of an oncogenic sequence of nucleotides in a cell can be enhanced or obscured by a preexisting proliferative abnormality. Our ability to insert oncogenes and proto-oncogenes into embryos and to study their effects in the normally differentiating cells of an intact organism has circumvented this problem, and results of such studies have made an enormous contribution to our understanding of neoplastic disease and its relationship to aberrant gene expression.

The large T antigen of the simian virus 40 (SV40) is a transforming gene that has been studied extensively in transgenic mice. The sequence corresponding to the T antigen, linked to its own enhancer-promoter, has been introduced into mice by microinjection (Brinster et al. 1984; Small et al. 1985). The gene was expressed preferentially in the central nervous system (CNS) and even produced choroid plexus papillomas. Recombinant molecules, consisting of the sequence for the T antigen linked to different promoters, caused tumors in those tissues that normally express the promoters that modulate transcription of the oncogene. Directed oncogenesis or hyperplasia mediated by T antigen has now been demonstrated in the endocrine pancreas (Hanahan 1985), the exocrine pancreas (Ornitz et al. 1985), the liver (Messing et al. 1985), the lens of the eye (Mahon et al. 1987), and even in heart muscle (Field 1988).

In addition to viral oncogenes, cellular proto-oncogenes and their transforming counterparts, modulated by a variety of promoters, have been introduced into mice. One series of transgenic mice produced by introduction of the c-*myc* oncogene under the influence of the promoter for mouse mammary tumor virus (MMTV) developed tumors in breast (Stewart et al. 1984), testis, and lymphatic tissues (Leder et al. 1986), demonstrating that aberrant regulation of proto-oncogenes can predispose an organism to malignancy.

One of the major contributions of the study of oncogenes in transgenic mice has been the support that it has provided for the hypothesis that oncogenesis is a multistep process that requires at least two transforming events. This concept is clearly suggested by the results of the experiments with c-*myc* and SV40, in which all cells in the target organ express the exogenous sequence and, therefore, are potential candidates for malignant transformation, but only a few of them develop into overt malignancies. Thus, it would appear that the oncogene initiates the first step and creates a situation in which a single additional event or "hit" precipitates oncogenesis. Whether the multiple-hit hypothesis will be borne out for all oncogenic sequences in animals remains an open question.

1.3.1.3 Transgenic Mice as Animal Models of Human Disease Animal models for human illnesses are useful for studying the pathogenesis of diseases as well as for developing and testing new therapies. However, many human diseases either do not exist in animals or are only developed by higher mammals, making models scarce and expensive to researchers. Human diseases can be induced in transgenic mice by expression of transferred genes, or by insertional disruption of endogenous sequences. Some examples of models created by transgene expression are listed below.

Hepatitis B is a human disease that lacks a readily workable animal model. Introduction of the HBsAg gene into mice results in transgenic mice that mimic the carrier state with production of HBsAg in the liver but with an absence of disease (Babinet et al. 1985; Chisari et al. 1985).

Progressive multifocal leukoencephalopathy can be recreated in mice by injection of the JC viral genome (Small et al. 1986). Another disease of the nervous system, neurofibromatosis, can be induced in mice by introduction of the gene for tyrosine aminotransferase (TAT) of HTLV-1 (Hinrichs et al. 1987).

Osteogenesis imperfecta is a dominant disorder that results from a mutation of the gene for $\alpha 1$ (I) collagen. The simple, resultant amino-acid substitution results in a change in the helical structure of mature collagen. This condition has been induced in mice by microinjection of a mutant gene for collagen $\alpha 1$ (I), created in the laboratory by site-directed mutagenesis (Stacey et al. 1988).

To date, the only disease model created by insertional mutagenesis has been Lesch-Nyhan disease (hypoxanthine phosphoribosyl transferase [HPRT] deficiency), created through the use of ES cells (Kuehn et al. 1987; Hooper et al. 1987). While mice with this enzyme deficiency do not manifest the pathology characteristic of the human disease, the production of such animals illustrates the principle that insertional mutagenesis can be exploited to create disease models.

1.3.1.4 Transgenic Mice as Models for Gene Therapy Genes can be inserted into transgenic animals and function to alleviate disease states. Such model systems can be of great importance in improving our understanding of the potential for gene transfer as an approach to treatment of disease.

Mice with growth-hormone deficiency are markedly reduced in size and males suffer from infertility. Introduction of the growth-hormone gene into these animals leads to growth which exceeds that of normal animals and restores male fertility (Hammer et al. 1984). However, the pattern of release of growth hormone that results from transgene function is apparently inconsistent with female fertility (Hammer et al. 1984).

Mice lacking genes at the MHC region are unable to respond to the synthetic antigen poly (Glu-Lys-Phe). When the gene for Ia is inserted into transgenic animals, transgene function restores immunoresponsiveness to antigens (LeMeur et al. 1985; Yamamura et al. 1985).

Insertion of either the mouse or human β-globin gene can reduce the severity of β-thalassemia in mice (Costantini et al. 1986). In these experiments, the product of the human globin gene was able to associate effectively with the mouse α chains, and it actually functioned better than the transferred mouse gene β-globin in palliating the thalassemic state.

Mice with a deficiency in gonadotropin-releasing hormone (GnRH) are infertile and exhibit profound perturbations of their reproductive endocrine functions. Cloning of the GnRH gene and its transfer into mice has resulted in restoration of normal endocrine function and in fertility (Mason et al. 1986).

1.3.1.5 Transgenic Mice as Models for Genetic Engineering Genetic engineering of resistance to chemotherapy is one technique that can assist in increasing the therapeutic margin of a drug. In cancer therapy the toxic doses of most antineoplastic drugs overlap with their therapeutic doses, so that the intensity and duration of drug regimens are, of necessity, limited. Production of transgenic animals that are systemically resistant to the effects of a drug (Isola and Gordon 1986) provides a test system with which to answer questions about whether intensive chemotherapy can improve the overall survival of tumor-bearing hosts or whether treatment would just select for highly resistant tumors.

Introduction of the gene for the receptor for low-density lipoprotein (LDL), under control of the MT-1 promoter, has resulted in inducible expression of the gene with a resultant decrease in levels of free LDLs within the circulation of the transgenic animals (Hoffman et al. 1988). This experiment suggests a potential strategy for a genetic approach to the control of high levels of LDL, which could eventually reduce the risk of atherosclerosis.

The successful production of transgenic rabbits, pigs, and sheep by microinjection (Hammer et al. 1985) suggests that, in the future, genetically engineered farm animals may play an important role in agriculture and medicine. The isolation of tissue plasminogen activator (TPA) from the milk of transgenic mice after linkage of the TPA gene to the promoter for whey acidic protein (Pittius et al. 1988) strongly suggests that transgenic animals might be exploited as "factories" for production of valuable products which, because of post-translational modification and other reasons, cannot readily be produced in bacteria.

1.3.1.6 Use of Promoter-Directed Expression of Genes for Ablation of Specific Lineages Complete cell lineages can be ablated in transgenic mice that carry constructs in which a cellular toxin is linked to a tissue-specific promoter. Such experiments with the diphtheria toxin A and the elastase I promoter have succeeded in ablating the exocrine pancreas (Palmiter et al. 1987), and with the toxin and the promoter for crystallin to have resulted in destruction of lens tissue (Breitman et al. 1987). This strategy may provide useful models for analyses of very specific interactions between cells.

1.3.1.7 Antisense Genes in Transgenic Mice Another method for negating gene function involves the use of antisense transcripts. When genes are cloned in reverse orientation with respect to the promoter, RNA may be produced from the noncoding strand. This RNA, presumably by forming a heteroduplex with the sense RNA, can block translation of cytoplasmic mRNA (Izant and Weintraub 1984). Thus, antisense genes can be used to obliterate production of proteins from specific genes in transgenic animals. The feasibility of this approach has recently been demonstrated by the transfer of an antisense construct of the gene for myelin basic protein (MBP) into mice. Interference with the production of MBP resulted in dysmyelination (Katsuki et al. 1988). Although this approach is still in its infancy, it has great potential for future experiments.

1.3.2 Insertional Mutagenesis

While efficient expression of microinjected genes can be exploited in a wide variety of basic and applied studies, gene transfer can also be used to disrupt the host's coding information, thereby inducing mutations. Such insertional mutations are particularly useful because the mutated gene is tagged with a defined sequence of nucleotides that can then be used for molecular cloning of the affected host locus. Some disagreement exists as to the frequency of occurrence of insertional mutations, which result from the fortuitous integration of foreign DNA into a functional region of the host genome. However, a safe guess is that about 5% of transgenic animals will sustain such mutations, a figure that corresponds roughly to the percentage of DNA in the genome devoted to coding sequences and their associated regulatory information. It is unknown at present whether integration occurs randomly throughout the genome, as suggested by the frequency of occurrence of insertional mutations. Evidence is now persuasive that retroviruses do not integrate at random but preferentially target the chromatin that is open for transcription at the perimplantation period, approximately the same time as that at which the gene transfer event occurs (Vijaya et al. 1986; Rohdewold et al. 1987). An important future goal for studies of microinjected genes is the characterization of the DNA that flanks the inserts, so that the mechanism of integration can be compared directly to that of retroviruses.

Insertional mutations must be distinguished from the effects of the expression of the foreign gene(s). When a mutant phenotype is inherited as a strict Mendelian recessive, linked inexorably to the transgene, strong evidence exists that insertional mutagenesis is responsible for the altered phenotype. However, such findings do not formally rule out the possibility that the additive effects of the expression of two transgene inserts in homozygotes are responsible for the observed effect. Some progress toward the resolution of this ambiguity is the demonstration that the transgene insertional mutation is allelic with a known spontaneous mutation with similar phenotypic effects. Such a demonstration was provided by Woychik et al. (1983), who discovered a recessively inherited limb deformity associated with the presence of a c-*myc* transgene. When the transgenic animals were crossed with mice that carry the spontaneous mutation limb deformity (*ld*), double heterozygotes for the transgene and the spontaneous mutation exhibited a mutant phenotype indistinguishable from *ld/ld* or transgenic homozygotes.

The ability of microinjected DNA to destroy host genes with extremely subtle and specific roles in development was demonstrated by Palmiter et al. (1983). They identified a transgenic line that carried pMK, the MT-1-Herpes *tk* fusion gene, in which males were fertile but failed to transmit the foreign sequence. Their eventual conclusion from analysis of this line was that insertion of the transgene inactivated a host gene that was active after meiotic segregation of the chromosomes, so that the sperm that inherited the mutation were destroyed (Palmiter et al. 1982).

In our own laboratory, we have identified two insertional mutations that affect neurologic function and male fertility. In the case of one of these mutations, transgenic homozygotes are ataxic and males are infertile. Histologically, the mutation is manifested as a profound degeneration of cerebellar Purkinje cells, less marked loss of retinal photoreceptors and olfactory bulb mitracells, and an absence of mature spermatozoa. We have found this mutation to be allelic with *pcd*, a gene in which mutations are associated with spontaneous cerebellar degeneration, which maps to mouse chromosome 13. Another mutant line also exhibits motor dysfunction and male infertility. In this case the transgene insert is allelic with *ho*, a gene in which mutations generate similar phenotypic manifestations, which maps to mouse chromosome 6. In this second case, we have independently demonstrated linkage of the transgene insert to Mi^{wh}, a semidominant mutation that also maps to chromosome 6. Analysis of male infertility in this line has revealed marked abnormalities in sperm morphology. We have employed another technique, originally developed in our laboratory, namely, zona drilling (Gordon and Talansky 1986), to restore fertility to these mutant males and genetically rescue the mutation. Our observations that grossly deformed sperm could fertilize eggs and give rise to normal offspring created the impetus for extending zona drilling to clinical trials. Thus, transgenic animals can be employed for the testing of new procedures destined for clinical use.

1.4 TRANSGENIC MICE IN DEVELOPMENTAL GENETICS

In addition to the specific uses outlined above, studies with transgenic mice have had a significant impact in the understanding of development and its genetic basis.

Although the parental contribution to every autosome is equal, it was clear from early experiments that both maternal and paternal alleles were indispensable for development of the fertilized oocyte into a mature individual (Markert and Petters 1977; Surani and Barton 1983; McGrath and Solter 1983). It has been proposed that maternal and paternal genes are imprinted with differential patterns of methylation that result in differences in activity of subsets of genes that are, in turn, important for development. In fact, such imprinting has been documented in the case of transgenes that are inherited maternally or paternally. HBsAg, a bacterial chloramphenicol acetyl transferase (CAT) construct, and a quail troponin gene showed extensive methylation and/or weak expression when inherited from the female parent, whereas the converse was demonstrated in the case of paternal inheritance (Reik et al. 1987; Hadchouel et al. 1987; Sapienza et al., 1987). A detailed study of similar findings with a Rous sarcoma virus construct has also been published by Swain et al. (1987).

Foreign DNA integrated into the X chromosome has provided unique insight into the mechanism of inactivation of X chromosomes. For instance, it has been observed that long concatamers of a transgene can escape the mechanism of X inactivation, perhaps by disturbing the geometry of GC-rich areas distributed along the X chromosome (Goldman et al. 1987).

1.5 SUMMARY AND CONCLUSIONS

In the relatively short period since its development, the technique for the production of transgenic animals by pronuclear microinjection, as established by Gordon et al. (1980), has resulted in studies that have clearly had a profound effect on our basic understanding of mammalian development and genetics, and upon our outlook for genetic engineering. The question now is: What remains to be done? Clearly, a greater understanding of the mechanism of integration could lead to more efficient transfer of genes and lessen the problem of rearrangement of host DNA. A better characterization of enhancers and their mode(s) of action would allow design of molecules with highly specific patterns of gene expression, as well as novel distributions of expression. When these goals are met, it may well become scientifically feasible to transfer genes into the human germline. Whether such an undertaking is advisable or even desirable is an issue requiring open discussion, but it is an issue that is clearly distinct from the question of technical feasibility. The most rational approach to this profound question lies in the continued pursuit of knowledge relating to transgenic technology, rather than self-imposed ignorance. Even if never used as a medical therapy, transfer of genes into the germline will reveal what is certainly one of nature's greatest secrets: the nature of the interaction of genes in the development of a multicellular eukaryotic organism. Through the contributions to this volume and in the years ahead, we will find ouselves in a position to unravel the details of the awesome and fascinating phenomenon that is mammalian development.

REFERENCES

Babinet, C., Farza, H., Morello, D., Hadchouel, M., and Pourcel, C. (1985) *Science* 230, 1160–1163.
Breitman, M.L., Clapoff, S., Rossant, J., et al. (1987) *Science* 238, 1563–1565.
Brinster, R.L., Chen, H.Y., Messing, A., et al. (1984) *Cell* 37, 367–379.
Brinster, R.L., Chen, H.Y., Trumbauer, M., et al. (1981) *Cell* 27, 223–231.
Brinster, R.L., Chen, H.Y., Trumbauer, M.E., Yagle, M.K., and Palmiter, R.D. (1985) *Proc. Natl. Acad. Sci. USA* 82, 4438–4442.
Brinster, R.L., Ritchie, K.A., Hammer, R.E., et al. (1983) *Nature* 306, 332–336.
Burki, K., and Ullrich, A. (1981) *EMBO* 1, 127–131.
Chada, K., Magram, J., and Costantini, F. (1986) *Nature* 319, 685–688.
Chada, K., Magram, J., Raphael, K., et al. (1985) *Nature* 314, 377–380.
Chisari, F.V., Pinkert, C.A., Milich, D.R., et al. (1985) *Science* 230, 1157–1160.
Costantini, F., Chada, K., and Magram, J. (1986) *Science* 233, 1192–1194.
Costantini, F., and Lacy, E. (1981) *Nature* 294, 92–94.
Covarrubias, L., Hishida, Y., and Mintz, B. (1986) *Proc. Natl. Acad. Sci. USA* 83, 6020–6024.
Davis, R.L., Weintraub, H., and Lassar, A.B. (1987) *Cell* 51, 987–1000.
Evans, M.J., and Kaufman, M.H. (1981) *Nature* 292, 154–156.
Field, L.J. (1988) *Science* 239, 1029–1032.
Goldman, M.A., Stokes, K.R., Idzerda, R.L., et al. (1987) *Science* 236, 593–595.
Gordon, J.W. (1989) *Int. Rev. Cytol.*, 115, 171–230.
Gordon, J.W., Chesa, P.G., Nishimura, H., et al. (1987) *Cell* 50, 445–452.
Gordon, J.W., and Ruddle, F.H. (1983) *Methods in Enzymol. Recombinant DNA, Part C.* 101, 411–433
Gordon, J.W., Scangos, G.A., Plotkin, D.J., Barbosa, J.A., and Ruddle, F.H. (1980) *Proc. Natl. Acad. Sci. USA* 77, 7380–7384.
Gordon, J.W., and Talansky, B.E. (1986) *J. Exp. Zool.* 239, 347–354.
Grosschedl, R., Weaver, D., Baltimore, D., and Costantini, F. (1984) *Cell* 38, 647–658.
Grosveld, F., van Assendelft, G.B., Greaves, D.R., and Kollias, G. (1987) *Cell* 51, 975–985.
Hadchouel, M., Farza, H., Simon, D., Tiollais, P., and Pourcel, C. (1987) *Nature* 329, 454–456.
Hammer, R.E., Palmiter, R.D., and Brinster, R.L. (1984) *Nature* 311, 65–67.
Hammer, R.E., Pursel, V.G., Rexroad, C.E. Jr., et al. (1985) *Nature* 315, 680–683.
Hanahan, D. (1985) *Nature* 315, 115–122.
Hinrichs, S.L., Nerenberg, M., Reynolds, R.K., Khoury, G., and Jay, G. (1987) *Science* 247, 1340–1343.
Hoffmann, S.L., Russell, D.W., Brown, M.S., Goldstein, J.L., and Hammer, R.E. (1988) *Science* 239, 1277–1281.
Hogan, B., Costantini, F., and Lacy, E. (1986) *Manipulating the Mouse Embryo*, Cold Spring Harbor Press, Cold Spring Harbor, NY.
Hooper, M., Hardy, K., Handyside, A., Hunter, S., and Monk, M. (1987) *Nature* 326, 292–295.
Isola, L.M., and Gordon, J.W. (1986) *Proc. Natl. Acad. Sci. USA* 83, 9621–9625.
Izant, J.G., and Weintraub, H. (1985) *Science* 229, 345–352.

Jahner, D., Haase, K., Mulligan, R., and Jaenisch, R. (1985) *Proc. Natl. Acad. Sci. USA* 82, 6927–6931.
Jaenisch, R. (1979) *Virology*. 93, 80–90.
Jaenisch, R. (1976) *Proc. Natl. Acad. Sci. USA* 73, 1260–1264.
Katuski, M., Sato, M., Kimura, M., et al. (1988) *Science* 241, 593–595.
Kelly, K.A., Chamberlain, J., Nolan, J., et al. (1988) *Mol. Cell. Biol.* 8, 1821–1825.
Krumlauf, R., Hammer, R.E., Tilghman, S.M., and Brinster, R.L. (1985) *Mol. Cell. Biol.* 5, 1639–1648.
Kuehn, M.R., Bradley, A., Robertson, E.J., and Evans, M.J. (1987) *Nature* 326, 295–298.
Lacy, E., Roberts, S., Evans, E.P., Burtenshaw, M.D., and Costantini, F. (1983) *Cell* 34, 343–358.
Leder, A., Pattengale, P.K., Kuo, A., Stewart, T.A., and Leder, P. (1986) *Cell* 45, 485–495.
LeMeur, M., Gerlinger, P., Benoist, C., and Mathis, D. (1985) *Nature* 316, 38–42.
Mahon, K.A., Chepelinsky, A.B., Khillan, J.S., et al. (1987) *Science* 235, 1622–1628.
Markert, C.L. and Petters, R.M. (1977) *J. Exp. Zool.* 201, 295–302.
Martin, G.R. (1981) *Proc. Natl. Acad. Sci. USA* 78, 7634–7638.
Mason, A.J., Pitts, S.L., Nikolics, K., et al. (1986) *Science* 234, 1372–1378.
McGrath, J., and Solter, D. (1983) *Cell* 37, 179–183.
Messing, A., Chen, H.Y., Palmiter, R.D., and Brinster, R.L. (1985) *Nature* 316, 461–463.
Mintz, B. (1962) *Amer. Zool.* 2, 541–542 (Abstract).
Mintz, B., and Cronmiller, C. (1981). *Somat. Cell Genet.* 7, 489–515.
Mintz, B., and Illmensee, K. (1975) *Proc. Natl. Acad. Sci. USA* 72, 3585–3589.
Nussenzweig, M.C., Shaw, A.C., Sinn, E., et al. (1987) *Science* 236, 816–819.
O'Brien, R.L., Brinster, R.L., and Storb, U. (1987) *Nature* 326, 405–409.
Ornitz, D.M., Palmiter, R.D., and Messing, A. (1985) *Cold Spring Harbor Symp. Quant. Biol.* 50, 399–409.
Overbeek, P.A., Lai, S.-P., Van Quill, K.R., and Westphal, H. (1986) *Science* 231, 1574–1577.
Palmiter, R.D., Behringer, R.R., Quaife, C.J., et al. (1987) *Cell* 50, 435–443.
Palmiter, R.D., Chen, H.Y., and Brinster, R.L. (1982) *Cell* 29, 701–710.
Palmiter, R.D., Wilkie, T.M., Chen, H.Y., and Brinster, R.L. (1983) *Cell* 36, 869–877.
Papaiaonnou, V.E., McBurney, M.E., Gardner, R.L., and Evans, M.J. (1975) *Nature* 258, 70–73.
Pellicer, A., Wagner, E.F., El Kareh, A., et al. (1980) *Proc. Natl. Acad. Sci. USA* 77, 2098–2101.
Pittius, C.W., Henninghausen, L., Lee, E., et al. (1989) *Proc. Natl. Acad. Sci. USA* 85, 5874–5878.
Rassoulzadegan, M., Leopold, P., Vailly, J., and Cuzin, F. (1986) *Cell* 46, 513–519.
Reik,, W., Collick, A., Norris, M.L., Barton, S.C., and Surani, M.A. (1987) *Nature* 328, 248–251.
Ritchie, K.A., Brinster, R.L., and Storb, U. (1984) *Nature* 312, 517–520.
Rohdewold, H., Weiher, H., Reik, W., Jaenisch, R., and Briendl, M. (1987) *J. Virol.* 61, 336–343.
Rubinstein, W.S., and Gordon, J.W. (1987) *Develop. Genet.* 8, 233–247.
Sapienza, C., Peterson, A.C., Rossant, J., and Balling, R. (1987) *Nature* 328, 251–254.

Scangos, G.A., and Ruddle. F.H. (1981) *Gene* 14, 1–10.
Shani, M. (1986) *Mol. Cell. Biol.* 6, 2624–2631.
Small, J.A., Blair, D.G., Showalter, S.D., and Scangos, G.A. (1985) *Mol. Cell. Biol.* 5, 642–648.
Small, J.A., Scangos, G.A., Cork, L., Jay, G., and Khoury, G. (1986) *Cell* 46, 13–18.
Stacey, A., Bateman, J., Choi, T. et al. (1988) *Nature* 332, 131–136.
Stewart, T.A., Pattengale, P.K., and Leder, P. (1984) *Cell* 38, 627–637.
Storb, U. (1987) *Ann. Rev. Immunol.* 5, 151–174.
Storb, U., Pinkert, C., Arp, B., et al. (1986) *J. Exp. Med.*
164, 627–641.
Stout, J.T., Chen, H.Y., Brennand, J., Caskey, C.T., and Brinster R.L. (1985) *Nature* 317, 250–252.
Surani, M.A.H., and Barton, S.C. (1983) *Science* 222, 1034–1036.
Swain, J.L., Stewart, T.A., and Leder, P. (1987) *Cell* 50, 719–727.
Swift, G.H., Hammer, R.E., MacDonald, R.J., and Brinster, R.L. (1984) *Cell* 38, 639–646.
Tarkowski, A.K. (1961) *Nature* (Lond.) 190, 857–860.
Townes, T.M., Lingrel, J.B., Chen, H.Y., Brinster, R.L., and Palmiter, R.D. (1985) *EMBO J.* 4, 1715–1723.
Vijaya, S., Steffen, D.L., and Robinson, H.L. (1986) *J. Virol.* 60, 683–692.
Widera, G., Burkly, L.C., Pinkert, C.A., et al. (1987) *Cell* 51, 175–187.
Woychik, R.P., Stewart, T.A., Davis, L.G., D'Eustachio, P., and Leder, P. (1983) *Nature* 318, 36–40.
Yamamura, K., Kikutani, H., Folsom, V., et al. (1985) *Nature* 316, 67–69.

CHAPTER 2

Determination of Retroviral Mutation Rates Using Spleen Necrosis Virus-Based Vectors and Helper Cells

Joseph P. Dougherty
Howard M. Temin

It is known that retroviruses have a high frequency of mutation (Chen et al. 1981). For example, it has been estimated that the frequency of nucleotide substitution per site per year for replication of HIV-1 is 1×10^{-3} (Yokoyama and Gojobori 1987). In cell-free systems, the nucleotide mutation frequency per site, using purified reverse transcriptase from avian myeloblastosis virus (AMV), was found to be 1×10^{-3} (Gopinathan et al. 1979). More recent studies have estimated the mutation frequency per site for purified AMV reverse transcriptase to be 1.3×10^{-5} per nucleotide (Takeuchi et al. 1988). Given the discrepancies in cell-free systems and the potential for in vitro artifacts, it is advantageous to measure mutation rates during viral replication. Moreover, during retroviral replication there is an RNA transcription step as well as a reverse transcription step. Reverse transcriptase-dependent mutation rates, as measured in cell-free systems, do not take into account errors occurring during RNA transcription, RNA-directed synthesis of DNA, or possible effects of other virion proteins.

22   Determination of Retroviral Mutation Rates

Until recently, retroviral mutation rates, that is, the frequency of mutation per cycle of viral replication, had not been determined for any replicating retrovirus. Using a vector based on spleen necrosis virus (SNV), which contains a gene that is suppressed because of a splicing defect, and the protocol described at the beginning of the Results section, we showed that the rate of mutation that led to expression of the suppressed gene was 5×10^{-3} per replication cycle in D17 cells (a dog osteosarcoma cell line; Dougherty and Temin 1986). Even though the exact nature of the mutations that resulted in expression of the suppressed gene was not characterized, this initial experiment indicated that retroviral mutation rates are high enough to be measured and that determination of retroviral mutation rates for defined mutations is feasible.

In this chapter we describe the determination of a base-substitution mutation rate at a defined locus in an SNV-based vector in D17 cells (Dougherty and Temin 1988). The mutation rate obtained was 2×10^{-5} per base pair per replication cycle. Moreover, our results indicated that the insertion-mutation rate in an SNV-based vector was 1×10^{-7} per base pair per replication cycle.

A mutation rate has also been determined for Rous sarcoma virus (RSV) (Leider et al. 1988). RSV viral RNA, derived from a single clone, was isolated after one and one-half cycles of replication and examined for mutations by denaturing gradient-gel electrophoresis. The rate of mutation for RSV was calculated to be 1.4×10^{-4} mutations per nucleotide per replication cycle, although the exact nature of the mutations was not reported.

2.1   MATERIALS AND METHODS

2.1.1   Nomenclature

The mutation rate means the frequency of mutations per single cycle of viral replication. *hygro* and *neo* refer to genes, while hygro^r and neo^r refer to resistant phenotypes. Plasmids have a small p before their names, while viruses derived from these plasmids do not.

2.1.2   Plasmid Constructions

The construction of pJD216NeoHy has been described elsewhere (Dougherty and Temin 1986; Dougherty and Temin 1987). pJD216Neo(Am)Hy was derived from pJD216NeoHy by substitution of a *Pvu*II fragment that contained the amber codon for the wild-type *Pvu*II *neo* fragment (Folger et al. 1985).

2.1.3   Cells

The D17 cell line is an osteosarcoma-derived line of canine cells, which is permissive for infection by SNV. The .2G cell lines are lines of helper cells derived from D17 cells. D17 and helper cells were grown as previously described (Watanabe and Temin 1983). Selection for hygromycin-resistant cells was performed in the presence of of hygromycin at a concentration of 50–100 µg/ml. Selection for G418-resistant cells was performed in the presence of G418 at a concentration of 400 µg/ml. Selection for methotrexate-resistant cells was performed in the presence of methotrexate (90 ng/ml).

2.1.4 Transfections and Viral Infections

Transfections were performed by the calcium phosphate method (Graham and van der Eb 1973; Wigler et al. 1979). Infections were performed as described previously by Dougherty and Temin (1986). Virus collected from helper cells was clarified by centrifugation which was followed by immediate use or by storage at −70°C. Virus titers were determined by infecting 2×10^5 D17 cells in 60-mm (diameter) petri dishes with 10-fold serial dilutions of virus from each clone of helper cells in the presence of Polybrene at a concentration of 100 µg/ml in 0.4 ml of medium. Infection was followed by selection for resistance to hygromycin or G418. Titers are given in hygro^r or neo^r transforming units (TU) per 0.2 ml of virus stock. TU are equivalent to the number of colonies of phenotypically transformed cells formed after infection and selection, multiplied by the dilution.

2.1.5 Amplification of DNA by Polymerase Chain Reaction and Sequencing of DNA

Amplification of DNA by the polymerase chain reaction was performed with 1.5 µg of genomic DNA in 75-µl aliquots that contained 67 mM Tris hydrochloride (pH 8.8), 6.7 mM $MgCl_2$, 16.6 mM ammonium sulfate, 10 mM β-mercaptoethanol, 6.7 mM EDTA, 0.4 mM deoxynucleoside triphosphates, 4.5 µg of each primer, 4.5 U of *Taq* polymerase, and 10% dimethyl sulfoxide. The first step was to heat the samples for the polymerase chain reaction (PCR) to 90°C for 2 min, followed by 30 cycles of incubation at 70°C for 5 min, followed by incubation at 90°C for 1 min. The samples were then electrophoresed in an 8% acrylamide gel and stained (Maniatis et al. 1982). After electrophoresis, amplified DNA was electroeluted from the acrylamide gel, and then it was end-labeled by T4 polynucleotide kinase and [gamma-^{32}P]ATP. The DNA was then digested with *Bgl*II, subjected to electrophoresis in an 8% acrylamide gel, and the larger end-labeled fragment was electroeluted from the gel. The purified end-labeled fragment was then sequenced by the method of Maxam and Gilbert (1980).

2.2 RESULTS

2.2.1 Single Cycle of Retrovirus Replication

The genomes of retroviruses can be functionally divided into *cis* and *trans* sequences (Wei et al. 1981; Weiss et al. 1982). *trans* sequences encode viral proteins while *cis* sequences must be present in the viral genome for efficient replication of the virus. Retroviral vectors contain the *cis*-acting sequences required for viral replication, while the *trans* sequences encoding the viral proteins are replaced with other genes for their introduction into the cells of choice (Coffin et al. 1980; Shimotohno and Temin 1981; Wei et al. 1981; Cepko et al. 1984). Retrovirus helper cells provide *trans*-acting functions required for replication of the virus vector (Watanabe and Temin 1983) without the production of replication-competent helper virus. Superinfection of such helper cells is blocked by a factor of over 100 as a result of virus interference, and vector virus cannot spread in the target cells because it is defective and there is no helper virus to supply viral proteins. Therefore, going from a vector provirus in a helper cell to a vector provirus in a target cell represents a single cycle of replication. Such a single cycle involves one round of RNA transcription and one round of reverse transcription (Figure 2–1A). Growth of a clone of cells then involves multiple rounds of cell replication.

To determine rates of retroviral mutation, helper cell clones are established by infection that contain a single provirus with two marker genes, one that is not expressed because of a defect and another that is expressed normally. Virus is harvested from helper cell clones and is used to infect target cells. Selection is then applied for each marker. The titer of the gene that was defective in the helper cell gives the frequency of mutation because a mutation has to occur in order to obtain revertants, and the titer of a nondefective gene gives the overall titer of the virus. The ratio of the titers represents the mutation rate. It should be noted that cellular replication of proviral DNA is relatively error-free, so that the mutations that do arise occur during the RNA transcription step or the reverse transcription step of viral replication.

Procedures in which exogenous genes are introduced into humans or animals via infection with retrovirus vectors involve essentially a single cycle of viral replication. In such procedures virus particles are harvested from helper cells and are then used to infect human or animal cells in the absence of replication-competent helper virus. The study described below and other studies using the procedure described here should shed light on the frequency with which mutations arise in vector proviruses introduced into humans and animals and should help us to predict the predominant types of mutation that can be expected. It should be noted that we find transfection procedures to be much more mutagenic than infection.

2.2.2 JD216NeoHy Expresses Both the *neo* and *hygro* Genes

For initial studies of the rates of retroviral mutation, SNV-based splicing vectors, pJD216NeoHy and pJD216Neo(Am)Hy (Figure 2–1B), were used. A splicing vector is a vector in which two genes can be expressed from a single, long terminal repeat promoter (LTR), one gene from unspliced viral

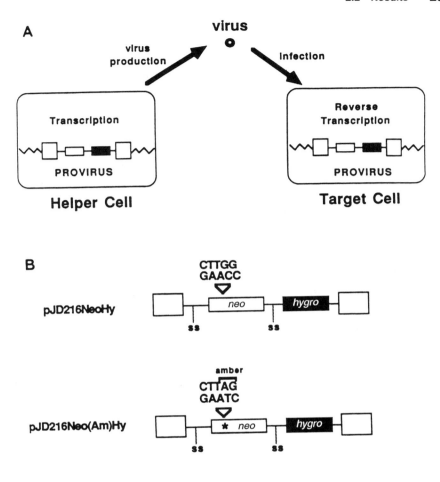

FIGURE 2-1 Single cycle of retroviral replication and vectors. (A) Single round of retroviral replication. The open rectangles represent LTRs, the horizontal lines represent viral sequences, the open and filled boxes represent inserted genes, and the jagged lines represent chromosomal sequences. (B) Vectors used. ss, Splice site. The solid boxes represent coding sequences of the *hygro* gene and the open boxes represent coding sequences of the *neo* gene. The asterisk represents an amber codon introduced into the *neo* gene (Folger et al. 1985). The amber codon creates a *Dde*I restriction site (CTTAG). The inverted triangle and the sequence above the *neo* gene of pJD216NeoHy represents the wild-type *neo* sequence in which the point mutation was introduced to create an amber codon and a *Dde*I cleavage site. The inverted triangle over the *neo* sequence of pJD216Neo(Am)Hy shows the change that was made.

RNA and the other from spliced viral RNA. JD216NeoHy contains both the neomycin gene (*neo*), which confers resistance to G418 (Jorgensen et al. 1979), and the hygromycin gene (*hygro*), which confers resistance to hygromycin B (Gritz and Davies 1983). To provide viral proteins without production of helper virus, a D17-based SNV helper cell line named .2G was used. As a positive control, ten .2G helper cell lines, each containing a JD216NeoHy provirus, were established by infection. Virus was harvested from each cell line, fresh D17 cells were infected, selection for colonies of G418-resistant (neo^r) or hygromycin-resistant (hygro^r) cells was applied, and *neo* and *hygro* titers, expressed as transforming units (TU; Table 2–1), were obtained. JD216NeoHy produced equal numbers of clones of neo^r and hygro^r cells.

2.2.3 Rate of Base-Pair Substitutions

To determine the rate of base-pair substitutions, we constructed pJD216Neo(Am)Hy, which differs from pJD216NeoHy by a single base pair, a change that results in the introduction of an amber codon into the 5′ coding region of *neo* (Figure 2–1B) (Folger et al. 1985). Forty-three .2G helper cell lines, each containing a JD216Neo(Am)Hy provirus, were es-

TABLE 2–1 *neo* and *hygro* Titers Produced by JD16NeoHy Proviruses in .2G or C321 Helper Cells[a]

	Titer (TU) of [b]	
Clone	neo	hygro
.2G clones		
1	110×10^2	130×10^2
2	51×10^2	47×10^2
3	34×10^2	39×10^2
4	31×10^2	27×10^2
5	17×10^2	18×10^2
6	47×10^1	83×10^1
7	75×10^1	61×10^1
8	65×10^1	56×10^1
9	51×10^1	55×10^1
10	36×10^1	43×10^1

[a]To establish .2G helper cells with a single JD216NeoHy provirus, helper cells were transfected with pJD216NeoHy, selected for hygromycin resistance, and virus was harvested from hygro^r helper cells and used to infect .2G cells at a low multiplicity of infection. Infected .2G cells were selected for hygromycin resistance, and individual cell clones were picked and grown. Virus titers for each clone of helper cells that harbored a JD216NeoHy provirus were determined as described in the Materials and Methods section after freezing and thawing at least once before use.
[b]For .2G clones, the overall titers of *neo* and *hygro* were 27×10^3 and 29×10^3 TU.

tablished. Again, virus was harvested from each cell line, fresh D17 cells were infected, with subsequent selection for colonies of neo^r or hygro^r cells, and *neo* and *hygro* titers were obtained and expressed as TU. The neo^r/hygro^r ratio obtained with the 10 clones that gave the highest overall titers was 2×10^{-5} (Table 2-2). The ratio of neo^r/hygro^r colonies obtained for all 43 clones was $129/5916 \times 10^3$ or 2.2×10^{-5}. The number of neo^r colonies represents the number of vector mutants that express the *neo* gene, while that of the hygro^r colonies represents the overall vector titer. The mutation rate is the frequency of mutation per replication cycle. Since only one round of viral replication occurred, 2×10^{-5} is the mutation rate per replication cycle.

To prevent any possible spread of the virus during the growth of the cell clones, the same experiment was performed with 12 .2G helper cell lines, each harboring a JD216Neo(Am)Hy provirus, but in this case the cell clones were grown in the presence of neutralizing antibodies against SNV proteins (Chen et al. 1981). The antibodies were removed from the cell cultures 24 hours before harvesting of the virus. The ratio of neo^r/hygro^r colonies obtained was $28/1146 \times 10^3$ or 2.4×10^{-5} and was the same as that obtained when the .2G clones were grown in the absence of neutralizing antibodies.

TABLE 2-2 Rate of Base-Pair Substitutions for Virus from .2G cells[a]

	Titer (TU) of [b]	
Clone	neo	hygro(10^4)
1	9	68
2	17	67
3	10	44
4	4	43
5	1	23
6	4	22
7	7	19
8	6	15
9	2	15
10	3	14

[a] .2G cell clones containing a JD216Neo(Am)Hy provirus were established in the same manner as the cell lines described in Table 2-1, except that the original transfections were performed with pJD216Neo(Am)Hy. Virus titers were determined as described in Table 2-1, footnote a, except that the virus stocks from the cell clones were not frozen and thawed but were only clarified by centrifugation. The titers in this experiment are higher than those in the experiment recorded in Table 2-1 because this virus was not frozen and thawed before assay.

[b] The overall titers of *neo* and *hygro* were 63 and 330×10^4, respectively, and the mutation rate was determined as follows: *neo* TU/*hygro* TU = $63/330 \times 10^4 = 2 \times 10^{-5}$.

28 Determination of Retroviral Mutation Rates

2.2.4 The *Dde*I Restriction Site was Lost in Most of the Cases Tested

Introduction of the amber codon into pJD216Neo(Am)Hy also resulted in creation of a *Dde*I restriction site with sequence CTNAG (Figure 2–1B). A base change in position 2 or 3 of this amber codon results in loss of the *Dde*I restriction site (Figure 2–1B). If the *neo* revertants obtained in the experiment described in the legend to Table 2–2 were the result of base-pair substitutions at position 2 or 3 of this amber codon, then the *neo* revertants should have contained proviruses that had lost the *Dde*I restriction site. To test this hypothesis, we grew 17 neo^r D17 cell clones, obtained as described in Table 2–2, footnote *a*. We isolated genomic DNA from these clones, digested the genomic DNA with *Dde*I, electrophoresed the DNA in a 1.2% agarose gel, blotted the DNA onto nitrocellulose, hybridized it with a *neo*-specific probe, and then autoradiographed the filters (Figure 2–2). In the three cases shown, the *Dde*I site was lost. In all, we found that 15 of 17 clones had lost the *Dde*I site.

2.2.5 Analysis of the Genomic Sequence of *neo* Revertants by PCR Amplification

The exact nature of the changes that give the neo^r phenotype was analyzed by direct sequencing of PCR-amplified genomic DNA from seven revertants (McMahon et al. 1987; Saiki et al. 1985; Wong et al. 1987). Two 20-nucleotide oligomers were synthesized and used as primers to amplify a 265-bp region that spans the amber codon. Thirty sequential cycles of primer annealing, DNA polymerase extension (with DNA polymerase from *Thermus aquaticus*), and denaturation were performed. Amplified DNA was purified in an 8% polyacrylamide gel (Figure 2–3A) and sequenced by the

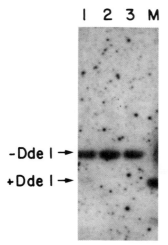

FIGURE 2–2 Analysis of JD216Neo(Am)Hy proviruses in neo^r cells. Three individual clones of neo^r D17 cells, described in Table 2–2, were grown. Genomic DNA was isolated, and 10 μg of each sample was digested with *Dde*I, with subsequent electrophoresis in a 1.2% agarose gel and blotting of the gel to nitrocellulose. The blot was then hybridized with a ^{32}P-labeled *neo*-specific probe, with subsequent autoradiography (Southern and Berg 1982). Genomic DNAs from the clones were run in lanes 1 to 3. Lane M shows a control that consisted of plasmid DNA in which the *Dde*I site in *neo* was present. DNA from such a clone and from one without the *Dde*I site were included in all gels.

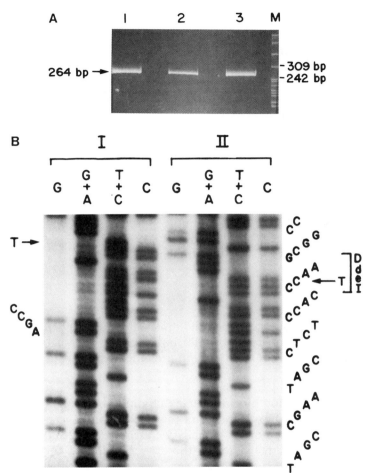

FIGURE 2-3 Genomic sequencing of *neo* revertants. (A) Electrophoretic pattern on acrylamide gel of amplified DNA obtained with three separate clones of neo^r D17 cells infected with virus from .2G helper cell clones. M refers to molecular weight markers (MspI-digested pBR322). (B) Autoradiograms obtained from DNA sequencing of two clones, I and II. The bracket labeled *Dde*I indicates the position of the *Dde*I site in the original JD216Neo(Am)Hy vector. In I, there was a 4-bp insertion, CCGA, indicated at the left.

protocol of Maxam and Gilbert (Figure 2-3B) (Maxam and Gilbert 1980). Figure 2-3A shows amplified genomic DNA from three neo^r clones of D17 cells originally infected with JD216Neo(Am)Hy virus harvested from .2G helper cell clones. Typically, we obtained 1-5 μg of amplified DNA from 1.5 μg of genomic DNA, indicative of approximately 10^8-fold amplification.

We performed this amplification on DNAs from seven neo^r clones of D17 cells that were originally infected with JD216Neo(Am)Hy virus from

.2G helper cell clones (Figure 2–2). In all cases in which the *Dde*I site was lost, base pair 4 of the *Dde*I site (corresponding to base 2 of the amber codon) was converted from A–T to G–C (Figure 2–3B, II).

Two clones of neo^r D17 cells originally infected with JD216Neo(Am)Hy virus from .2G helper cell clones did not lose their *Dde*I sites. DNAs from these two clones were used as templates for amplification and sequencing. The sequence obtained for the DNA from one clone is shown in Figure 2–3B, I. As expected, the *Dde*I site was maintained. There was no change from T to C. However, there was an insertion of four bases (CCGA) seven bases downstream from the amber codon. The insertion is a duplication of the adjacent CCGA. Thomas and Capecchi have shown that +1, +4, +7, etc., frameshift mutations just downstream from this amber codon can restore *neo* function (see Discussion) (Thomas and Capecchi 1986). Sequencing of the other *neo* revertant that retained the *Dde*I site confirmed that the *Dde*I site was maintained without change, but we were not able to find any compensating mutations within 40 bases that surrounded the amber codon.

2.3 DISCUSSION

In this chapter we have described the direct measurement of rates of mutation during retroviral replication for both a single base-pair substitution at a defined locus and for insertions in a defined region.

2.3.1 Rate of Mutation by Base-Pair Substitution

The rate of substitution of base pairs that we obtained was 2×10^{-5} per base pair per replication cycle. Initially we assumed that mutations would occur at any of the three bases of the amber codon (TAG) unless they produced another stop codon. However, genomic sequencing of PCR-amplified DNA showed that substitutions were limited to a single change at the second base of the amber codon, namely a transition from A–T to G–C, which restores the original wild-type sequence. The wild-type codon is TGG, which is the tryptophan codon. All other base substitutions at the amber codon would result in changes to codons for different amino acids. It seems possible that we were limited to this change because when the 5' end of the wild-type *neo* gene product is present, there may be an absolute requirement for tryptophan at the locus we were studying. Thus, the value we obtained for the rate of base-pair substitution is a minimal one, because there may have been selection against other mutations. Experiments with other mutations (in progress) will test this hypothesis.

The mutation rate we observe is similar to that calculated for two other RNA viruses, poliovirus type 1 (less than 10^{-5} for the VP1 gene) and influenza virus (8×10^{-5} for the NS segment), and higher than that for a third, Sindbis virus (2×10^{-7}) (Durbin and Stollar 1986; Parvin et al. 1986; Southern and Berg 1982). These other RNA viruses utilize different types of polymerase than those utilized by retroviruses.

Mutations in our system could have occurred either at the RNA transcription step or at the reverse transcription step. We are not sure at which step the mutations occurred. It is possible to separate these two steps physically, and such experiments may allow determination of the mutation rate at each step.

2.3.2 Insertion Mutation Rate

Two *neo* revertant clones of D17 cells originally infected with JD216Neo(Am)Hy virus from .2G helper cells did not lose the *Dde*I site. Genomic sequencing with PCR-amplified DNA revealed that one clone retained the *Dde*I site and the amber codon (Figure 2–3B, I), but there was a 4-bp insertion 7 bp downstream from the amber codon. Thomas and Capecchi (1986) have found that +1 frameshift mutations within 11 bp downstream from the amber codon can compensate for the stop codon. Translation can be initiated at an AUG in the −1 reading frame upstream from the amber codon, allowing readthrough of the amber codon, which is in the 0 reading frame. The +1 frame shift then allows the ribosome to regain the proper phase. The target size for selectable insertions encompasses about 11 bp downstream from the amber codon (Thomas and Capecchi 1986). Since the only insertions that score as revertants are +1 frameshift mutations, our effective target is the equivalent of a single codon. The insertion mutation rate per codon is 1/17 of the base-substitution mutation rate. Therefore, the insertion mutation rate at this site is about 10^{-7} per base pair per replication cycle.

The other clone that retained the *Dde*I site also retained the amber codon. We discovered no compensating mutations (examining 20 bases 5' or 20 bases 3' to the amber codon) to account for its neo^r phenotype. It is possible that there is a compensating mutation outside the area that we sequenced or that there is a cellular mechanism by which cells of this clone can suppress the amber codon. (We have not been able to recover neo^r-transforming virus by superinfection of cells that carry this provirus, even though the superinfecting virus replicated well [unpublished data]).

If the retroviral mutation rates we measured hold true for the entire genome, we can estimate roughly the number of replication cycles undergone by different retroviral isolates since divergence from a cell or other viral isolate. Reticuloendotheliosis virus, strain T (REV-T), contains the oncogene v-*rel*. Comparing v-*rel* from REV-T coordinates 4290 to 4675 with its proto-oncogene, c-*rel*, we found that v-*rel* and c-*rel* differ by 6 bp and an insertion of 6 bp in v-*rel* (Wilhelmsen et al. 1984). The changes in this area of v-*rel* are not important for the transforming ability of v-*rel* (Sylla and Temin 1986), and so we consider them to be neutral. Therefore, from the time the c-*rel* sequence was transduced into REV-T until the REV-T provirus was cloned (Chen et al. 1981), REV-T underwent approximately 800 cyles of replication (number of replication cycles = 6-bp substitutions/2 × 10^{-5} substitutions per base pair per replication cycle × 385 bp). A similar calculation can be made for other retroviruses, if we assume that the mutation rates we measured also apply to them and that the observed differences in nucleotide sequence are primarily neutral. Thus, we can calculate that the observed substitution rate for lentiviruses (visna and human immunodeficiency viruses) of approximately 10^{-3} per nucleotide per year (Braun et al. 1987; Yokoyama and Gojobori 1987) indicates approximately 50 cyles of replication per year since the divergence of different isolates.

REFERENCES

Braun, M.J., Clements, J.E., and Gonda, M.A. (1987) *J. Virol.* 61, 4046–4054.
Cepko, C.L., Roberts, B.E., and Mulligan, R.C. (1984) *Cell* 37, 1053–1062.
Coffin, J.M., Tsichlis, P.N., Barker, C.S., et al., X.X. (1980) *Ann. N.Y. Acad. Sci.* 354, 410–425.
Chen, I.S.Y., Mak, T.W., O'Rear, J.J., and Temin, H.M. (1981) *J. Virol.* 40, 800–811.
Dougherty, J.P., and Temin, H.M. (1986) *Mol. Cell. Biol* 6, 4387–4395.
Dougherty, J.P., and Temin, H.M. (1987) in *Gene Transfer Vectors for Mammalian Cells*, pp. 18–23, Cold Spring Harbor Laboratory, Cold Spring Harbor, NY.
Dougherty, J.P., and Temin, H.M. (1988) *J. Virol.* 62, 2817–2822.
Durbin, R.K., and Stollar, V. (1986) *Virology* 154, 135–143.
Folger, K.R., Thomas, K., and Capecchi, M.R. (1985) *Mol. Cell. Biol.* 5, 59–69.
Gopinathan, K.P., Weymouth, L.A., Kunkel, T.A., and Loeb, L.A. (1979) *Nature* 278, 857–859.
Graham, F.L., and van der Eb, A.J. (1973) *Virology* 52, 456–467.
Gritz, L., and Davies, J. (1983) *Gene* 25, 179–188.
Jorgensen, R.A., Rothstein, S.J., and Reznikoff, W.S. (1979) *Mol. Gen. Genet.* 177, 65–72.
Leider, J.M., Palese, P., and Smith, F.I. (1988) *J. Virol.* 62, 3084–3091.
Maniatis, T., Fritsch, E.F., and Sambrook, J. (1982) in *Molecular Cloning*, Cold Spring Harbor Laboratory, Cold Spring Harbor, NY.
Maxam, A.M., and Gilbert, W. (1980) *Methods Enzymol.* 65, 499–560.
McMahon, G., Davis, E., and Wogan, G.N. (1987) *Proc. Natl. Acad. Sci. USA* 84, 4974–4978.
Parvin, J.D., Moscona, A., Pan, W.T., Leider, J.M., and Palese, P. (1986) *J. Virol.* 59, 377–383.
Saiki, R.K., Scharf, S.J., Faloona, F., et al. (1985) *Science* 230, 1350–1354.
Shimotohno, K., and Temin. H.M. (1981) *Cell* 26, 67–77.
Southern, P.J., and Berg, P. (1982) *J. Mol. Appl. Genet.* 1, 327–341.
Sylla, B.S., and Temin, H.M. (1986) *Mol. Cell. Biol.* 6, 4709–4716
Takeuchi, Y., Haumo, T., and Hoshino, H. (1988) *J. Virol.* 62, 3900–3902.
Thomas, K.R., and Capecchi, M.R. (1986) *Nature* (London) 324, 34–38.
Watanabe, S., and Temin, H.M. (1983) *Mol. Cell. Biol.* 3, 2241–2249.
Wei, C., Gibson, M., Spear, P.G., and Scolnick, E.M. (1981). *J. Virol.* 39, 935–944.

Weiss, R., Teich, N., Varmus, H., and Coffin, J., eds. (1982) *Molecular Biology of Tumor Viruses. RNA Tumor Viruses, 1st ed. 1/Supplements and Appendices.* Cold Spring Harbor Press, Cold Spring Harbor, NY.
Wigler, M., Sweet, R., Sim, G.K., et al. (1979) *Cell* 16, 777–785.
Wilhelmsen, K.C., Eggleton, K., and Temin., H.M. (1984) *J. Virol.* 52, 172–182.
Wong, C., Dowling, C.E., Saiki, R.K., et al. (1987) *Nature* 330, 384–386.
Yokoyama, S., and Gojobori, T.J. (1987) *J. Mol. Evol.* 24, 330–336.

CHAPTER 3

Steroid Hormone Receptors as Transactivators of Gene Expression

Bert W. O'Malley

Since the early 1970s, a great deal of evidence has accumulated in favor of the hypothesis that steroid hormones act at the level of nuclear DNA to regulate gene expression (Jensen et al. 1968; Gorski et al. 1968; O'Malley and Means 1974; O'Malley et al. 1979). The earliest studies were qualitative and involved experiments which showed that steroid hormones: (1) caused accumulation of new species of hybridizable RNAs that did not exist prior to stimulation; (2) caused stimulation of synthesis of new and specific proteins de novo; (3) caused a corresponding increase in the cellular levels of specific mRNAs; and (4) stimulated the rate of transcription of certain nuclear genes (O'Malley et al. 1969). At that time, the primary pathway for the action of steroid hormones was defined as follows: steroid → (steroid-receptor) → (steroid-receptor-DNA) → mRNA → protein → functional response (O'Malley et al. 1979). Figure 3–1 contains a diagram of the pathway elucidated for the action of progesterone in the chick oviduct. The steroid enters cells by passive diffusion and allosterically activates receptors in either the cytoplasm or the nucleus. The activated receptor binds, usually at the 5'-flanking region of genes, such as the gene for ovalbumin, and stimulates transcription and protein synthesis.

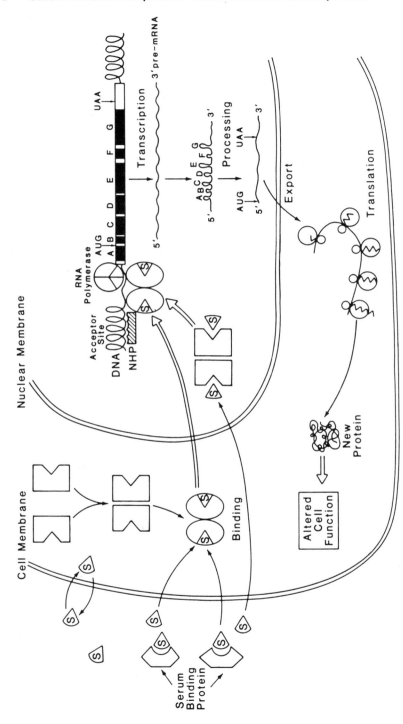

FIGURE 3-1 Pathway for regulation by a steroid hormone of a target gene (Ovalbumin). S, steroid hormone.

The first receptors were subsequently purified to near homogeneity and characterized as to size, charge, etc. Antibodies were raised against them, structural domains were postulated after proteolytic analyses, and assays of receptors for sex steroids became commonplace in the diagnosis and assignment of therapy for breast cancer. Investigators isolated specific target genes for steroid hormones, defined their structures and proved that *cis*-acting regulatory sequences were located near such genes, usually in the 5'-flanking sequences. When such sequences, termed steroid response elements (SREs), are occupied by receptors, these genes come under the control of the relevant hormones.

Molecular biologists became interested in these receptors for steroid hormones as they came to realize that these receptors were the most intensively studied and highly purified transactivation factors for control of eucaryotic transcription, and that they were the specific activators of an emerging and fascinating genetic *cis*-element, the enhancer. In the past five years, a great deal more has been learned about the structure-function relationships of steroid receptors and the mechanisms by which they interact with DNA. Biochemical studies in the late 1970s suggested that steroid receptors, thyroid receptors, and receptors for vitamins, such as vitamin D, belonged to a family of gene-regulatory proteins. Furthermore, it was suggested that these proteins were organized into domains that participated in the functions of (1) specific and high-affinity binding of ligands, (2) specific binding to DNA, and (3) "transcriptional modulation." Molecular cloning of the receptors confirmed the "superfamily" concept. Molecular cloning and sequence analyses not only substantiated the existence of these domains but showed that they could be rearranged as independent cassettes within their own molecules or as hybrid molecules with other regulatory peptides (Evans 1988).

Perhaps the most intensively studied domain has been that responsible for binding to DNA. This domain has been shown to be a cysteine-rich region, which is capable of binding zinc in such a manner that two peptide projections, referred to as "zinc fingers," are formed. These zinc fingers promote the interactions of receptors with target enhancers and clearly mark each protein as a member of this evolutionarily conserved family. Each zinc finger is important for high-affinity binding to target DNA sequences, although certain experiments indicate that the first (N-terminal) finger plays a greater role in the recognition of specific sequences. A surprising observation has been that certain oncogenes, such as v-*erb* A, are members of this family of receptor genes. The avian erythroblastosis virus appears to have captured the cellular gene that encodes the thyroid receptor and this retrovirus uses this mutated molecule for its own oncogenic purposes.

One of the most revealing observations to emanate from the molecular analyses of the receptor superfamily of genes, using recombinant DNA methods, is the existence of many new members of this receptor family, whose functions are as yet undetermined (Evans 1988). As evidenced by cloning of their cellular cDNAs, more receptors with unknown function appear to exist than the total known number of ligand-activated members of the family. This result means that many new hormones, many of which may be nonsteroidal or nutritional in nature, await discovery.

The sequences of the target enhancer sequences, referred to as SREs and regulated by steroid hormones, have been determined for most members of this family to date (Evans 1988; Payvar et al. 1983; Renkawitz et al. 1984; Jantzen et al. 1987). A summary of such regulatory sequences (termed steroid response elements or SRE) for a series of genes responsive to glucocorticoid and progesterone (termed GRE/PRE), estrogen (ERE), and throid hormone (TRE) is given in Figure 3-2. In general, there are 15 bp per sequence, composed of two half-sites of 6 bp arranged on a dyad axis of symmetry (inverted repeats) around a few central base pairs of random composition. The SREs for various receptors share similarities in sequence and, in fact, the identical sequence allows activation by glucocorticoid, progesterone, and androgen receptors. One copy of such an SRE is usually sufficient to bring a promoter under moderate hormonal control and two copies often provide a synergistic response to the cognate hormone.

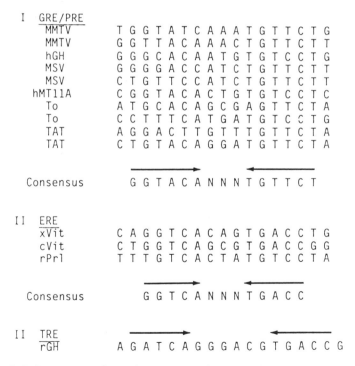

FIGURE 3-2 Sequences of steroid response elements. GRE/PRE, glucocorticoid/progesterone response element; ERE, estrogen response element; TRE, thyroid hormone response element.

The precise mechanism of interaction of receptors with their target SREs has come under close scrutiny of late. After synthesis in the cytoplasm, many steroid receptors form transient complexes with heat-shock proteins, such as hsp90. Receptors for glucocorticoid, progesterone, and estrogen are prepared from extracts of cells in vitro in the form of such aggregates. In cells, this interaction may promote proper folding and stability of the molecule; in complexes with hsp90, receptors cannot bind to DNA. When a receptor is complexed with a heat-shock protein, binding of hormone drives the dissociation of the complex in vitro. The exact meaning of this putative association of receptors with heat-shock proteins is unclear at present but is the subject of intensive study.

Recent evidence shows that receptors for glucocorticoid, progesterone, and estrogen bind to their SREs as dimers, one molecule to each half-site. This interaction appears to be cooperative, at least for receptors of glucocorticoid and progesterone and is shown schematically in Figure 3-3. In this manner, receptor dimers bind with greater affinity and stability to their SREs than do monomers (Tsai et al. 1988). Interactions between receptor dimers at separate SREs allow a higher-order cooperative interaction, which stabilizes the two dimers into a tetrameric structure with a 100-fold greater affinity for its SRE sequences than that of a single dimer (Tsai et al. 1989). Such protein-protein interactions may occur among homologous receptor complexes, heterologous receptor complexes, or receptor-promoter/TATA complexes. These interactions are thought to stabilize binding of transcription factors to their specific sites on DNA and to promote a high degree of efficiency of initiation of transcription at nearby genes (see Figure 3-3). Conversely, incompatible complexes or proteins that disrupt such interaction should promote negative control by decreasing the chance that RNA polymerase will initiate transcription of a gene. Protein-protein interaction is the basis of the currently popular hypothesis for formation of stable transcription complexes on regulated genes. However, a great deal more needs to be learned about the structure of the chromatin in regions of genetic control. Recent experiments suggest that specific phasing of nucleosomes around select SREs may allow recognition by receptors. Some studies indicate that hormonal stimulation may lead to structural arrangements of nucleosomes that promote transcriptional changes at target genes.

The definitive role of the ligand in activation of the receptor remains something of a mystery. Certainly receptors are inactive in cells in the absence of hormone. Administration of hormone in vivo leads to formation of a bound protein complex at SREs of hormone-regulated genes, as demonstrated by footprint analyses in vivo. Receptors in crude extracts of cells bind DNA poorly until they undergo a temperature- and salt-aided "activation," which is driven by hormone. This activation is accompanied by disaggregation of receptors from heat-shock and/or other proteins. Antihormones promote such disaggregation but do not activate target genes, indicating an additional level of ligand-induced allosteric control. Purified

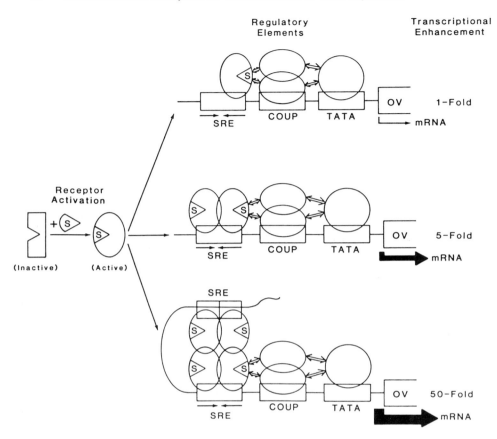

FIGURE 3-3 Cooperative enhancement of transcription proceeds from the stable occupation of SRE by steroid receptors. COUP, chicken ovalbumin upstream response element; SRE, steroid response element; TATA, downstream promoter element; OV, ovalbumin gene.

receptors, by contrast, show a clear preference for their SREs whether or not they are complexed with hormone.

Obviously, the hormone plays a pivotal role in activating the intracellular receptor but, to date, its exact role has been difficult to define from studies in vitro.

Finally, the role of steroid receptors has been defined now for a genetic disease that involves resistance to a hormone. Defects in receptors have been suspected to be the cause of testicular feminization, vitamin D-resistant rickets, and hypercortisolism without Cushing's syndrome. Recent familial studies have proven that vitamin D-resistance can result from a mutation in a single amino acid at the tip of either "zinc finger" of the DNA-binding domain of its receptor (Hughes et al. 1988). These observations are the first to prove that mutations in human steroid receptors, or indeed in any transcription factor, can cause a human genetic disease. Many recent reports show that androgen receptor mutations are a frequent cause of the androgen insensitivity syndrome, commonly called testicular feminization.

Studies of the molecular mechanisms of the actions of steroid hormones have had an immense impact on the field of endocrinology, advancing it as a legitimate discipline, wherein technologies inherent to biochemistry, molecular biology, biophysics, pharmacology, and immunology can be unified toward the study of cellular physiology and regulatory biology. Much excitement has been generated to date but research in this area in the near future holds promise for a greater understanding of fertility, early embryonic development, genetic disease, stress, eating and nutritional disorders, emotional and depressive disorders, cancers, and aging. Our results to date should not only lead to the discovery of a series of new hormones but also to the rational design and synthesis of new agonists and antagonists for therapy of the above mentioned disorders.

REFERENCES

Evans, R.M. (1988) *Science* 240, 889–895.
Gorski, J., Toft, D., Shyamala, G., Smith, D., and Notides, A. (1968) *Rec. Prog. Horm. Res.* 24, 45–80.
Hughes, M.R., Malloy, P.J., Kieback, D.G., et al. (1988) *Science* 242, 1702–1705.
Jantzen, H.-M., Strahle, U., Glass, B., et al. (1987) *Cell* 49, 29–38.
Jensen, E.V., Suzuki, T., Kawashima, T., et al. (1968) *Proc. Natl. Acad. Sci. USA* 59, 632–638.
O'Malley, B.W., and Means, A.R. (1974) *Science* 183, 610–620.
O'Malley, B.W., McGuire, W.L., Kohler, P.O., and Korenman, S.G. (1969) *Rec. Prog. Hormone Res.* 25:105.
O'Malley, B.W., Roop, D.R., Lai, E.C., et al. (1979) *Rec. Prog. Horm. Res.* 35, 1–42.
Payvar, F., DeFranco, D., Fivestone, G.L., et al. (1983) *Cell* 35, 381–392.
Renkawitz, R., Schutz, G., Von der Ahe, D., and Beato, M. (1984) *Cell* 37, 503–510.
Tsai, S.Y., Carlstedt-Duke, J., Weigel, N.L., et al. (1988) *Cell* 55, 361–369.
Tsai, S.Y., Tsai, M.-J., and O'Malley, B.W. (1989) *Cell* 57, 443–448.

PART II

Gene Expression

CHAPTER 4

Targeted Mutagenesis in Embryo-Derived Stem Cells
Kirk Thomas

The generation of transgenic animals depends upon the propensity of eukaryotic cells to form stable recombinants between introduced, exogenous DNA and endogenous, chromosomal DNA. Although the majority of such recombinational events are nonhomologous reactions, many cell types also possess the enzymatic machinery required for homologous recombination. The homology-dependent recombination between chromosomal DNA and exogenous sequences is referred to as gene targeting, and it offers an additional dimension to transgenic technology. Gene targeting permits the transfer of genetic alterations created in vitro to precise sites within the cellular genome. If the host cells are pluripotent, embryo-derived stem (ES) cells, such alterations can then be transferred to the germ line of a living organism. This strategy of creating animals with specific genomic changes has immense potential in medicine and agriculture, and in furthering our understanding of the genetic control of mammalian development.

In this chapter, a discussion is presented of our studies of gene targeting in cultured mouse cells, with emphasis on the ways in which our under-

---

This work was done in the laboratory of Dr. Mario Capecchi with his collaboration and that of Suzanne Mansour and Dusan Kostic. All tissue culture was performed by Laurie Fraser, Susan Tamowski, Carol Lenz, and Marjorie Allen.

standing of homologous recombination in somatic cells has allowed us to target mutations to a number of different loci within the genome of mouse ES cells. This technology should now permit the systematic modification of any murine gene with the eventual creation of mice with predetermined genotypes.

4.1 HOMOLOGOUS RECOMBINATION IN CULTURED MOUSE FIBROBLASTS

Cultured mammalian cells have been shown to possess the machinery required for catalysis of homologous recombination. This machinery is quite active, for example, in promoting genetic exchange between shared regions of homology on co-introduced fragments of exogenous DNA (Folger et al. 1982, Miller and Temin 1983; Robert de Saint Vincent and Wahl 1983; Shapira et al. 1983; Small and Scangos 1983; Subramani and Berg 1983). Nevertheless, integration of these exogenous sequences into the host genome generally occurs through a nonhomologous reaction, such that the pattern of integration of introduced DNA appears to be random (Robins et al. 1981; Kato et al. 1986).

To determine whether there could be recombination through regions of homology between exogenous and endogenous DNA sequences, we designed a model system in mouse L cells that would allow us to detect even rare gene-targeting events (Thomas et al. 1986). We first established a number of cell lines that contained mutant alleles of the bacterial neomycin phosphotransferase gene (neo^r) integrated at different genomic positions. Each cell line was then transfected by microinjection with DNA that contained a different, nonoverlapping, mutant allele of the neo^r gene. Gene-targeting events could, therefore, be identified by the generation of a wild-type neo^r gene which would render the cells resistant to the drug G418. From this system we made the following observations:

1. The targeting vector must be linearized prior to its introduction into the host cells; no targeting was seen using closed-circular DNA.
2. The absolute frequency of targeting in mouse L cells was one targeting event per 1000 cells injected with DNA.
3. The relative frequency of targeting was one homologous event per 100 nonhomologous integration events.
4. The targeting frequency was independent of the copy number of the two reactants. Neither variation of the number of exogenous molecules from 5 to 100 copies per cell, nor variation of the number of target sites from 1 to 5 copies per cell affected the frequency of gene targeting.

4.2 TARGETED MUTAGENESIS OF THE *HPRT* LOCUS IN MOUSE ES CELLS

The unexpectedly high frequency of gene targeting in the L-cell system encouraged us to pursue a similar analysis in ES cells (Thomas and Capecchi 1987). As discussed in other chapters of this volume, it is possible to isolate lines of pluripotent stem cells from mouse blastocysts. After passage and even selection in culture, these cells can be reintroduced into blastocysts where they may contribute to a variety of tissue types, including the germ line, in the resultant chimeric animal (Bradley et al. 1984; Gossler et al. 1986; Robertson et al. 1986).

We chose to begin our studies of gene targeting in ES cells by targeting null mutations into the hypoxanthine phosphoribosyl transferase (*hprt*) locus. The *hprt* gene is located on the X-chromosome, so that in male ES cells only a single copy of the gene needs to be inactivated in order to generate a mutant phenotype. Moreover, cells containing null mutations of the *hprt* locus can be identified by their ability to grow in the presence of the toxic purine analog, 6-thioguanine (6-TG).

The protocol for targeted mutagenesis of the *hprt* gene is illustrated in Figure 4–1A. A plasmid vector containing sequences from the mouse *hprt* gene was modified to contain an insertion mutation in the eighth exon of the gene. This insertion was a copy of the bacterial neo^r gene (engineered to function in ES cells) and was designed to perform two functions. First, it disrupted the protein-coding sequences of the *hprt* gene. Second, it provided a positive selectable marker, resistance to G418, for those cells transformed by the vector DNA. Homologous recombination between this targeting vector and the chromosomal target would transfer the neo^r insertion cassette into the endogenous *hprt* gene. Cells containing such an insertion would be resistant not only to G418 but also to 6-TG. Figure 4–1B illustrates three different vectors that were used to inactivate the *hprt* locus. All of the vectors contain the neo^r gene inserted into the eighth exon, and they all have a common 3' end point. They differ from one another, however, with respect to the amount of *hprt* sequences present on the 5' side of the neo^r gene. The total amounts of homology to the *hprt* sequence in the three vectors are 4.0 kb in pRV4.0, 5.4 kb in pRV 5.4, and 9.1 kb in pRV9.1.

Each of these vectors was linearized by restriction endonuclease digestion and then introduced by electroporation into ES cells. Aliquots of the cells were then placed in three separate growth media: nonselective medium, to assess the number of cells that survived electroporation; G418 medium, to assess the number of vector-transformed cells; and G418 plus 6-TG medium, to assess the number of transformed cells that lacked *hprt* activity. Cells in the final category were then characterized by Southern transfer analysis to determine whether the absence of *hprt* activity was due to the presence of the neo^r gene within the endogenous *hprt* locus. The results of these experiments are shown in Table 4–1. Although the transforming ca-

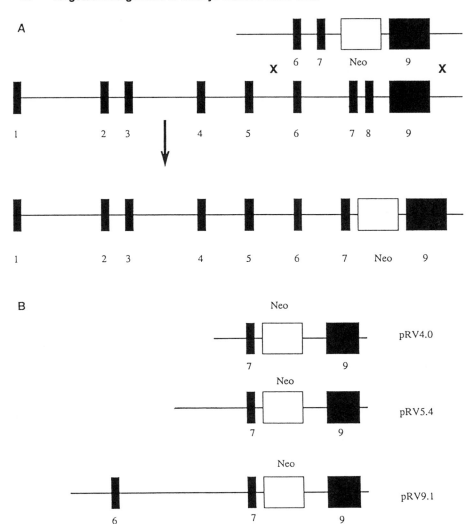

FIGURE 4-1 Gene Targeting at the *hprt* Locus. (A) Targeting reaction. The linearized targeting vector (top line) contains sequences homologous to the *hprt* gene with an insertion of the *neo*r gene in the eighth exon. After homologous pairing with the endogenous *hprt* locus (middle line), recombination between the vector and the target results in the transfer of the *neo*r gene to the endogenous *hprt* gene (bottom line). (B) The three vectors used to target the *neo*r gene to the *hprt* locus are shown in linear form after their removal from bacterial plasmid sequences. Thin lines represent noncoding sequences (introns and flanking DNA). Thick lines represent the *hprt* exons. Open box represents the *neo*r gene. Each exon is numbered.

TABLE 4-1 Targeting Frequency at the *hprt* Locus

Vector	Transformation Frequency[a]	Absolute Targeting Frequency[b]	Relative Targeting Frequency[c]
pRV4.0	10×10^{-4}	5×10^{-8}	5×10^{-5}
pRV5.4	6×10^{-4}	9×10^{-8}	15×10^{-5}
pRV9.1	4×10^{-4}	40×10^{-8}	100×10^{-5}

Conditions for electroporation and selection have been described elsewhere (Thomas and Capecchi 1987). 10^7 ES cells were electroporated with 25 µg of linearized vector in 1 ml of buffer. Eight separate electroporations were performed with each vector. Aliquots of cells were plated under one of three conditions: in nonselective medium; in G418 medium; or in G418 + 6-TG medium. After ten days of growth, colonies were counted.
[a] The number of colonies in G418 medium/number of colonies in nonselective medium.
[b] The number of colonies in G418 + 6-TG medium/number of colonies in nonselective medium.
[c] The number of colonies in G418 + 6-TG medium/number of colonies in G418 medium.

pabilities of each of the three vectors were similar (approximately one transformant per 1,000 electroporated cells), the targeting frequencies were quite different: when the length of *hprt* sequences in the vector increased, so did the targeting frequency.

These experiments demonstrated that mouse ES cells possess enzymatic machinery that enables them to participate in gene targeting. The ratio of homologous recombination to nonhomologous recombination is, however, strongly dependent upon the extent of homology between the vector and its target. A twofold increase in the extent of homology resulted in a 20-fold increase in the frequency of gene targeting.

4.3 TARGETED MUTAGENESIS OF NONSELECTABLE GENES

The *hprt* locus has proven to be an excellent model system for studies of gene targeting (Thomas and Capecchi 1987; Doetschman et al. 1987; Mansour et al. 1988). Its presence as a hemizygous locus in male ES cells and the availability of a selection scheme for mutant *hprt* alleles have permitted the detection of even rare gene-targeting events (1 in less than 2×10^7 electroporated cells [see Table 4-1, line 1]). The luxury of a selection system is, however, not available for most genes, so identification of the products of gene targeting becomes a serious problem. Although use of the *neo*[r] insertion casette as the mutagenic element permits the identification of transfected cells from a pool of nontransfected cells, it is still necessary to identify the homologous recombinant from the larger pool of nonhomologous, random integrants.

The most straightforward solution to this problem is merely to screen recombinants by Southern transfer analysis for the desired cell. If such a cell were present at a frequency of 1 per 1,000 cells (as in the case of targeting the *hprt* locus with the vector pRV9.1), this task would be tedious, but not impossible. If the targeting event were much rarer, however, the effort required to identify it could prove prohibitive, although polymerase chain reaction (PCR) technology might increase the sensitivity of this screening procedure by several orders of magnitude (Kim and Smithies 1988). Another approach is specifically to design the targeting vectors such that the expression of the positive selectable marker is activated only after the appropriate recombination event (Jasin and Berg 1988). For example, a neo^r gene lacking a transcriptional enhancer may function only upon its insertion with an appropriately expressed gene. Such a manipulation could effectively reduce the number of cells that contain randomly inserted vectors and must be screened. Of course, such a protocol is restricted to those loci that are active in the target cell type and the design of the vector may require extensive knowledge of the control elements in the target locus.

Faced with the limitations imposed by the aforementioned protocols for screening and/or enrichment, we have developed a selection scheme for identification of cells modified by gene targeting (Mansour et al. 1988) This scheme is called positive-negative selection (PNS). The targeting vectors used in the PNS protocol are characterized by two components: a positive selectable marker for selection of transfected cells (as a result of both the homologous recombination and the random integration of the vector), and a negative selectable marker for selection against transfected cells that contain random integrations of the targeting vector.

The PNS protocol takes advantage of two observations that we have made in our studies of homologous recombination. First, although linear vectors are required for gene targeting, the terminal sequences of the vectors need not share homology with the target sequence; that is, genetic exchange can occur at internal regions of homology. Second, when a linear DNA molecule is integrated by the nonhomologous reaction, it does so via its ends and very little, if any, terminal sequence information is lost during the insertion of DNA into the chromosome. Thus, it is possible to position a negative selectable marker at the end of the targeting vector, outside the region of vector/target homology. A homologous recombinational reaction will not transfer the marker to the chromosome; a nonhomologous recombinational event will result in the integration of the negative selectable marker. Application of the appropriate selection will thus eliminate cells in which random integration of the vector has occurred.

An example of the PNS procedure is illustrated in Figure 4–2. The procedure was used to enrich for gene-targeting events at the *hprt* locus in mouse ES cells. The replacement vector, pRV9.1 (see Figure 4–1B), containing the neo^r gene as the positive selectable marker, was modified to include the thymidine kinase gene from *Herpes simplex* virus (HSV-*tk*) at

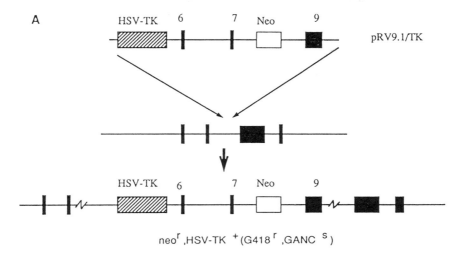

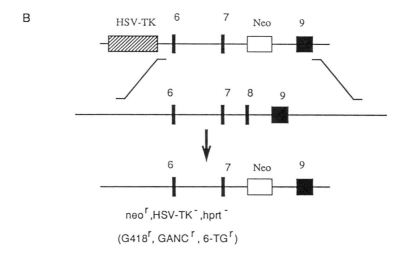

FIGURE 4-2 Gene targeting at the *hprt* locus, using the PNS procedure. (A) Random integration. The vector, pRV9.1/TK, contains 9.1 kb of DNA from the mouse *hprt* gene, with the *neo*^r gene inserted into the eighth exon and the HSV-*tk* at the 5′ end. When the vector integrates via nonhomologous recombination at some random chromosomal position through its ends, the *neo*^r gene and the HSV-*tk* gene are both inserted into the genome, and the recipient cell becomes G418-resistant and GANC-sensitive. (B) Gene targeting. When the vector, pRV9.1/TK, recombines through regions of shared homology with the endogenous *hprt* gene, gene targeting will transfer the *neo*^r, but not the HSV-*tk* gene, to the *hprt* locus. The recipient cell becomes G418-resistant, GANC-resistant and *hprt*−.

its 5' end. The HSV-*tk* gene serves as a negative selectable marker: when it is expressed in mammalian cells it confers sensitivity to a number of nucleoside analogs, for example, gancyclovir (GANC). If the *neo*^{r} gene is transferred via homologous recombination to the *hprt* locus, the resulting cell will become *G418*^{r}, *GANC*^{r}, and *hprt*$^-$. By contrast, if the vector inserts randomly, the host cell will become *G418*^{r}, but GANC-sensitive (HSV-*tk*$^+$) and will remain *hprt*$^+$.

To test this premise, the vector pRV9.1/TK was introduced into ES cells by electroporation. The cells surviving electroporation were then subjected to a variety of growth conditions, as summarized in Table 4–2. Of the cells surviving electroporation, 1 in 100 were *G418*^{r}; of the *G418*^{r} cells, 1 in 3,000 were both *G418*^{r} and *GANC*^{r}. Those cells resistant to both G418 and GANC were candidates for *hprt* mutants and were, therefore, tested for resistance to 6-TG and subjected to Southern transfer analysis. The vast majority of these cells, 19 out of 24, were resistant to 6-TG and, of those 19, all contained the insertion of the *neo*^{r} gene within the endogenous *hprt* locus. Therefore, the PNS procedure was successful; the majority of *G418*^{r} + *GANC*^{r} cells contained targeted mutations of the *hprt* gene, although they were not selected directly for that phenotype.

In the experiment described above, the PNS procedure gave more than a 2,000-fold enrichment in the ratio of detectable homologous to nonhomologous events. Such an enrichment is sufficient to permit the isolation of rare recombinants, and this protocol has been successfully used to inactivate several loci in mouse ES cells, including *int-2*, *hox1.2*, and *hox1.3*. Cell lines containing these mutations have all been introduced into mouse blastocysts and chimeric progeny have been produced. These loci were chosen because of their deduced importance to the mammalian developmental program. The *int-2* gene, for example, is a proto-oncogene that codes for a protein related to the family of fibroblast growth factors (Dickson and Peters 1987). Expression of the gene follows a highly restricted pattern during normal embryogenesis, yet no role(s) for the gene product has yet been defined (Jakobovits et al. 1986; Smith et al. 1988; Wilkinson et al. 1988). Our understanding the function of this gene may be enhanced by examination of

TABLE 4–2 Gene Targeting Using PNS

No. of Cells Surviving Electroporation	No. of G418^r Cells	No. of G418^r + GANC^r Cells	6-TG^r Cells/ G418^r + GANC^r Cells
1.3 × 10^7	1.5 × 10^5	48	19/24

Conditions for selection have been described elsewhere (Mansour et al. 1988). 10^7 ES cells were electroporated with 25 μg of the linearized vector, pRV9.1/TK. Eight separate electroporations were performed. Aliquots of cells were plated under the following conditions: in nonselection medium; in G418 medium; or in G418 + GANC medium. Those colonies surviving in G418 + GANC medium were transferred to G418 + 6-TG medium.

the consequences of homozygous null mutations at the *int-2* locus. The existence of ES cells that contain a mutant allele represents a first step in this process.

4.4 CONCLUSION

The gene-targeting strategy discussed above has been used to transfer mutations created in vitro to specific loci in mouse ES cells. This technology, specifically the PNS procedure, should permit investigators to generate mice with defined mutations at virtually any genetic locus. As presented here, the PNS procedure was used only to generate insertion-type mutations but its use need not be limited to such experiments. For example, by positioning the positive selection element in an intron or 3' untranslated region of the cloned gene, this element can serve as a selection marker in the co-conversion of linked mutations, such as point mutations, deletions, or even gene corrections, when targeting mutant loci. The ability to create mice with any desired genotype should prove invaluable in studies of mammalian development and should provide a route towards creation of animal models of human genetic disorders.

REFERENCES

Bradley, A., Evans, M., Kaufman, M.H., and Robertson, E. (1984) *Nature* 309, 255–256.
Dickson, C., and Peters, G. (1987) *Nature* 326, 833.
Doetschman, T., Maeda, N., and Smithies, O. (1987) *Nature* 330, 576–578.
Folger, K.R., Wong, E.A., Wahl, G., and Capecchi, M.R. (1982) *Mol. Cell. Biol.* 2, 1372–1287.
Gossler, A., Doetschman, T., Korn, R., Serfling, E., and Kemler, R. (1986) *Proc. Natl. Acad. Sci. USA* 83, 9065–9069.
Jakobovits, A., Shackleford, G.M., Varmus, H.E., and Martin, G.R. (1986) *Proc. Nat. Acad. Sci. USA* 83, 7806–7810.
Jasin, M., and Berg, P. (1988) *Genes Dev.* 2, 1353–1363.
Kato, S., Anderson, R.A., and Camerini-Otero, R.D. (1986) *Mol. Cell. Biol.* 6, 1787–1795.
Kim, H.S. and Smithies, O. (1988) *Nucleic Acids Res.* 16, 8887–8903.
Mansour, S.L., Thomas, K.R., and Capecchi, M.R. (1988) *Nature* 336, 348–352.
Miller, C.K., and Temin, H.M. (1983) *Science* 220, 606–609.
Robert de Saint Vincent, B. and Wahl, G. (1983) *Proc. Natl. Acad. Sci. USA* 80, 2002–2006.
Robertson, E., Bradley, A., Kuehn, M., and Evans, M. (1986) *Nature* 323, 445–447.
Robins, D.M., Ripley, S., Henderson, A.S., and Axel, R. (1981) *Cell* 23, 29–39.
Shapira, G.J., Stachelek, J.L., Letsou, A., Soodak, L., and Liskay, R.M. (1983) *Proc. Natl. Acad. Sci. USA* 80, 4827–4831.
Small, J., and Scangos, G. (1983) *Science* 219, 174–176.

Smith, R., Peters, G., and Dickson, C. (1988) *EMBO J.* 7, 1013–1022.
Subramani, S., and Berg, P. (1983) *Mol. Cell. Biol.* 3, 1040–1052.
Thomas, K.R., and Capechi, M.R. (1987) *Cell* 51, 503–512.
Thomas, K.R., Folger, K.R., and Capecchi, M.R. (1986) *Cell* 44, 419–428.
Wilkinson, D.G., Peters, G., Dickson, C., and McMahon, A.P. (1988) *EMBO J.* 7, 691–695.

CHAPTER 5

Application of Germline Transformation to the Study of Myogenesis

Moshe Shani

In an attempt to define the control elements involved in the activation of muscle-specific genes during development, we are taking three different gene-transfer approaches, using transgenic mice, multipotent embryonic stem (ES) cells and retroviral vectors. The results obtained so far indicate the following:

1. A fragment of 210 bp from the 5' region of the actin gene from rat skeletal muscle confers the capacity for preferential expression in skeletal and cardiac muscles of transgenic mice.
2. The presence of sequences from the first intron of the rat gene eliminates the great variability in the tissue-specificity of chloramphenicol acetyl transferase (CAT) constructs.

I thank my colleagues David Yaffe, Doron Shinar, Paz Einat, Itzhak Dekel, Olla Yoffe, Iris Ben-Dror, Yonat Magal, and Amotz Nehushtan for their contributions to the studies described here. This work was supported in part by the Muscular Dystrophy Association (USA), the Israel Academy of Sciences and Humanities, and the United States-Israel Binational Agricultural Research and Development Fund (BARD).

3. The expression of transgenes may lead to a significant suppression of the corresponding endogenous genes.
4. A cloned muscle-specific gene introduced into ES cells is expressed specifically in myogenic cells derived from the transfected ES cells, provided that it is inserted without the vector DNA.
5. The promoter of the actin gene from rat skeletal muscle can function properly in close proximity to the control elements of a viral, long terminal repeat (LTR).
6. Retroviral vectors probably integrate into active chromosomal domains.

Terminal differentiation of muscle cells involves major changes at the morphological as well as the molecular level (for reviews, see Buckingham 1985; Blau et al. 1985). At the morphological level, mononucleated myoblasts fuse into multinucleated myotubes. At the molecular level, a large battery of muscle-specific genes is activated. Among them are genes involved in the formation of the contractile apparatus (the sarcomere), such as those for actin, myosin, tropomyosin, troponin, and genes involved in the provision of energy for muscle contraction, such as the gene for creatine phosphokinase. It is generally assumed that the activation of such a large number of muscle-specific genes is mediated by specific *trans*-acting factors that recognize common DNA sequences within and/or in the immediate vicinity of these genes. To date, no muscle-specific, *trans*-acting regulatory factors have been isolated by either transfection of cloned genes or of specifically modified genes into myogenic cells. A genetic approach has recently been successful in demonstrating that the myogenic program is controlled by a surprisingly small number of regulatory genes, which may also serve as the putative *trans*-acting factors (Konieczny and Emerson 1984; Lassar et al. 1986; Davis et al. 1987; Pinney et al. 1988; Tapscott et al. 1988). To define the control mechanisms involved in the tissue-specificity and the developmental regulation of skeletal muscle genes, we are taking three separate gene-transfer approaches, all of which involve insertion of genes into the germline, using transgenic mice, ES cells, and retroviral vectors. In this chapter our recent results with these biological systems are reviewed.

5.1 EXPRESSION OF RAT MUSCLE-SPECIFIC GENES IN TRANSGENIC MICE

5.1.1 *Cis*-Acting DNA Sequences in the Actin Gene From Rat Skeletal Muscle

We previously reported that a chimeric actin/globin gene is expressed in a tissue-specific and developmentally regulated manner in transgenic mice (Shani 1986). In this chimeric gene, the 3' end of the actin gene was replaced by the corresponding region of the human embryonic globin gene (Melloul et al. 1984). Thus, it appears that the highly conserved 3' region of the rat

5.1 Expression of Rat Muscle-Specific Genes in Transgenic Mice

gene is not essential for its proper expression. To determine whether the regulatory elements required for the tissue-specificity of this gene reside at the 5' end, we microinjected into fertilized mouse eggs the three fusion genes shown in Figure 5-1, each of which contains a different length of the 5' region of the actin gene from rat skeletal muscle fused to the bacterial CAT gene. We analyzed CAT activity in different tissues of the resultant transgenic mice. Our results, summarized in Table 5-1, demonstrate that 145 bp of the 5' flanking region of the rat gene are sufficient to permit preferential expression in striated muscles. However, it should be emphasized that, unlike the case with mice containing the chimeric actin/globin gene, low levels of CAT activity were also detected in nonmyogenic tissues, such as the liver, spleen, thymus, brain, and lung. This ectopic expression was found in different strains of mice that carried the same DNA construct and sometimes in different individuals within the same transgenic strain. This variation may indicate that additional elements are required for a more stringent control of transcription of this gene.

It is surprising that no such variability was observed in the six transgenic strains in which the entire first intron was included. These results may imply that the first intron contains elements essential for the correct activation of the actin gene from rat skeletal muscle. Thus far, there is no evidence for the existence of such elements. Alternatively, splicing may play an important role in the expression of transgenes, as suggested recently by Brinster et al. (1988). However, since the three CAT constructs already contained an intron (derived from the SV40 small T-antigen) at the 3' end, it appears that heterologous introns cannot always substitute for the homologous one.

5.1.2. The Interrelationship between Exogenous and Endogenous Genes

Although a large number of transgenic mice has already been produced and analyzed, very little is known about the interplay between the expression of endogenous and exogenous genes. To address this question, we have

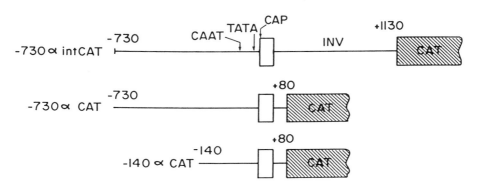

FIGURE 5-1 The structure of the constructs used for microinjection.

TABLE 5-1 CAT Activity in Different Tissues from Transgenic Mice That Carry the Three Gene Constructs

Tissue	−730 CAT1 G_o	−730 CAT2 G_o	−730 CAT2a G_1	−730 CAT2b G_2	−730 CAT4 G_2	−140 CAT1 G_o	−140 CAT1a G_1	−140 CAT1a G_1	−140 CAT1b G_1	−140 CAT3 G_o	−INT CAT1 G_o	−INT CAT2 G_o	−INT CAT3 G_o	−INT CAT4 G_o
Skeletal muscle	100	100	100	100	100	55	1	<1	100	22	100	100	100	100
Heart	100	14	11	0	7	100	17	100	3	100	75	10	56	77
Tongue	0	ND[a]	11	0	<1	0	ND	0	0	0	1	18	4	38
Liver	0	18	11	31	<1	3	<1	<1	<1	0	<0.1	0	<0.1	<0.1
Spleen	0	0	0	0	<1	0	100	0	0	3	<0.1	<0.1	<0.1	<0.1
Brain	0	0	40	0	<1	0	3	0	0	0	ND	ND	ND	ND
Lung	0	0	0	0	<1	0	5	0	0	2	<0.1	<1	<0.1	<1
Kidney	0	0	0	0	<1	0	<1	0	<1	0	<0.1	0	<0.1	<0.1
Thymus	0	0	0	0	<1	0	28	0	0	3	<0.1	<0.1	<0.1	<1
Stomach	0	0	0	0	<1	0	ND	0	0	0	<1	30	<0.1	<1
Testes	0	0	0	0	<1	ND	<1	0	1	0	0	ND	<0.1	ND

[a]ND, not determined.

analyzed the levels of transcripts of both types of gene in a number of our transgenic mice. In two strains, one of which carried the myosin light-chain 2 (MLC2) gene from rat (Nudel et al. 1984; Shani 1985; Einat et al. 1987) and the second which carried a modified actin gene from skeletal muscle (Einat et al. unpublished data), we found a profound effect on the expression of the endogenous gene. In the strain carrying the rat MLC2 gene, the activation of the exogenous gene in skeletal muscle was correlated with a significant suppression of the corresponding endogenous gene (Shani et al. 1988). Similarly, in the second strain, the expression of the endogenous actin gene from skeletal muscle was down-regulated in the tongue. However, in this strain, the expression of the endogenous gene for skeletal muscle actin in other striated muscles (skeletal and cardiac) was not affected.

These results indicate that a transgene can functionally replace, at least to some extent, the endogenous gene without homologous recombination. The suppression of the endogenous gene could be due either to competition in vivo for limited amounts of tissue-specific, *trans*-acting factors or to feedback inhibition as a result of overproduction of transcripts or protein. Unfortunately, the strain carrying the rat MLC2 gene contains about 15 copies per cell of the transgene, and expresses the rat gene at an exceptionally high level (about five to ten times the level of the endogenous gene). Therefore, we could not distinguish between the two models. However, the modified actin gene from rat skeletal muscle in the second strain is expressed at low levels in skeletal and cardiac muscle (about 20% of the level of the endogenous gene), while it is expressed at relatively higher levels in the tongue. These results, therefore, favor the model involving competition rather than that involving feedback inhibition. To better distinguish between the two possibilities, we are trying to develop a system of transfected myogenic cells in vitro that will mimic this effect. Preliminary results indicate that it is possible to suppress the expression of the endogenous gene by incorporation of a transfected gene.

To the best of our knowledge, there are only two other reports in which the expression of a transgene has been correlated with the suppression of the corresponding endogenous gene (Adams et al. 1985; Alexander et al. 1987; Grosveld et al. 1987). In the case of mice carrying a fusion *myc* gene, the suppression of the endogenous gene was very similar to the exclusive expression of the translocated allele in conventional plasmacytomas (Adams et al. 1985; Alexander et al. 1987). In another case, in two of nine transgenic strains carrying the human β-globin "minilocus," the expression of the mouse gene was suppressed, presumably as a result of a competition for transcription factors (Grosveld et al. 1987).

5.2 EXPRESSION OF MUSCLE-SPECIFIC GENES INTRODUCED INTO ES CELLS

We have established a number of lines of ES cells that retain their multipotency both in vivo and in vitro. These cells were transfected by electroporation with a fragment of DNA that carried the muscle-specific actin/

globin gene and the *SV2neo* gene, which confers resistance to the neomycin analog G418. ES cells resistant to G418 were injected subcutaneously into nude mice. This treatment resulted in the formation of teratocarcinoma-like tumors, which contained islands of many differentiated tissues of all three germ layers. Forty to sixty days later, the tumors were excised, lines of myogenic and nonmyogenic cells were established, and the presence of transcripts of transfected genes was determined. The results demonstrated the cell-specific expression of the chimeric actin/globin gene introduced into undifferentiated multipotent ES cells. The exogenous gene was not expressed at the ES cell stage; low levels of its transcripts were found in nonmyogenic cells, whereas a significant increase in the level of these transcripts was detected during the transition to multinucleated fibers (Shinar et al., 1989). The low but detectable level of transcripts in cultures of nonmyogenic cells contrasted with the stringent control of expression of the same chimeric gene in six independent, transgenic strains (Shani 1986). This difference may be due either to the proximity of the SV40 enhancer (of the *SV2neo* gene) to the actin/globin sequences, or to the fact that by selecting for resistance to neomycin, we may have selected for cells in which the exogenous gene has integrated into constitutively active domains. Two recent reports have also described tissue-specific expression of genes transfected into ES cells. Lovell-Badge et al. (1987) reported the expression of a gene for collagen II in the carcass of chimeric mice generated from transfected ES cells. However, in this case, some activity of the transfected gene was found at the stem-cell stage.

Takahashi et al. (1988) have reported the expression of the δ-crystallin gene from the chick in the lenses of chimeric mice. They also reported that the expression of the transfected gene in teratocarcinomas produced by the transfected ES cells was less stringently controlled than it was in chimeric mice.

It is of interest that in most of the ES cells transfected with entire plasmids (including the vector DNA), transcripts of the chimeric gene could not be detected in the myogenic or in the nonmyogenic cells derived from these cells. Moreover, in the presence of prokaryotic DNA sequences, the expression of the selectable gene declined rapidly, even at the stem-cell stage, and could not be detected in the differentiated derivatives. Removal of the vector DNA resulted in the stable maintenance of expression of resistance to neomycin. The inactivation of the inserted genes, in the presence of prokaryotic DNA sequences, was correlated with methylation de novo. No such methylation was observed in clones transfected without the pBR sequences. These results demonstrate the involvement of methylation of DNA in the inactivation of genes inserted into ES cells. Methylation de novo was rarely observed in somatic cells transfected with entire plasmids, regardless of the state of expression of the introduced gene.

Methylation of foreign DNA has frequently been implicated in the inactivation of viruses in embryonal carcinoma cells (Jahner and Jaenisch

1984). However, it is probably not the primary event in the inactivation of genes inserted into ES cells. There are several reports indicating the presence in early embryonic cells of negative *trans*-acting factors (Gorman et al. 1985; Shinar et al., 1989). Thus, methylation of DNA may be involved in the establishment and maintenance of the inactive state.

5.3 GENE TRANSFER MEDIATED BY RETROVIRAL VECTORS: EXPRESSION OF AN INTERNAL MUSCLE-SPECIFIC PROMOTER

A number of serious problems are encountered in attempts to adapt the microinjection approach to farm animals: (a) it is difficult to visualize the pronuclei; (b) the number of one-cell embryos available for manipulation is low; (c) the number of both donor and recipient animals is limited; (d) the microinjected embryos have reduced viability; and (e) the overall efficiency of transgenesis is low. It is therefore important to develop alternative approaches to overcome these difficulties. Germline transformation via retroviral vectors could offer a solution to some of these problems (for reviews, see Temin 1986; Gilboa 1987; Grindley et al. 1987).

Retroviruses are suitable for gene transfer because integration is an obligatory part of their life cycle. After replacement of the viral genes (*gag*, *pol*, *env*) with the genes of interest and taking advantage of the efficient process of viral infection, genes can be inserted into the target cells as if they were viral genes.

The advantages of the use of retroviral vectors, as compared to the microinjection approach, are high efficiency of infection, single-copy integration, no gross rearrangements at the site of integration, and predictable structure of the proviral DNA. In most cases, the vectors are harmless to the host cells, and they have broad specificity with respect to cell type and host range. However, there are also several limitations to this method: it requires extra manipulation; there is a limit to the size of the inserted genes; and there may be interference by the strong LTR elements, with the expression of internal, tissue-specific and developmentally regulated promoters.

To determine whether the expression of an internal, tissue-specific promoter would be affected by the presence of the viral control elements, we constructed the N2-based (Keller et al. 1985) recombinant retroviruses shown in Figure 5-2. In these vectors the bacterial CAT gene, fused to fragments of various sizes from the 5' end of the actin gene from rat skeletal muscle, was inserted in both orientations with respect to viral transcription. Surprisingly, no infectious viruses were recovered when 730 bp of 5' flanking sequences of the rat gene were included. Removal of the distal 300 bp resulted in the release of virus at high titer ($>10^5$ colony forming units (cfu)/ ml). These viruses were first used to infect cultures of the rat myogenic cell line L8. Clones carrying the viral genome were then isolated either after

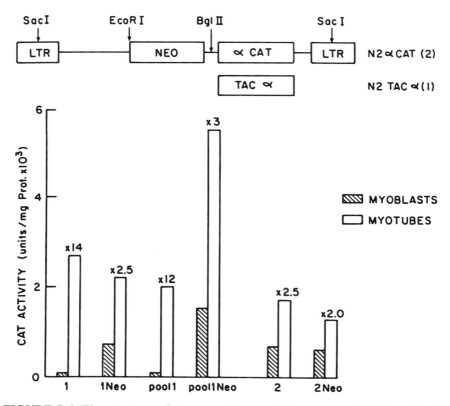

FIGURE 5-2 The structure of the recombinant viral vectors, and CAT activity in extracts of undifferentiated (myoblasts) and differentiated (myotubes) muscle cells. After infection with the two viruses, individual or pooled clones were isolated either with or without selection for neomycin. For example, clone 1 was infected with the vector N2TACα(1) and not selected for resistance to neomycin, while clone 1Neo was selected for resistance to neomycin. The factor by which the CAT activity increased during differentiation is indicated above the bars.

selection with G418 or without selection. CAT activity was determined in extracts prepared from cultures that contained mainly mononucleated myoblasts or myotubes (see Figure 5-2). CAT activity was found with all constructs, regardless of the orientation of the inserted gene with respect to viral transcription. However, there was a marked difference between the stringency of expression with regard to terminal differentiation of the myogenic cells and the orientation of the inserted gene. Expression was best regulated in clones that were not selected for resistance to G418, in which the inserted gene was in the opposite orientation to the transcription of the LTR. This result held true in isolated as well as in pooled clones (see Figure 5-2), and it is in agreement with a number of previous reports (Emerman and Temin 1984; Joyner and Bernstein 1983). Placing the two genes in the same orientation may lead to inhibition by steric hindrance (Garner et al. 1986).

An important question related to transfer of genes into somatic cells is whether, by selection for neomycin-resistance, we select for sites of integration into chromosomal domains that favor expression. The fact that the CAT gene was expressed in all clones, both selected and unselected, suggests that retroviruses apparently integrate in a nonrandom manner (Vijaya et al. 1986; Rohdewohld et al. 1987; Shih et al. 1988). It remains to be determined whether a similar tendency with respect to integration also applies to transfected or microinjected DNA.

REFERENCES

Adams, J.M., Harris, A.W., and Pinkert, C.A., et al. (1985) *Nature* 318, 533-538.
Alexander, W.S., Schrader, J.W., and Adams, J.M. (1987) *Mol. Cell Biol.* 7, 1436-1444.
Blau, H., Pavlath, G.C., Hardeman, E.C., et al. (1985) *Science* 230, 758-766.
Brinster, R.L., Allen, J.M., Behringer, R.R., Gelinas, R.E., and Palmiter, R.D. (1988) *Proc. Natl. Acad. Sci. USA* 85, 836-840.
Buckingham, M.E. (1985) *Essays in Biochem.* 20, 77-109.
Davis, R.L., Weintraub, H., and Lassar, A.B. (1987) *Cell* 51, 987-1000.
Einat, P., Bergman, Y., Yaffe, D., and Shani, M. (1987) *Gene Devel.* 1, 1075-1084.
Emerman, M., and Temin, H.M. (1984) *Cell* 39, 459-467.
Garner, I., Minty, A., Alonso, S., Barton, P., and Buckingham, M.. (1986) *EMBO J.* 5, 2559-2567.
Gilboa, E. (1987) *BioAssays* 5, 252-257.
Gorman, C., Rigby, W.J., and Lane, D.P. (1985) *Cell* 42, 519-626.
Grindley, T., Soriano, P., and Jaenisch, R. (1987) *Trends in Genet.* 3, 162-166.
Grosveld, F., Van Assendelft, G.B., Greaves, D.R., and Kollias, G. (1987) *Cell* 51, 975-985.
Jahner, D. and Jaenisch, R. (1984) in *DNA Methylation; Biochemical and Biological Significance* (Razin, A., Cedar, H., and Riggs, A.D., eds.), pp. 189-220, Springer Verlag, New York.
Joyner, A.L., and Bernstein, A. (1983) *Mol. Cell Biol.* 3, 2191-2202.
Keller, G., Paige, P., Gilboa, E., and Wagner, E. (1985) *Nature* 318, 149-154.
Konieczny, S.F., and Emerson, C.P. (1984) *Cell* 38, 791-800.
Lassar, A.B., Paterson, B.M., and Weintraub, H. (1986) *Cell* 47, 649-656.
Lovell-Badge, R.H., Byrgrave, A., Bradly, A., Tilly, R., and Cheah, K.S.E. (1987) *Proc. Natl. Acad. Sci. USA* 78, 7634-7638.
Melloul, D., Aloni, B., Calvo, J., Yaffe, D., and Nudel, U. (1984) *EMBO J.* 3, 983-990.
Nudel, U., Calvo, J., Shani, M., and Levy, Z. (1984) *Nucl. Acids Res.* 12, 7175-7186.
Pinney, D.F., Pearson-White, S.H., Konieczny, S.F., Latham, K.E., and Emerson, C.P. (1988) *Cell* 53, 781-793.

Rohdewohld, H., Weiher, H., Reik, W., Jaenisch, R., and Breindl, M. (1987) *J. Virol.* 61, 336–343.
Shani, M. (1985) *Nature* 314, 283–286.
Shani, M. (1986) *Mol. Cell Biol.* 6, 2624–2631.
Shani, M., Dekel, I., and Yoffe, O. (1988) *Mol. Cell Biol.* 8, 1006–1009.
Shih, C.C., Stoye, J.P., and Coffin, J.M. (1988) *Cell* 53, 531–537.
Shinar, D., Yoffe, O., Shani, M., and Yaffe, D. (1989) *Differentiation*, 41, 116–126.
Takahashi, Y., Hanaoka, K., Hayasaka, M., et al. (1988) *Development* 102, 259–269.
Tapscott, S.J., Davis, R.L., Thayer, M.J. et al. (1988) *Science* 242, 405–411.
Temin, H.M. (1986) in *Gene Transfer* (Kucherlapati R., ed.), pp. 149–187, Plenum Press, New York.
Vijaya, S., Steffen, D.L., and Robinson, H.L. (1986) *J. Virol.* 60, 683–692.

CHAPTER 6

Regulation of Expression of Genes for Milk Proteins

Lothar Hennighausen
Christoph Westphal
Lakshmanan Sankaran
Christoph W. Pittius

The tissue-specific and stage-specific regulation of expression of genes for milk proteins is influenced by a variety of factors, including peptide and steroid hormones (Topper and Freeman 1980; Hobbs et al. 1982). The whey acidic protein (WAP) and β-casein are abundant proteins in the milk of mice and rats (Hobbs et al. 1982; Hennighausen and Sippel 1982b) and the corresponding mRNAs comprise approximately 30% of the poly(A)$^+$-RNA in the mammary glands of lactating animals (Hobbs et al. 1982; Hennighausen and Sippel 1982a and 1982b; Hennighausen et al. 1982). To elucidate the mechanisms by which expression of the mouse WAP and β-casein genes is regulated in a tissue-, stage-, and hormone-specific fashion, corresponding cDNA and genomic sequences were isolated and characterized (Hennighausen and Sippel 1982a and 1982b; Campbell et al. 1984). Using hybrid-

---

C.W.P. was supported by the Deutscher Akademischer Austauschdienst and C.W. was supported by a special program from the NIH. We also thank William Jakoby for his most generous support and Peter McPhie for comments on the manuscript.

ization probes specific for the WAP and β-casein genes, we analyzed steady-state levels of the corresponding mRNAs in a variety of tissues, both during mammary development and upon hormonal stimulation of mammary tissue from pregnant animals.

Sequences in the promoter regions of several genes for milk proteins, including the WAP (Lubon and Hennighausen 1987) and α-lactalbumin genes (Hall et al. 1987; Lubon and Hennighausen 1988), are conserved and recognized by nuclear proteins from mammary glands of lactating animals (Lubon and Hennighausen 1987 and 1988; Hennighausen and Lubon 1987), a result that suggests that the upstream sequences include regulatory elements. Functional mammary epithelial cell lines do not exist, but transgenic mice (Palmiter and Brinster 1986) provide a system for the study of DNA control elements in vivo that regulate gene expression in the mammary gland. We have recently described the generation of several lines of transgenic mice that carry a hybrid gene composed of the promoter of the mouse WAP gene and a cDNA that encodes human tissue plasminogen activator (tPA) (Gordon et al. 1987; Pittius et al. 1988a). Human tPA has been found in the milk of lactating animals from most of these transgenic lines (Gordon et al. 1987) and expression of the hybrid gene was specific to the mammary gland (Pittius et al. 1988b), suggesting that the WAP promoter contains mammary regulatory elements.

In this chapter we compare expression of the WAP-tPA hybrid gene in transgenic mice with that of the endogenous genes for WAP and β-casein.

6.1 EXPRESSION OF THE WAP AND β-CASEIN GENES DURING MAMMARY DEVELOPMENT

To investigate the developmentally regulated expression of the WAP and β-casein genes, we measured steady-state levels of the corresponding mRNAs in mammary glands from mice at different stages of development. Low levels of WAP RNA were found up to day 14 of gestation and these low levels increased about 50-fold around day 16, with additional 2- to 4-fold increases both at parturition and on day 7 of lactation (Figure 6–1). With an RNA probe that spans the first exon of the WAP gene (Pittius et al. 1988a and 1988b), the RNA levels measured at midlactation were found to be about 10^4-fold higher than in virgin animals (Pittius et al. 1988b, see Figure 6–1). Levels of β-casein RNA during mammary development were measured with an RNA probe that spanned β-casein mRNA sequences between nucleotides 441 and 634 (Hennighausen and Sippel 1982a). Induction of β-casein during development followed similar kinetics to those observed for the WAP RNA but the increase between days 14 and 16 was less pronounced (see Figure 6–1).

In addition to analyzing steady-state levels of WAP mRNA during mammary development with RNA probes that corresponded to the first exon,

6.1 Expression of the WAP and β-Casein Genes

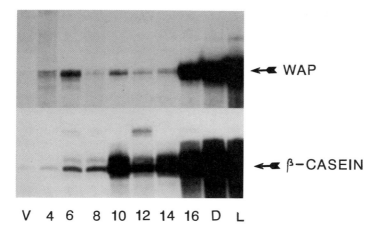

FIGURE 6-1 Accumulation of WAP and β-casein mRNAs during mammary development. Total RNA was isolated from mammary glands of virgin, pregnant, and lactating animals. Aliquots of 100 ng RNA were analyzed with 200,000 cpm of RNA probe for the presence of WAP and β-casein RNA, respectively, and the reaction products were separated in a sequencing gel. The WAP-specific probe is indicative of the first exon of the gene (Pittius et al. 1988a; Pittius et al. 1988b) and the β-casein probe corresponds to nucleotides 441–634 of the β-casein mRNAs. The arrow labeled WAP indicates a protected fragment of 112 nucleotides which encodes the first exon of the WAP gene (Pittius et al. 1988a; Pittius et al. 1988b) and the arrow labeled β-casein indicates a protected fragment of 193 nucleotides. V, virgin; D, delivery; L, midlactation. The numbers refer to the day of gestation at which RNA from the mammary gland was analyzed.

we also used probes specific for the fourth exon of the WAP gene. An analysis of RNA from various stages of mammary development with an RNA probe that spanned parts of the third intron and fourth exon (Figure 6-2) resulted in two interesting observations. First, two protected fragments were detected and, second, the increase in levels of WAP RNA during mammary development appeared to be less pronounced than appeared to be the case with the probe that spanned the first exon (see Figure 6-1). The protected fragment of 95 nucleotides (upper arrow in Figure 6-2) is indicative of the presence of sequences from exon IV, as predicted from cDNA and genomic sequences (Campbell et al. 1984), and the slightly smaller fragment (lower arrow in Figure 6-2) could be the result of the involvement of a cryptic splice site, as predicted for the corresponding rat gene (Campbell et al. 1984). The less pronounced increase during mammary development of levels of WAP RNA sequences that correspond to the fourth exon can be interpreted as being the result of premature termination of transcription within the WAP gene. Premature termination of transcription within the β-casein gene has been reported recently by Rosen et al. (1988). However, we cannot rule

68 Regulation of Expression of Genes for Milk Proteins

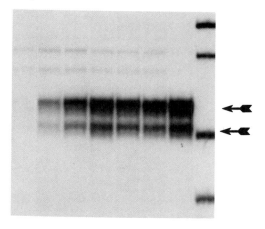

V 14 15 16 17 18 L M

FIGURE 6-2 Accumulation of WAP RNA during mammary development. Total RNA was isolated from mammary glands and analyzed as described in the legend to Figure 6-1 (Pittius et al. 1988a; Pittius et al. 1988b). The RNA probe was derived from a 600-bp *BglII-BamHI* fragment (Campbell et al. 1984) spanning sequences from the third intron and fourth exon. The upper arrow indicates a 95-bp fragment that corresponds to sequences from the fourth exon (Campbell et al. 1984) and the lower band represents a potential alternative splicing product. V, virgin; L, midlactation. The numbers refer to the day of gestation at which RNA from the mammary gland was analyzed. M, pBR322 cut with *MspI*.

out the possibility that some sequences in the transcript of the WAP gene are more stable than others during turnover.

6.2 TISSUE-SPECIFIC EXPRESSION OF THE GENES FOR WAP AND β-CASEIN

To assess the tissue-specific expression of the genes for WAP and β-casein we analyzed, by the RNA protection assay, RNA from 15 tissues taken from lactating females (Pittius et al. 1988a). High levels of WAP RNA were found in mammary gland and 10^4- to 10^6-fold lower levels were found in tongue, thymus, liver, and pancreas (Figure 6-3, upper panel). The tissue-specific distribution of β-casein RNA was similar to that of WAP RNA with the exception that the concentrations in tongue, thymus, liver, and pancreas were about 10-fold higher than those of WAP RNA (see Figure 6-3, middle panel). Neither WAP nor β-casein RNA was detected in brain, heart ventricle, lung, stomach, intestine, spleen, kidney, or uterus (see Figure 6-3). The presence of WAP and β-casein RNAs in the same set of nonmammary

6.2 Tissue-Specific Expression of the Genes for WAP and β-Casein

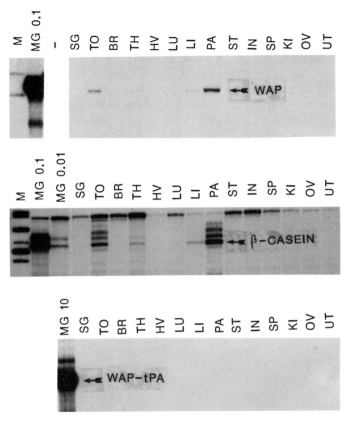

FIGURE 6-3 Distribution in tissues of endogenous WAP RNA, β-casein RNA, and transgenic WAP-tPA hybrid RNA. RNA from 15 tissues from lactating animals was hybridized with radioactively labeled RNA complementary to the first exon of the WAP gene (top panel), to the β-casein gene (middle panel) or to the WAP-tPA hybrid gene (lower panel) (Pittius et al. 1988a; Pittius et al. 1988b) and then digested with RNAase; the reaction products were separated in sequencing gels. The arrows indicate the 112-nucleotide fragment that encodes the first exon of the WAP gene (top panel), the 193-nucleotide fragment specific for mouse β-casein mRNA (middle panel) and the 132-nucleotide fragment specific for the WAP-tPA hybrid RNA (lower panel). The top band in the middle panel reflects the presence of undigested probe and the other bands reflect the presence of partially digested products. The appearance of undigested probe is inherent in the properties of the β-casein gene and is probably due to the high GC content of the RNA. M, lambda DNA cut with *Msp*I. Ten micrograms of RNA from each tissue was analyzed. The amount of mammary gland RNA analyzed is indicated above the respective lanes. MG, mammary gland; SG, salivary gland; TO, tongue; BR, brain; TH, thymus; HV, heart ventricle; LU, lung; LI, liver; PA, pancreas; ST, stomach; IN, intestine; SP, spleen; KI, kidney; OV, ovary; UT, uterus.

tissues suggests that the expression of the corresponding genes in these tissues is specific. However, specific functions for milk proteins in nonmammary tissues have not yet been demonstrated.

The whey acidic protein is a member of the "four-disulfide-core" family of proteins and bears a striking structural similarity to neurophysin (Hennighausen and Sippel 1982a). Neurophysin is part of the precursor to oxytocin and acts as a carrier for oxytocin from the hypothalamus to the neural lobe of the pituitary gland. Since the mammary gland is a target tissue for oxytocin, it has been speculated that WAP might be a receptor or a carrier for oxytocin (Hennighausen and Sippel 1982a). Therefore, we analyzed pituitary tissue for the presence of WAP RNA. It is noteworthy that WAP RNA was found in the neurointermediate lobe but not in the anterior lobe of the pituitary gland (Figure 6–4). However, β-casein RNA was also found predominantly in the intermediate lobe (see Figure 6–4). By contrast, higher levels of c-myc RNA were found in the anterior lobe (see Figure 6–4). Taken together, these results suggest that genes for milk proteins are preferentially transcribed in cells associated with the neural lobe of the pituitary gland. The significance of this observation remains to be determined.

To assess whether the tissue specificity of the expression of the WAP gene is conferred by promoter/upstream sequences, we analyzed a transgenic mouse that carried the WAP-tPA hybrid gene (Gordon et al. 1987; Pittius et al. 1988a; Pittius et al. 1988b). The levels of transcript of the WAP-tPA hybrid gene found in the mammary gland were about 100-fold lower than those of WAP RNA but expression was essentially confined to the mammary gland (see Figure 6–3, lower panel). This result suggests that regulatory

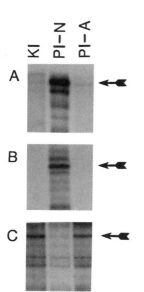

FIGURE 6–4 Expression of WAP, β-casein, and c-myc RNA in the anterior and neurointermediate lobes of the pituitary gland. The two lobes of the pituitary gland were physically separated, total RNA was prepared, and 10 µg of RNA were analyzed for the presence of WAP (A), β-casein (B) and c-myc (C) RNAs by an RNAase protection assay. The c-myc probe spanning the second exon of the gene was generously provided by Kathleen Kelly of the National Cancer Institute. The arrows point to the protected fragments of predicted sizes. KI, kidney; PI-N, neural lobe of the pituitary gland; PI-A, anterior lobe of the pituitary gland.

elements that confer the capacity for mammary-specific gene expression are located within the promoter/upstream region of the WAP gene.

6.3 HORMONE-INDUCED ACCUMULATION OF WAP AND β-CASEIN RNA IN MAMMARY GLANDS OF PREGNANT MICE

Hormonal requirements for the accumulation of endogenous WAP and β-casein RNAs were analyzed in explant cultures of mammary glands obtained from pregnant animals on day 8 to 10 of gestation. The explants were cultured for 48 hours in serum-free medium in the absence or presence of different combinations of hormones and were analyzed for the respective RNAs by RNAase mapping. After tissue was cultured for two days without insulin, hydrocortisone or prolactin, levels of WAP, and β-casein RNA dropped sharply (Figure 6–5). Similarly, insulin alone or in combination with either hydrocortisone or prolactin was not sufficient to sustain high levels of WAP or β-casein RNA (see Figure 6–5). However, the combination of all three hormones resulted in an approximately 50-fold increase in levels of both RNAs above starting levels (see Figure 6–5). The extent of induction was similar to that observed during midgestation (see Figure 6–1).

To investigate whether promoter/upstream sequences of the WAP gene participate in the hormonally responsive accumulation of RNA (probably as a result of stimulation of transcription), we analyzed the levels of WAP-tPA RNA in explant cultures from transgenic mice (Pittius et al. 1988b). Whereas insulin alone or in combination with either hydrocortisone or prolactin was not sufficient to sustain the level seen at the initiation of the culture, the combination of all three hormones resulted only in a small net increase of WAP-tPA RNA (see Figure 6–5C).

6.4 SUMMARY

Our results demonstrate that mRNAs that encode two major milk proteins in the mouse, namely, WAP and β-casein, are found predominantly in the mammary glands of lactating animals. During the course of mammary development from the virgin to the fully lactating animal the steady-state levels of the two RNAs increase about 10^4-fold, with the most pronounced increase occurring around midpregnancy. Low levels of WAP and β-casein RNAs are also found in some nonmammary tissues such as tongue, pancreas, and pituitary gland, but not in others, for example, heart and brain. This variation suggests that the transcriptional machinery in some nonmammary cells is capable of recognizing milk-protein genes. The analysis of transgenic animals that carry a hybrid gene composed of the WAP promoter plus the

72 Regulation of Expression of Genes for Milk Proteins

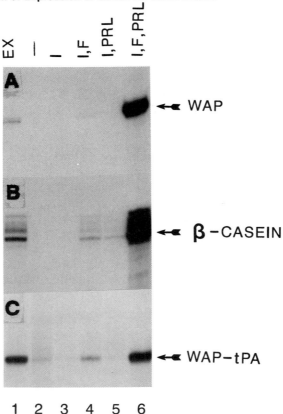

FIGURE 6-5 Hormonally induced accumulation of endogenous WAP, β-casein, and WAP-tPA hybrid RNA in organ cultures of mammary glands. Mammary gland tissue pooled from 10 animals at days 8 to 10 of gestation was cultured for two days in the absence or presence of different combinations of hormones. Total RNA was isolated and 1 μg was analyzed in an RNAase protection assay for the presence of WAP (A), β-casein (B), or WAP-tPA (C) transcripts. Lane 1 (EX) shows WAP RNA at day 8 to 10 of gestation. Lanes 2 through 6 show WAP transcripts after two days in culture in the presence of hormones (I, insulin at 0.1 μg/ml; F, hydrocortisone at 0.1 μg/ml; PRL, prolactin at 1 μg/ml). Lane 2 (−), no hormones; lane 3, I; lane 4, I and F; lane 5, I and PRL, lane 6, I, F, and PRL. The arrows indicate the respective transcripts of predicted sizes.

tPA gene strongly suggests that mammary specificity of gene expression is encoded in the promoter/upstream region of the WAP gene.

Induction of the genes for WAP and β-casein during development of mammary glands requires the synergistic action of insulin, hydrocortisone, and prolactin. The increase in levels of the two mRNAs in mammary tissue from midpregnant animals was about 50-fold upon incubation with all three of these hormones, suggesting that no additional hormones may be required in the intact animal. The characteristics of the hormonal induction of the WAP-tPA hybrid gene in transgenic animals differed from those of the induction of the endogenous milk-protein genes in that the presence of all three hormones was not sufficient to increase the levels of the hybrid RNA. This difference suggests that some elements involved in the hormone-dependent accumulation of WAP RNA are located outside the promoter/upstream region, (Pittius et al. 1988a and 1988b). Similar conclusions have been drawn by Groner and coworkers, who analyzed expression of the WAP promoter in two additional hybrid genes (Andres et al. 1988). In addition to the WAP gene, the β-casein gene also appears to contain hormone-sensitive regulatory elements within the transcribed region. Hybrid genes containing the rat β-casein promoter (Lee et al. 1989a and 1989b) are expressed in a mammary-specific fashion in transgenic animals but have lost their ability to be regulated by steroid and peptide hormones, a property of their endogenous counterpart.

Aside from providing vital information about the nature and location of elements that confer the capacity for mammary-specific gene expression, the transgenic mice carrying the WAP-tPA gene were the first animal model for the production of human pharmaceuticals in milk (Gordon et al. 1987; Pittius et al. 1988a). The conversion of the mammary gland into a bioreactor could revolutionize biotechnology and provide ample amounts of pharmaceuticals and other valuable proteins that cannot be produced in sufficient quantities by current methods.

REFERENCES

Andres, A-C., van der Valk, M.A., Schonenberger, C-A., et al. (1988) *Genes and Dev.* 2, 1486–1495.
Campbell, S.M., Rosen, J.M., Hennighausen, L.G., Strech-Jurk, U., and Sippel, A.E. (1984) *Nucleic Acids Res.* 12, 8685–8697.
Gordon, K., Lee, E., Vitale, J.A., et al. (1987) *Bio/Technol.* 5, 1183–1187.
Hall, L., Emery, D.C., Davies, M.S., Parker, D., and Craig, R.K. (1987) *Biochem. J.* 242, 735–742.
Hennighausen, L., and Lubon, H. (1987) *Methods Enzymol.* 152, 721–734.
Hennighausen, L.G., and Sippel, A.E. (1982a) *Nucleic Acids Res.* 10, 2677–2684.
Hennighausen, L.G., and Sippel, A.E. (1982b) *Eur. J. Biochem.* 125, 131–141.
Hennighausen, L.G., Sippel, A.E., Hobbs, A.A., and Rosen, J.M. (1982) *Nucleic Acids Res.* 10, 3733–3744.
Hobbs, A.A., Richards, D.A., Kessler, D.J., and Rosen, J.M. (1982) *J. Biol. Chem.* 257, 3598–3605.
Lee, K-F., Atiee, S.H., Henning, S.J., and Rosen, J.M. (1989a) *Mol. Endocrinology* 3, 447–453.
Lee, K-F., Atiee, S.H., and Rosen, J.M. (1989b) *Mol. Cell. Biol.* 9, 560–565.
Lubon, H., and Hennighausen, L. (1987) *Nucleic Acids Res.* 15, 2103–2121.
Lubon, H., and Hennighausen, L. (1988) *Biochem. J.* 356, 391–396.

Palmiter, R.D., and Brinster, R.L. (1986) *Annu. Rev. Genet.* 20:465–499.
Pittius, C.W., Hennighausen, L., Lee, E., et al. (1988a) *Proc. Natl. Acad. Sci. USA* 85, 5874–5878.
Pittius, C.W., Sankaran, L., Topper, Y.J., and Hennighausen, L. (1988b) *Mol. Endocrinol.* 88, 1027–1032.
Rosen, J.M., Poyet, P., Goodman, H., and Lee, K-F. (1989) Biochemical Society, Symposium, *Gene Expression: Regulation at the RNA and Protein Levels*, Vol 55, 115–123.
Topper, Y.J., and Freeman, C.S. (1980) *Physiol. Rev.* 60, 1049–1106.

CHAPTER 7

Expression of a Silent MUP Gene in Transgenic Mice

H. Jin Son
Kate Shahan
Eva Derman
Frank Costantini

The mouse major urinary proteins (MUPs) are secreted proteins of low molecular weight that are synthesized in the liver as well as in the mammary, submaxillary, sublingual, parotid, and lachrymal glands (Shaw et al. 1983; Shahan and Derman 1984; Shahan et al. 1987a). Although the function of MUPs has yet to be established, they belong to a family of proteins that include retinol-binding protein, lactoglobulin and apolipoprotein D, all of which bind small hydrophobic ligands (Godovac-Zimmerman 1988). This familial relationship suggests that MUPs may serve as binding proteins for as yet unidentified small molecules.

The MUPS are encoded by a family of approximately 35 closely related genes, all of which are linked on chromosome 4 (Bennett et al. 1982; Bishop et al. 1982; Shi and Derman, submitted). It is not known precisely how

---

This research was supported by Grant AGO7538 from the Public Health Service and Grant CD-315 from the American Cancer Society to E.D., and by Grant CA22376 from the Public Health Service to F.C.

many of these genes are functional and, indeed, several of them (the "group 2" genes; Bishop et al. 1982) are known to be pseudogenes. At least six distinct MUP mRNA sequences, MUP I through MUP VI, have been identified and shown to account for the great majority of all MUP mRNAs (Shahan et al. 1987a and 1987b). MUPs I, II, and III are found to be predominant in the liver, while MUP II is also found in the mammary gland, MUP IV is expressed in the lachrymal and parotid glands, MUP V primarily in the submaxillary gland, and at a lower level in the sublingual gland, and MUP VI in the parotid gland. Genomic clones including several of these genes have been isolated (Bishop et al. 1982; Clark et al. 1985; Held et al. 1987; Shahan et al. 1987b; Shi and Derman, submitted).

Shi et al. (1989) have recently shown that the MUP V mRNA is encoded by one member of a pair of very closely related genes, designated *Mup*-1.5a and *Mup*-1.5b. The two genes appear to be the product of a recent gene duplication, and they differ at only three positions: an A to C transition at position 817 of the mRNA, a C to T transition at position 1029 in intron II, and a deletion of the sequence $(GT)_3$ within a tract of $(GT)_{15}$ in intron III. In addition, in the sequences in the 5'- and 3'-flanking regions (500 nucleotides in each case), the two genes are identical, and the analysis of restriction sites in the flanking DNA suggests that the region of conservation extends over at least 35 kb. Nevertheless, in Balb/c mice, only the *Mup*-1.5a gene and not the *Mup*-1.5b gene is expressed (Shi et al. 1989).

In this chapter we shall describe our surprising discovery that the *Mup*-1.5b gene, although silent in its normal genomic position, is capable of high-level, tissue-specific expression in the submaxillary gland when introduced into new genomic loci in transgenic mice. We argue that the *Mup*-1.5b gene, represents a new class of *Mup* genes, the silent *Mup* genes, which are distinct from both the expressed genes and the pseudogenes, and we discuss possible mechanisms of silencing.

7.1 RESULTS

7.1.1 The *Mup*-1.5b Gene Is Expressed in the Submaxillary Gland of Transgenic Mice

None of the differences in sequence observed between the *Mup*-1.5a and 1.5b genes appeared sufficient to explain their differential expression (Shi et al. 1989). Therefore, we reasoned that a difference in sequence of a regulatory element outside the region examined was likely to be responsible for the inactivity of the *Mup*-1.5b gene. To begin to localize this putative regulatory element, we generated a series of transgenic mice that carried the silent *Mup*-1.5b gene on a 9.5-kb Hind III fragment, which included about 5 kb of 5'-flanking DNA. The transgene was marked by the insertion of an oligonucleotide into the 5'-untranslated region, so that the *Mup*-1.5b transgene and its mRNA could be distinguished from the endogenous *Mup*-1.5

genes and mRNA. The marked gene (construct "pE") was introduced into mouse zygotes by microinjection, and four transgenic founder mice were produced that carried intact copies of the gene. Transgenic offspring from all four pE founder mice were used for analysis of RNA at three weeks of age, the time at which endogenous MUP V mRNA is maximally expressed (Shi et al. 1989). RNA was isolated from various tissues and examined by RNAse protection analysis, using a riboprobe that distinguishes the marked transgenic mRNA from the endogenous MUP V mRNA (Figure 7-1).

In two of the four transgenic lines, the transgene was expressed at high levels in the submaxillary glands and was not detectable in any other tissues (see Figure 7-1). In the other two lines, no expression of the transgene was detected in any tissue (data not shown). The levels of expression of the transgene averaged 3.7 times the endogenous level of MUP V mRNA in one line and 2.9 times the endogenous level in the other line. We conclude that the *Mup*-1.5b gene, although normally silent, is capable of high-level, tissue-specific expression when introduced into transgenic mice as a 9.5-kb DNA fragment. Therefore, the absence of the expression of the endogenous *Mup*-1.5b gene must be due to the effects of sequences outside the 9.5-kb fragment, rather than to any inherent defect in the gene or in its nearby flanking sequences.

7.2 DISCUSSION

Shi et al. (1989) have described a pair of nonallelic mouse *Mup* genes; one member of the pair (*Mup*-1.5a) is active and the other (*Mup*-1.5b) is structurally similar but functionally silent. The *Mup*-1.5a and 1.5b genes are highly conserved, containing common restriction sites as far as 20 kb from the transcription unit and exhibiting complete sequence identity in the immediate 5'- and 3'-flanking regions. This high degree of homology appears to be the result of a relatively recent gene duplication (roughly 85,000 years ago), followed by coordinate evolution (Shi et al. 1989). While both *Mup*-1.5 genes are located on chromosome 4, closely linked to the rest of the *Mup* genes (Shi and Derman, submitted), their relative location within the *Mup* gene complex is not yet known.

The *Mup*-1.5a gene is expressed primarily in the submaxillary gland and at a much lower level in the sublingual gland (Shahan et al. 1987b). In contrast, the *Mup*-1.5b gene is apparently expressed in neither of these tissues (nor in other tissues known to synthesize MUPs) in normal mice. This absence of expression is not due to any obvious structural defect that would interfere with transcription or processing of RNA, as shown by analysis of the sequence of the *Mup*-1.5b gene and its immediate flanking regions (Shi et al. 1989). It could be explained if an enhancer outside the sequenced region were found to be defective. However, the tissue-specific and high-level expression of the *Mup*-1.5b gene in transgenic mice effectively rules

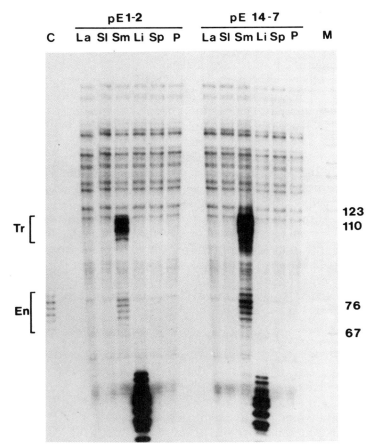

FIGURE 7-1 Tissue-specific expression of the *Mup*-1.5b gene in transgenic mice. Total RNA was isolated from various tissues of two male transgenic mice, one from line pEl and one from line pE14, and 10 μg of each RNA was examined by RNase protection analysis. C, Control RNA from submaxillary gland of a nontransgenic mouse (0.5 μg of polyA+ RNA). La, Lachrymal gland; Sl, sublingual gland; Sm, submaxillary gland; Li, liver; Sp, spleen; p, preputial gland; M, molecular weight markers (pBR322 DNA digested with *Hpa* II). Tr, Fragments protected by transgenic mRNA; En, fragments protected by endogenous MUP V mRNA. The intense bands with high mobility in the lanes loaded with samples from liver result from cross-hybridization to liver MUP mRNAs. The diagram (next page) represents a 240-bp DNA fragment from the 5' end of the transgene, used to generate the RNA probe. The arrow indicates the 5' terminus of the mRNA, and the hatched box a 10-bp oligonucleotide inserted in the first exon of the gene. Hybridization to mRNA from the transgene results in an RNase-resistant fragment of approximately 115 nucleotides, while hybridization to endogenous MUP V mRNA, which lacks the oligonucleotide, results in a shorter (approximately 72 nucleotides) RNase-resistant fragment. Reproduced from Shi et al. (1989).

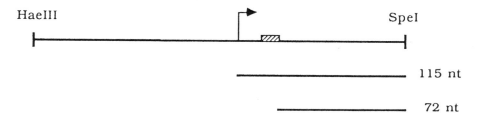

FIGURE 7-1 Continued

out this as well as two other explanations. First, the absence of expression of the *Mup*-1.5b gene cannot be due to instability of the corresponding mRNA, caused by the A to C transversion in the 3′-untranslated region because the mRNA accumulates to high levels in the transgenic mice. Second, although it is difficult to rule out the possibility that the endogenous *Mup*-1.5b gene is expressed in a different tissue or at a different developmental stage from expression of the 1.5a gene, the observed pattern of expression of the 1.5b gene in transgenic mice is inconsistent with this hypothesis. We are left with the conclusion that the *Mup*-1.5b gene is a potentially functional gene, complete with its own tissue-specific regulatory elements, that is silent in its normal chromosomal position. The *Mup*-1.5b gene, therefore, is not a pseudogene but represents a new class of "silent" *Mup* genes, which are potentially functional but normally inactivated by an as yet undefined mechanism.

The mechanism of silencing of the *Mup*-1.5b gene is not yet known. One possibility is that when the duplication of the *Mup*-1.5 gene occurred, or thereafter, the 1.5b gene was separated from a distant positive-activating element, as was the β-globin gene in the Dutch β-thalassemia (Kioussis et al. 1983; Grosveld et al. 1987). However, the high level of expression of the *Mup*-1.5b gene in transgenic mice argues against this possibility. A more likely possibility is that the *Mup*-1.5b gene is silenced by a negative position effect, which is escaped by the transgene. Indeed, the sensitivity of the *Mup*-1.5b gene to negative position effects is demonstrated by the complete absence of expression of this gene in two of the four transgenic lines that were analyzed. At the sites at which the gene has integrated in these two transgenic lines, as well as in its normal chromosomal location, the *Mup*-1.5b gene seems to be located in a chromosomal domain that prohibits its expression. It may be possible to distinguish between these two models by producing transgenic animals that carry the *Mup*-1.5b gene together with larger regions of flanking chromosomal DNA (i.e., on cosmid clones or yeast artificial-chromosome clones). The negative position effect model predicts that the *Mup*-1.5b gene would fail to be expressed when transferred together with enough DNA to include the putative "silencing" element.

It appears that the *Mup*-1.5b gene is not unique as a silent gene. Shi et al. (1989) have argued that several other members of the *Mup* gene family are likely to be silent genes. In addition, certain members of families of both the alpha-globin and growth hormone genes may represent additional examples of silent genes; they are not normally expressed in vivo, yet they are capable of expression after transfection into cultured cells and, thus, are not pseudogenes (Hill et al. 1985; Pavlakis et al. 1981; Seeburg 1982).

In addition to its implications for the mechanism of silencing, the expression of the *Mup*-1.5b gene in transgenic mice establishes an assay for the *cis*-acting DNA sequences that lead to expression in the submaxillary gland. By microinjecting the gene together with varying amounts of 5'-flanking DNA, we have shown that an upstream element is required for expression in the submaxillary gland, and we are in the process of localizing this element (unpublished data). Similar studies with other members of the family of *Mup* genes should allow us to identify the regulatory elements responsible for the differential tissue-specific expression of various *Mup* genes.

REFERENCES

Bennett, K.L., Lalley, P.A., Barth, R.K., and Hastie, N.E. (1982) *Proc. Natl. Acad. Sci. USA* 79, 1220–1224.

Bishop, J.O., Clark A.J., Clissold, P.M., Hainey, S., and Francke, U. (1982) *EMBO J.* 1, 615–620.

Clark, A.J., Chave-Cox, A., Ma, X., and Bishop, J.O. (1985) *EMBO J.* 4, 3167–3171.

Godovac-Zimmerman, J. (1988) *Trends in Biochemical Sciences* 13, 64–66.

Grosveld, F., Assendelft, G.B., Greaves, D.R., and Kollias, G. (1987) *Cell* 51, 975–985.

Held, W.A., Gallagher, J.F., Hohman, C.M., Kuhk, N.J., Sampsell, B.M., and Hughes Jr., R.G. (1987) *Mol. Cell. Biol.* 7, 3705–3712.

Hill, A.V.S., Nicholls, R.D., Thein, S.L., and Higgs, D.R. (1985) *Cell* 42, 809–819.

Kioussis, D.E., Vanin, E., Delange, T., Flavell, R.A., and Grosveld, F.G. (1983) *Nature* 306, 662–666.

Pavlakis, G.N., Hizuka, N., Gorden, P., Seeburg, P.H., and Hamer, D.H. (1981) *Proc. Natl. Acad. Sci. USA* 78, 7398–7402.

Seeburg, P.H. (1982) *DNA* 1, 239–249.

Shahan, K., Gilmartin, M., and Derman, E. (1987a) *Mol. Cell. Biol.* 7, 1938–1946.

Shahan, K., Denaro, M., Gilmartin, M., Shi, Y., and Derman, E. (1987b) *Mol. Cell. Biol.* 7, 1947–1954.

Shahan, K., and Derman, E. (1984) *Mol. Cell. Biol.* 4, 2259–2265.

Shaw, P.H., Held, W.A., and Hastie, N.D. (1983) *Cell* 32, 755–761.

Shi, Y., Son, H.J., Shahan, K., Rodriquez, M., Costantini, F., and Derman, E. (1989) *Proc. Natl. Acad. Sci. USA*, 86, 4584–4588.

CHAPTER 8

The Activation and Silencing of Gene Transcription in the Liver

Sally A. Camper
Shirley M. Tilghman

Members of multigene families often have programs for gene expression that differ developmentally or in terms of tissue specificity. We have focused our attention on the genes that encode albumin and α-fetoprotein (AFP), the major serum proteins in the developing mammalian fetus (reviewed by Tilghman 1985). In fetal liver, the activation of transcription of these two evolutionarily related genes is coordinated, and the tight linkage between the genes led to the proposal that they might share regulatory elements (Kioussis et al. 1981). However, after birth, these genes have independent programs for developmental regulation. In contrast to the continued high rate of transcription of the albumin gene in adults, levels of AFP mRNA are reduced to a basal value approximately 10,000 times lower than levels during late gestation (Tilghman and Belayew 1982). At birth and again at weaning the overall pattern of genes expressed in the liver changes dramatically (Greengard 1971). By studying the DNA sequences required for activation of transcription of the genes for albumin and AFP, as well as the silencing of transcription of the AFP gene at parturition, we hope to gain

---

This work was supported by grant CA44976 (S.M.T.) and fellowship GM10237 (S.A.C.) from the National Institutes of Health.

a better understanding of ways in which the changing patterns of gene expression in the liver are orchestrated.

Inactivation of the transcription of a gene can be envisioned to occur by many different mechanisms. For example, the inactive state might arise because certain positive regulatory factors are no longer available, perhaps as a result of competition by other genes for the same factors (Choi and Engel 1988), or as a result of a simple decline in the concentration of the various factors. Alternatively, in spite of the continued presence of a necessary factor or factors, their action is blocked by a dominant repressor. Such a repressor could act by displacing a positive factor (Barberis et al. 1987) or by displacing RNA polymerase itself (Ptashne 1986), or even by binding at a site different from those recognized by other positive factors. If inactivation is achieved via binding of a repressor molecule to DNA, complete inactivation might require that either a single or multiple binding sites be occupied (Ptashne 1986). In order to distinguish between and refine these models, an initial step is to identify the *cis*-acting sequences that are required for silencing and to determine whether they are distinct from those required for activation of the gene. There are numerous examples of negative factors that can also act as inducers, for example, Lambda repressor, ara C, and the glucocorticoid receptor (reviewed in Raibaud and Schwartz 1984; Yamamoto 1985). Some repressors, however, apparently act strictly as repressors (for example, cro; Ptashne 1986). The precedent of a dual role for many repressors as activators of gene transcription establishes the potential relevance of the postnatal silencing of transcription of the AFP gene to the simultaneous activation of other genes in the liver.

The regulatory elements required for activation and high-level expression of the AFP and albumin genes have been defined by experiments with cell cultures and transgenic animals. Both genes have tissue-specific, promoter-proximal elements located within 200 bp of the sites of inhibition of transcription (Gorski et al. 1986; Godbout et al. 1986). In addition, the albumin-AFP locus contains several enhancers. One of these lies approximately 10 kb upstream from the albumin gene (Pinkert et al. 1987), and three others are located within the 15-kb intergenic region, 6.5, 5.0, and 2.5 kb upstream from the AFP gene (Godbout et al. 1988). In order to understand the different developmental programs involved in transcription of these genes, and to explore the significance of the close linkage between the genes, the ability of each gene to function independently has been tested in transgenic mice. In addition, the developmental regulation of chimeric albumin and AFP transgenes has been tested. Our results indicate that the AFP gene functions appropriately when separated from the albumin gene (Hammer et al. 1987), while albumin transgenes are activated at low levels and with poor penetrance in the absence of the AFP-specific enhancer elements (Camper and Tilghman 1989). Analysis of chimeric genes, which consist of the albumin enhancer paired with the AFP transcription unit, plus the three AFP enhancers paired with the albumin gene, has shown that repression of the AFP gene is mediated by its promoter or by intragenic elements, but not by its enhancers (Camper and Tilghman 1989).

8.1 FUNCTION OF MULTIPLE AFP ENHANCER ELEMENTS IN TISSUE SPECIFICITY AND GENE ACTIVATION

The three enhancers located between the structural genes for AFP and albumin are redundant in transient expression assays (Godbout et al. 1986 and 1988). The existence of such multiple elements is probably more common than is currently appreciated because distal regions are often not examined. Furthermore, if their effects are nonadditive, multiple elements escape detection in studies that employ simple 5'-sequential deletions. We considered the possibility that the AFP enhancers are not redundant in vivo. Perhaps the diversity of tissues that express AFP is achieved through various combinations of multiple, nonredundant elements. When the tissue specificity of each of the enhancers was tested in conjunction with the AFP promoter by germline transformation of mice, it became apparent that the elements are not redundant. Rather, each has a different pattern of activity in each of the three fetal tissues that naturally express AFP. In the yolk sac, where endogenous expression of AFP is higher than in any other tissue, each of the enhancers exhibited equivalent activity. In the liver, the enhancers displayed additive behavior, although only the -5- and -2.5-kb enhancers had significant activity. In the gastrointestinal tract, where the AFP message is normally a minor mRNA, only the -2.5-kb enhancer exhibited marked activity (Hammer et al. 1987). Thus, the diverse tissue specificity of AFP arises not from the existence of multiple enhancers with unique specificity but rather from the combined action of enhancers with somewhat broader specificities and different strengths, similar to the action of the regulatory elements of the gene for alcohol dehydrogenase in *Drosophila* (Maniatis et al. 1987).

Another function of multiple AFP enhancer elements may be to ensure the establishment of an active transcription complex. Experiments both with transgenic mice and F9 teratocarcinoma stem cells demonstrated that the intergenic enhancer domain was a prerequisite for activation of transcription of the AFP gene (Scott and Tilghman 1983; Vogt et al. 1988; Hammer et al. 1987). When all three AFP enhancers were present in a transgene construction, but not when any one was present by itself, 100% of the transgenic mice expressed the linked gene at a high level (Hammer et al. 1987). In contrast, no detectable expression was found in approximately 80% of the transgenic mice that carried the albumin enhancer, and the 20% that did express the gene did so at low levels (Camper and Tilghman, 1989). The potency of the AFP enhancer domain in establishing the transcriptional activity of transgenes, regardless of its position of integration, is remarkable.

The AFP and albumin genes apparently arose by gene duplication (Kioussis et al. 1981). If this duplication occurred without duplication of the regulatory regions, evolutionary pressure would help maintain linkage between the two genes. Our experiments with transgenic mice have shown that AFP can function independently of the albumin gene in terms of activation of transcription, developmental regulation, and reinduction during regeneration of the liver (Hammer et al. 1987). The expression of the albumin gene in transgenic mice, by contrast, appears to be very sensitive to the site of integration (Pinkert et al. 1987), but this sensitivity can be overcome by inclusion of the three AFP enhancers (Camper and Tilghman, 1989), a result that implies that the intergenic enhancer domain may influence expression of both genes. Experiments are in progress to test this model.

8.2 ROLE OF THE AFP PROMOTER-PROXIMAL REGION

The tissue specificity of the AFP enhancers was first probed by transient expression of the gene in cultures of hepatoma cells. The activity of each of the elements, when coupled to a heterologous promoter, is 5- to 10-fold greater in hepatoma cells than in non-hepatoma cells, such as HeLa cells. The activity of the AFP enhancers in cells where the AFP gene is inactive indicates that the enhancers alone are not likely to be responsible for restriction of the expression of the AFP gene to the appropriate tissues (Godbout et al. 1986 and 1988). Rather, the promoter itself may be primarily responsible. One kilobase (kb) of 5'-flanking DNA, containing the AFP promoter, is transcriptionally active in hepatoma cells but not in heterologous cells, such as HeLa or L cells (Godbout et al. 1986). The absence of activity in HeLa cells can be overcome if a strong enhancer, such as the SV40 enhancer, is included (Scott and Tilghman 1983). In this case, the only sequence in the promoter that is required for activity is the TATAA sequence at -33 bp. However, for transcriptional activity in hepatoma cells, there is an absolute requirement for the region that spans -250 to -33 bp (Feuerman et al. 1989). The strict requirement for this proximal regulatory region in hepatoma cells and its repressive effects in heterologous cells, in the absence of the SV40 enhancer, suggest that the promoter-proximal regulatory region may play a crucial role in restricting the expression of the AFP gene to the appropriate tissues.

8.3 REGULATORY ELEMENTS OF THE ALBUMIN GENE

The albumin promoter has been the focus of considerable attention (Heard et al. 1987; Gorski et al. 1986, Babiss et al. 1987; Lichsteiner et al. 1987). A 400-bp region of the rat albumin promoter is sufficient for tissue specificity as assessed by transient expression assays in cell cultures (Ott et al. 1984).

The tissue specificity of the albumin promoter can be overcome by the inclusion of a strong SV40 enhancer (Heard et al. 1987), just as we observed in the case of the AFP promoter (Scott and Tilghman 1983). Thus, in spite of their dissimilar nucleotide sequences, the AFP and albumin promoters have some similar properties.

Transgenic mice have been used in assays of the transcriptional activity of the 5'-flanking region of the albumin gene (Pinkert et al. 1987). In spite of the high level of activity of the albumin promoter in cell cultures (Ott et al. 1984), the same promoter fused to a different reporter gene was essentially inactive in transgenic mice (Pinkert et al. 1987). The inclusion of a region that spans the region 8.5 to 12 kb upstream from the site of initiation of transcription of the albumin gene resulted in a dramatic increase in the percentage of transgenic animals that expressed the reporter gene in the adult liver. However, the absolute level of expression was variable. This element functions in an orientation- and position-independent manner but it failed to activate a heterologous promoter. There are two sites that are hypersensitive to DNase I at approximately -3.5 and -8 kb, which lie in the 5'-flanking region of the DNA between the promoter and the distal enhancer-containing domain. While no function has yet been attributed to these regions, they may indeed play some important role because their inclusion resulted in a progressive increase in the level of expression of the reporter gene (Pinkert et al. 1987).

In contrast to the predictably strong activation of AFP transgenes in yolk sac, liver, and gut (Hammer et al. 1987), a line of transgenic mice that carried the entire 12 kb of the 5'-flanking DNA of the albumin gene, plus an albumin minigene, was not activated in the yolk sac and expression was very weak in the fetal liver and gut (Camper and Tilghman 1989). This result will be pursued with more albumin-transgenic lines of mice. Nonetheless, the differences between the albumin and AFP enhancers in the tissue-specific activation of the genes are apparent.

8.4 DEVELOPMENTAL REGULATION OF EXPRESSION OF AFP AND ALBUMIN GENES

The developmental programs for transcription of AFP genes and albumin differ after birth. Expression of the albumin gene is maintained at a very high level, whereas transcription of the AFP gene declines dramatically until a low basal level of transcription is reached in adult animals (Tilghman and Belayew 1982). The developmental regulation of the transcription of the albumin gene remains to be examined in detail. Several lines of transgenic mice were constructed that contained the albumin enhancer and the entire 12-kb sequence of 5'-flanking DNA, coupled to the albumin promoter and minigene. One line of transgenic mice expressed the gene in the adult liver. When fetal livers and neonatal livers were examined, expression of the transgene was quite low relative to the levels noted in adults of the same line (Camper and Tilghman 1989). If this trend is reproduced in other transgenic lines, it may indicate that full activation of the albumin gene in fetal liver relies on the presence of transcriptional control elements located elsewhere in the locus. The AFP enhancers, lying more than 30 kb downstream from the site of initiation of transcription of the albumin gene, are obvious candidates to serve this function.

We have used transgenic mice to map the regions required for the silencing of transcription of the AFP gene (Krumlauf et al. 1985; Hammer et al. 1987; Camper and Tilghman 1989). To determine whether regulation during development results from stage-specific enhancers, we tested the capacity of the AFP enhancers to direct the postnatal repression of a heterologous gene. For this purpose, we constructed a chimeric gene that consisted of the AFP enhancers fused to the albumin promoter and minigene. In three separate lines of transgenic mice, expression in fetal livers was high, and this high level of expression persisted into adulthood (Camper and Tilghman 1989). In contrast, three transgenic mouse lines carrying an AFP minigene that relied on the albumin enhancer for activation, exhibited repression of the transgene after birth. These results prove that the AFP enhancer domain is not sufficient to direct the developmental program of AFP. In other words, the enhancers themselves are not stage-specific. The decline in transcription after birth is mediated by sequences in the proximal 1 kb of the 5′-flanking DNA, or by sequences within a limited portion of the AFP structural gene. Further experiments are in progress to localize the DNA sequences within this interesting domain and to allow us to develop a mechanistic model for repression. An understanding of the silencing of transcription of the AFP gene may be relevant to understanding changes in transcription of other liver-specific genes at birth.

REFERENCES

Babiss, L., Herbst, R., Bennett, A., and Darnell, J. (1987) *Genes Dev.* 1, 256–267.
Barberis, A., Superti-Furga, G., Busslinger, M. (1987) *Cell* 50, 347–359.
Camper, S., and Tilghman, S. (1989) *Genes Dev.* 3, 537–546.
Choi, O-R., and Engel, J. (1988) *Cell* 55, 17–26.
Feuerman, M., Godbout, R., Ingram, R., and Tilghman, S. (1989) *Mol. Cell. Biol.* 9, 4204–4212.
Godbout, R., Ingram, R., Tilghman, S. (1986) *Mol. Cell. Biol.* 6, 477–487.
Godbout, R., Ingram, R., and Tilghman, S. (1988) *Mol. Cell. Biol.* 8, 1169–1178.
Gorski, K., Carneiro, M., and Schibler, U. (1986) *Cell* 47, 767–776.
Greengard, O. (1971) *Essays in Biochemistry* 7, 159–205.
Hammer, R., Krumlauf, R., Camper, S., Brinster, R., and Tilghman, S. (1987) *Science* 235, 53–58.
Heard, J-M., Herbomel, P., Ott M-O., et al. (1987) *Mol. Cell. Biol.* 7, 2425–2434.

References

Kioussis, D., Eiferman, F., van de Rijn, P., et al. (1981) *J. Biol. Chem.* 256, 1960–1967.
Krumlauf, R., Hammer, R., Tilghman, S., and Brinster, R. (1985) *Mol. Cell. Biol.* 5, 1639–1648.
Lichsteiner, S., Wuarin, J., and Schibler, U. (1987) *Cell* 51, 963–973.
Maniatis, T., Goodbourn, S., and Fischer, J. (1987) *Science* 236, 1237–1245.
Ott, M-O., Sperling, L., Herbomel, P., Yaniv, M., and Weiss, M. (1984) *EMBO J.* 3, 2505–2510.
Pinkert, C., Ornitz, D., Brinster, R., and Palmiter, R. (1987) *Genes Dev.* 1, 268–276.
Ptashne, M. (1986) *A Genetic Switch*, Blackwell Scientific Publications, Palo Alto, CA.
Raibaud, O., and Schwartz, M. (1984) *Ann. Rev. Genet.* 18, 173–206.
Scott R., and Tilghman, S. (1983) *Mol. Cell. Biol.* 3, 1295–1309.
Tilghman, S. (1985) in *Oxford Surveys in Eukaryotic Genes* (Dawkins, R. and Ridley, M., eds.), vol 2, pp. 160–206, Oxford University Press, Oxford.
Tilghman, S., and Belayew, A. (1982) *Proc. Natl. Acad. Sci. USA* 79, 5254–5257.
Vogt, T., Compton, R., Scott, R., and Tilghman, S. (1988) *Nucl. Acids Res.* 16, 487–500.
Yamamoto, K. (1985) *Annu. Rev. Genet.* 19, 209–252.

CHAPTER 9

Gene Targeting in Embryonic Stem Cells

Thomas C. Doetschman

Mouse blastocyst-derived embryonic stem (ES) cells provide us with the opportunity to investigate gene function in the context of the developing and adult animal, and to create animal models for human genetic diseases. The characteristics of ES cells that make this possible are as follows: 1) they can differentiate in vitro and produce large quantities of complex embryonic structures; 2) they can be genetically modified in a site-specific manner by gene targeting; and 3) they can reconstitute a mouse.

This chapter includes an introduction to ES cells and their experimental capabilities. Experiments are reviewed in which transgenic mice were made with ES cells, gene targeting experiments are described in which homologous recombinational events occurred between exogenous DNA sequences and a specific chromosomal locus, and future prospects for gene targeting are discussed.

I would like to acknowledge the support and assistance of Dr. Rolf Kemler, Tubingen, FRG; Dr. Oliver Smithies, Chapel Hill, NC; and the members of their laboratories in which this work was carried out.

9.1 ES CELLS IN VITRO

Lines of mouse and hamster ES cells have been established directly from cultured blastocysts. In both cases, the ES cells are pluripotent and spontaneously differentiate in vitro into complex embryonic structures. In this section, the various procedures for establishing and maintaining ES cells in culture will be outlined, the developmental processes that they undergo in culture will be described, and the usefulness of ES cells as a model for embryogenesis will be discussed.

9.1.1 The Establishment and Maintenance of ES Cell Lines

ES cell lines are derived from the cells of the inner cell mass of mouse and hamster blastocysts. The former can be established from many different strains of mice. In the case of both mouse and hamster, ES cells are pluripotent in culture; and, in the case of the mouse, tumors produced by subcutaneous injection of ES cells into syngeneic animals contain many differentiated structures (Evans and Kaufman 1981; Martin 1981; Wobus et al. 1984; Axelrod 1984; Doetschman et al. 1985 and 1988a). ES cells are maintained in the stem-cell state by growth on a feeder layer of primary embryonic fibroblasts (Wobus et al. 1984; Doetschman et al. 1985 and 1988a) or on a layer of the embryonic fibroblastic cell line, STO (Evans and Kaufman 1981; Martin 1981; Axelrod 1984). When carefully maintained in culture in the undifferentiated state, ES cells can be kept in culture for as long as three months to one year and can be frozen and thawed several times without any apparent loss in developmental potential (Doetschman et al. 1985).

Mouse and hamster blastocysts attached to feeder layers of mouse embryonic fibroblasts are shown in Figures 9–1A and 9–1C. Much as the trophectoderm cells of the blastocyst invade the uterine epithelium, the large trophectoderm cells in culture invade the feeder layer and push the fibroblasts to the periphery. A cluster of cells forming the inner cell mass can be seen in the center. After several mechanical passages with a flame-drawn, glass capillary tube, and further passages in a solution of trypsin and EDTA, lines of ES cells can be established. Figures 9–1B and 9–1D show aggregates of undifferentiated mouse and hamster ES cells cultured on feeder layers of mouse embryonic fibroblasts.

9.1.2 In Vitro Model for Postimplantation Embryogenesis

When separated from the feeder cells, ES cells differentiate into embryoid bodies of increasing complexity. Figure 9–2 shows a series of such embryoid bodies from differentiating mouse (A–C) and hamster (D–F) ES cells. Simple embryoid bodies (A and D) consist of a peripheral layer of endoderm cells that surround undifferentiated ES cells. The endoderm cells contain cyto-

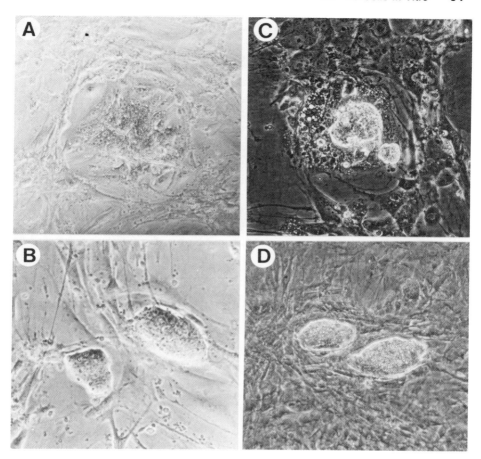

FIGURE 9-1 Establishment of mouse and hamster ES cells. Mouse (A) and hamster (B) blastocysts were flushed from the uterus of superovulated females 3 days and 9 hours (mouse) and 3 days and 5 hours (hamster) after fertilization and placed in culture on mitomycin-C-treated (10 µg/ul for 2 hours) or irradiated (3,000 rads) mouse embryonic fibroblasts with standard ES cell culture medium (15% fetal bovine serum in high-glucose Dulbecco's Modified Eagle's Medium (DMEM) and 10^{-4}M β-mercaptoethanol). The attached blastocysts take on a fried egg appearance with an outgrowth of cells of the inner cell mass surrounded by trophectoderm cells. Aggregates of mouse ES-D3 (C) and hamster ES-Ma1 (D) ES cells are shown after they have been established as cell lines. Note the morphological similarity of the aggregates of ES cells to the cells of the inner cell mass of the attached blastocyst. A, D, ×80; B, C, ×100. Reprinted with permission from Doetschman et al. (1985 and 1988a).

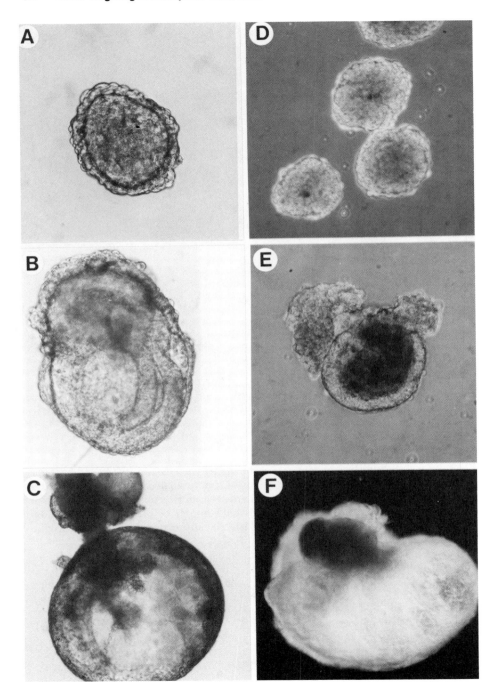

FIGURE 9-2 Development in vitro of ES cells into embryoid bodies. Undifferentiated ES cells were separated from the fibroblastic feeder cells by differential sedimentation (fibroblasts reattach after trypsinization much more rapidly than do ES cells). The ES cells were placed in suspension culture and grown in standard ES cell culture medium for 4 days, after which the concentration of serum was raised to 20%. (A) A simple embryoid body from ES-D3 cells of a mouse after 3 days in suspension culture. (B) A complex embryoid body from mouse ES cells after 6 days in suspension culture. (C) A cystic embryoid body formed from mouse ES cells after 10 days in suspension culture. (D–F) Similar embryoid structures from the hamster ES-Ma2 cell line after 3, 4 and 16 days in suspension culture, respectively. A, ×120; B, ×60; C, F, ×40; D, E, ×100. Reprinted with permission of the publishers from Doetschman et al. (1987a and 1988a).

keratins, form junctional complexes with each other, and secrete an underlying basal lamina. These characteristics suggest that the simple embryoid body can control its inner environment, a condition necessary for further development of the inner cells into the embryonic ectoderm, which can be seen in complex embryoid bodies as bands of columnar epithelial cells below the endoderm layer (B and E). A few days after the appearance of embryonic ectoderm cells, the embryoid bodies become cystic (C and F). In the mouse these cysts have many of the characteristics of the visceral yolk sac. They have blood islands with macrophages and embryonic erythrocytes, and the cystic fluid contains alpha-fetoprotein, transferrin, and several lipoproteins characteristic of the visceral yolk sac (Doetschman et al. 1985). From morphological criteria and the staging of the developing embryonic structures, it is assumed that hamster cystic embryoid bodies are also visceral yolk sac-like structures. Both mouse and hamster cystic embryoid bodies can support cardiogenesis in the form of the development of clusters of synchronously beating myocytes.

In studies of cultured embryoid bodies from ES cells, we have demonstrated that they can be used to assay for factor(s) that induce embryonic hematopoiesis (Doetschman et al. 1985) and perhaps also to assay for factors that induce vasculogenesis (Risau et al. 1987 and 1988). The embryoid bodies can be produced in sufficient quantities for investigations of growth factors that play a role in embryogenesis, for example, the acidic fibroblast growth factor (Risau et al. 1988). The availability of such large quantities of embryonic material has also made it possible to produce cDNA libraries from embryoid bodies in various developmental states (Eistetter 1988). Such libraries should contain cDNAs from low- as well as medium- and high-abundance messages, and they should yield developmentally significant molecules upon differential screening.

9.2 ES CELLS IN VIVO

Mouse ES cells are totipotent in vivo because they can reconstitute a mouse when removed from culture and are introduced into the normal embryonic environment of a blastocyst. Hamster ES cells have not yet been tested for their ability to colonize the hamster germline after introduction into the blastocyst.

9.2.1 Transgenesis by Means of ES Cells

When ES cells are injected back into blastocysts, they can colonize most if not all tissues of the animal, including the germline (Robertson et al. 1986). Genetically modified ES cells can maintain karyotypic stability throughout transfection and selection procedures, and they can stably transmit an expressing transgene through several generations of transgenic mice (Gossler et al. 1986). In the latter case, experimental transgenic mice were generated from blastocyst injections of ES-D3 cells (black, agouti male), which had been transfected with a neo^r gene and selected with the antibiotic G418. The cells were tested for their ability to make embryoid bodies in culture and for a normal chromosome count before injection into blastocysts. G418-resistant cell lines were tested and all had a modal number of 40 chromosomes and all but two of the lines were able to form large quantities of embryoid bodies of the types mentioned above. Injection of two of the most pluripotent cell lines into blastocysts from a black mouse each produced germline chimeras. Figure 9-3 shows a mouse family in which one of these germline chimeras (mostly agouti with a few small black patches on the head) was mated with a black female. The one agouti pup was the product of fertilization by a sperm derived from a neo^r ES cell. In the case of the remaining pups, the sperm were from the blastocyst into which the ES cells had been injected.

The ability to make transgenic mice with ES cells has made it possible to investigate the effects of insertional mutations in transgenic offspring (Bradley and Robertson 1986). In addition, mouse models for human genetic diseases can now be made. For example, mice with a deficiency in hypoxanthine-guanine phosphoribosyl transferase (Hprt), which in man produces the Lesch-Nyhan syndrome, have been made by the introduction of Hprt-deficient ES cells at the blastocyst stage (Hooper et al. 1987; Kuehn et al. 1987). The advantage of this approach to transgenesis is that one can manipulate the mouse genome in culture, check that the manipulation fits the experimental plan, and then make a transgenic mouse with the desired genetic modification.

9.2.2 Nongermline Chimeric Embryos from ES Cells

Valuable information can be gained by studying chimeric embryos derived from genetically altered ES cells. The regulation of the crystallin gene during embryogenesis has been investigated in chimeric embryos made from ES cells transfected with the gene for crystallin (Takahashi et al. 1988). In a study designed to analyze the effects of expression of the middle T antigen of polyoma virus on mouse development, chimeric embryos derived from

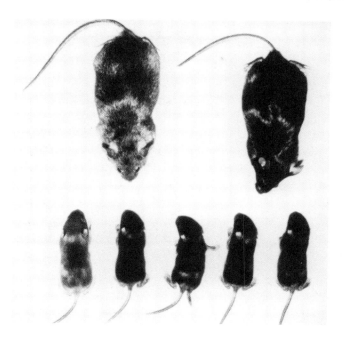

FIGURE 9-3 Transgenic animals produced by use of ES cells. ES-D3 cells (strain 129/Sv-CP; black, agouti male) that had been transfected with the *neo*^{r} gene were injected into C57Bl/6 blastocysts to produce chimeric animals (Gossler et al. 1986). A male that was heavily chimeric with respect to coat color (top left; note patches of black on head) was crossed with a C57Bl/6 (black, nonagouti) female (top right). This particular litter has one black, agouti (bottom left) and four black, nonagouti offspring, proving that the chimeric father was a germline chimera. Reprinted with permission of the publishers from Doetschman et al. (1987a).

polyoma-infected ES cells yielded information on oncogenesis of endothelial cells (Williams et al. 1988). These studies demonstrate that important information can be obtained from chimeric embryos and in somatic chimeric adults without germline chimerism.

9.3 GENE TARGETING

Random integration of foreign DNA is the preferred pathway for the integration of DNA into mammalian genomes, whereas in yeast, homologous recombination is the preferred pathway (Hinnen et al. 1978; Orr-Weaver et al. 1983). The existence of the enzymatic machinery for homologous recombination in mammalian cells was demonstrated in plasmid × plasmid

96 Gene Targeting in Embryonic Stem Cells

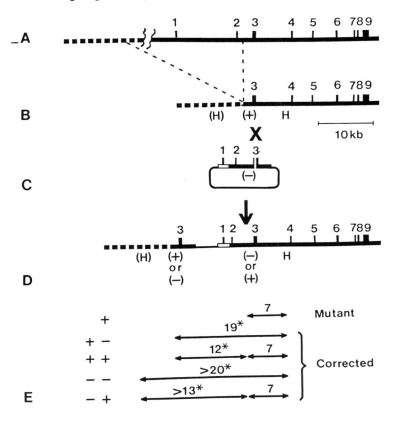

experiments (Folger et al. 1982; de Saint Vincent and Wahl 1983; Miller and Temin 1983). Later, modification of integrated plasmid sequences by exogenous plasmid DNA (plasmid × chromosomal plasmid experiments: Smith and Berg 1984; Smithies et al. 1984; Lin et al. 1985) demonstrated that chromosomal loci could also be targeted by homologous recombination. In general, these experiments were designed in such a way that a selectable function resulted only in the event of homologous recombination between the plasmid sequences.

9.3.1 Gene Targeting in Somatic Cells

Genetic modification of a natural chromosomal locus by gene targeting was first demonstrated in mouse erythroleukemia cells (containing a human chromosome 11), in the human beta-globin locus by first selecting for a marker gene (neo^r) and then sib-selecting with a gene-rescue assay (Smithies et al. 1985). Recently, deletions in IgM genes of mutant hybridoma cell lines have been corrected by homologous gene targeting (Baker et al. 1988; Baker

FIGURE 9-4 Targeted correction of a deletion mutation in the *Hprt* gene of ES cells. (A) The normal mouse *Hprt* gene shows its nine exons (not to scale) and the spontaneous deletion (dashed, thin lines) that produced the *Hprt* mutation in the ES cell line, E14TG2a (see Hooper et al. 1987). The interrupted heavy line represents DNA at an undetermined distance 5' to the *Hprt* locus. (B) The *Hprt* locus in the deletion mutant E14TG21. H indicates a Hind III site present in the chromosome but absent in the correcting plasmid. (C) The 12.4-kb plasmid (pNMR133), used to correct the deletion by homologous recombination, was cut in exon 3 as shown before introduction into the ES cells. The open box represents human sequences; the heavy line, mouse sequences; and the continuous thin line, pAT153 sequences (not to scale). The (−) sign shows where the (+) Hind III site was removed by the insertion of 4 bp during construction of the plasmid. (D) The *Hprt* locus after functional correction by homologous recombination. The corrected locus can have any combination of the two (+) or (−) sites, depending on the occurrence of branch migration, heteroduplex repair, and exolytic degradation with gap repair during the recombination, and on the sites where crossing over is resolved. (E) The predicted sizes and locations of Hind III fragments from the Hprt⁻ mutant and from corrected Hprt⁺ cells having the indicated combinations of the (+) or (−) sites and expected to hybridize to a probe specific for exon 3. The asterisks indicate fragments that should also hybridize to a probe specific for the plasmid vector. The uncertainty in sizes of two of the fragments stemmed from lack of knowledge of the location of the (H) site. Subsequent Southern analysis revealed the lengths of these fragments to be 22 and 15 kb, respectively. Procedure: ES-E14TG2a cells ($4-10 \times 10^7$) were electroporated in the presence of 5 nM linearized targeting fragment in standard ES cell culture medium. Selection was applied after 1 day by growth in HAT medium for 14 days. Reprinted with permission from Doetschman et al. (1988a).

and Shulman 1988). In such experiments, the targeted cells were ultimately isolated by sib-selection using a hapten-based complement assay system.

9.3.2 Gene Targeting in ES Cells

Gene targeting in ES cells combines the advantages of site-specific genetic modification with the developmental potential of ES cells. The genetic modification occurs in the natural chromosomal environment with all the regulatory machinery in its proper place. The cells provide experimental versatility because they provide a model in vitro system for embryogenesis, in which targeted modification of developmentally significant genes can be investigated and the targeted cells can be used to reconstitute mice.

Targeted gene modification in ES cells was first performed at the X-linked *Hprt* locus in male ES cells (Thomas and Capecchi 1987; Doetschman et al. 1987b). In these experiments, the ultimate selection of the targeted cells was based upon the state of expression of the *Hprt* gene. In the latter experiments, a mutant *Hprt* gene, with the upstream region and first two exons deleted, was corrected by homologous recombination between an exogenous vector that contained the deleted sequences (including the promoter region) and the endogenous *Hprt* gene. The gene targeting scheme is depicted in Figure 9-4. The targeting vector was designed to produce an insertional recombination event in which the exogenous sequences would become inserted at the *Xho*I site within the third exon of the gene, resulting in restoration of a functional *Hprt* gene. Eighteen HAT-resistant colonies of targeted cells were analyzed. Twelve were of the (+)(+) pattern, indicating that a cross-over event with exolytic degradation and gap repair had occurred; seven were of the (+)(−) pattern, indicating a simple cross-over event; and one was a mixed colony with several patterns. No (−)(−) pattern was found. The frequency of the gene targeting event with respect to cells that survived the electroporation treatment was about three per million. Transgenic mice have recently been made from these cells (Koller et al. 1989) and from cells targeted in a similar manner by Thompson et al. (1989).

In another set of gene targeting experiments (shown in Figure 9-5) the *Hprt* gene was mutated by an homologous recombinational event in which a 1.3-kb fragment that contained the third exon and flanking intronic DNA was replaced by homologous sequences, which had a promoterless neo^r gene inserted into the *Xho*I site of the third exon (Doetschman et al. 1988b). This event resulted in the disruption of the *Hprt* gene at the third exon by the neo^r gene. The neo^r gene was expressed from the *Hprt* promoter and conferred resistance to about 150 µg/ml of G418 on the ES cells. The second to fourth bases of the coding region (TGA) were placed in reading frame with the *Hprt* gene causing the termination of translational of the *Hprt* gene at the beginning of the neo^r gene. It is assumed that the ribosome can "step back" 1 bp to the initiation codon, ATG, of the neo^r gene and begin translation anew (see Peabody and Berg 1986). In most of the targeted cells analyzed, the recombination occurred as predicted. However, in two of the six targeted recombinants, the *Eco*RI site at the 5' cross-over point was missing. Amplification by the polymerase chain reaction (PCR) of a fragment that spanned this cross-over point revealed small deletions, suggesting that sequencing of critical regions of a recombinant may be required to confirm that the modification occurred as planned.

Recently, schemes have been developed to isolate cells in which nonselectable genes have been targeted. In one such experiment, the *int-2* gene has been targeted and the cells isolated by a double-selection scheme built into the targeting vector (Mansour et al. 1988), as described in Chapter 4. This targeting scheme will be very useful, provided that the promoters for the marker genes are active in the target locus.

A second scheme has been designed by Kim and Smithies (1988) and involves amplification by the PCR in a screening procedure designed to identify ES cells with the desired targeted modification. This type of targeting scheme will be useful in cases where a selection scheme will not work because the promoters for the selectable markers are not active in the target locus. In such an experiment, five targeted ES cells (those produced by the scheme shown in Figure 9-5) were mixed with 50,000 of the parental, nontargeted

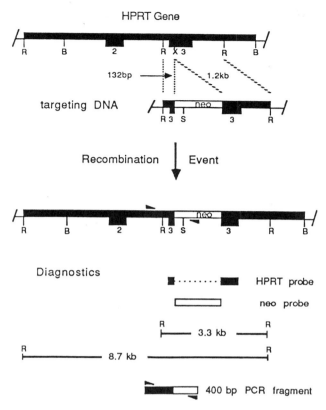

FIGURE 9-5 Targeted mutation of the mouse *Hprt* gene of ES cells. The homologous recombination event requires the replacement of the endogenous 1.3-kb *Eco*RI fragment, which includes the third exon of the *Hprt* gene, with a 3.3-kb fragment from pTDA139 in which the 2-kb promoterless *neo*^{r} gene has been inserted into the *Xho*I site of the exon. Dashed lines define the extents of homology between the target *Hprt* locus and the targeting DNA. Diagnostic restriction fragments are shown to scale underneath the recombination scheme. For purposes of clarity the sizes of exons are exaggerated. Filled narrow bar, Hprt-encoding DNA; thick closed bar, DNA exons; open bar, *neo*^{r} gene; R, B, and X, *Eco*RI, *Bam*HI, and *Xho*I sites, respectively; S, translational stop-stepback-start ATGA sequence. Procedure: ES-D3 cells (2×10^7) were electroporated at 800 Volts/cm and 200 μF capacitance in the presence of 1.3 nM targeting fragment in standard ES cell culture medium. Two minutes later the cells were replated. Targeted colonies were obtained by selection with either 150 μg/ml G418 for 5 days, then 10 μM 6-thioguanine for 11 days, or, alternatively, no selection for 4 days followed by treatment with 6-thioguanine for 11 days. Reprinted from Doetschman et al. (1987b).

ES-D3 cells. Sequences were amplified by the PCR, with one primer specific for sequences unique to the target DNA and another specific for sequences unique to the targeting DNA (see Figure 9–5). The method allowed detection of the five targeted cells, demonstrating that PCR amplification will enable us to isolate targeted ES cells by dilution cloning when selection schemes fail.

9.4 FUTURE PROSPECTS FOR ANIMAL MODELING VIA GENE TARGETING

The experiments discussed here have demonstrated that it will be possible to make predetermined genetic alterations in the mouse germline. It remains, however, to be resolved whether or not silent genetic loci are accessible to targeted gene modification. Fortunately, there is evidence that this will not be a widespread problem. In the report referred to above, in which the human β-globin locus was targeted by homologous recombination in mouse erythroleukemia cells that contained the human chromosome 11 (Smithies et al. 1985), experiments to target the human gene for β-globin in human bladder carcinoma cells were also carried out. Although no targeted cell lines were isolated, chromosomal DNA containing targeted β-globin sequences was rescued from a pool of electroporated cells in two independent experiments. These results strongly suggest that the silent β-globin locus in these cells was accessible to targeted gene modification.

In summary, combining gene targeting with the potential of ES cells to provide a model for embryogenesis in vitro and with their potential to reconstitute mice, we will be able to make animal models for genetic diseases and to study the function of regulatory and structural sequences in the context of embryonic and whole-animal models.

REFERENCES

Axelrod, H.R. (1984) *Dev. Biol.* 91, 227–234.
Baker, M.D., Pennell, N., Bosnoyan, L., and Shulman, M.J. (1988) *Proc. Natl. Acad. Sci. USA* 85, 6432–6436.
Baker, M.D., and Shulman, M.J. (1988) *Mol. Cell. Biol.* 8, 4041–4047.
Bradley, A., and Robertson, E. (1986) *Curr. Topics Dev. Biol.* 20, 357–371.
de Saint Vincent, B.R., and Wahl, G.M. (1983) *Proc. Natl. Acad. Sci. USA* 80, 2002–2006.
Doetschman, T.C., Eistetter, H., Katz, M., Schmidt, W., and Kemler, R. (1985) *J. Embryol. Exp. Morphol.* 87, 27–45.
Doetschman, T., Gossler, A., and Kemler, R. (1987a) in *Future Aspects In Human In Vitro Fertilization* (Feichtinger, W. and Kemeter, P., eds.) pp. 187–195, Springer Verlag, Heidelberg.
Doetschman, T., Gregg, R.G., and Maeda, N., et al. (1987b) *Nature* 330, 576–578.

Doetschman, T., Maeda, N., and Smithies, O. (1988b) *Proc. Natl. Acad. Sci. USA* 85, 8583–8587.
Doetschman, T., Williams, P., and Maeda, N. (1988a) *Dev. Biol.* 127, 224–227.
Eistetter, H.R. (1988) *Eur. J. Cell Biol.* 45, 315–321.
Evans, M.J., and Kaufman, M. (1981) *Nature* 292, 154–156.
Folger, K.R., Wang, E.A., Wahl, G., and Capecchi, M.R. (1982) *Mol. Cell. Biol.* 2, 1372–1387.
Gossler, A., Doetschman, T., Korn, R., Serfling, E., and Kemler, R. (1986) *Proc. Natl. Acad. Sci. USA* 83, 9065–9069.
Hinnen, A., Hicks, H.B., and Fink, G.R. (1978) *Proc. Natl. Acad. Sci. USA* 75, 1929–1933.
Hooper, M., Hardy, K., Handyside, A., Hunter, S., and Monk, M. (1987) *Nature* 326, 292–295.
Kim, H.-S., and Smithies, O. (1988) *Nucl. Acids Res.* 16, 8887–8904.
Koller, B.H., Hageman, L., Doetschman, T., et al. (1989) *Proc. Natl. Acad. Sci. USA*, 86, 8972–8931.
Kuehn, M.R., Bradley, A., Robertson, E.J., and Evans, M.J. (1987) *Nature* 326, 295–298.
Lin, F.-L., Sperle, K., and Sternberg, N. (1985) *Proc. Natl. Acad. Sci. USA* 82, 1391–1395.
Mansour, S.L., Thomas, K.R., and Capecchi, M.R. (1988) *Nature* 336, 338–352.
Martin, G. (1981) *Proc. Natl. Acad. Sci. USA* 78, 7634–7638.
Miller, C.K., and Temin, H.M. (1983) *Science* 220, 606–608.
Orr-Weaver, T.L., Szostak, J.W., and Rothstein, R.J. (1983) *Methods in Enzymology* 101, 228–245.
Peabody, D.S., and Berg, P. (1986) *Mol. Cell. Biol.* 6, 2695–2703.
Risau, W., Hallman, R., and Sariola, H et al. (1987) in *Current Communications in Molecular Biology. Angiogenesis: Mechanism and Pathology* (Rifkin, D.B. and Klagsburg, M., eds.), pp. 134–138, Cold Spring Harbor.
Risau, W., Sariola, H., and Zerwes, H.-G., et al. (1988) *Development* 102, 471–478.
Robertson, E., Bradley, A., Kuehn, M., and Evans, M. (1986) *Nature* 323, 445–448.
Smith, A.J.H. and Berg, P. (1984) *Cold Spring Harbor Symp. On Quant. Biol.* 49, 171–181.
Smithies, O., Gregg, R.G., Boggs, S.S., Koralewski, M.A., and Kucherlapati, R.S. (1985) *Nature* 317, 230–234.
Smithies, O., Korelewski, M.A., Song, K.-Y., and Kucherlapati, R.S. (1984) *Cold Spring Harbor Symp. on Quant. Biol.* 49, 161–170.
Takahashi, Y., Hanaoka, K., and Hayasaka, M., et al. (1988) *Development* 102, 259–269.
Thomas, K.R., and Capecchi, M.R. (1987) *Cell* 51, 503–512.
Thompson, S., Clarke, A.R., Pow, A.M., Hooper, M.L., and Melton, D.W. (1989) *Cell* 56, 313–321.
Williams, R.L., Courtneidge, S.A., and Wagner, E.F. (1988) *Cell* 52, 121–131.
Wobus, A.M., Holzhausen, H., Jakel, P., and Schoneich, J. (1984) *Exp. Cell Res.* 152, 212–219.

CHAPTER 10

Expression of a Human Multidrug-Resistance cDNA (*MDR*1) under the Control of a β-Actin Promoter in Transgenic Mice

Hanan Galski
Glenn T. Merlino
Michael M. Gottesman
Ira Pastan

One of the major mechanisms by which cultured cells become resistant to chemotherapeutic drugs, such as *Vinca* alkaloids, doxorubicin (Adriamycin), colchicine, and actinomycin D (Akiyama et al. 1985; Beck et al. 1979; Biedler and Riehm 1970; Dano 1973; Juliano and Ling 1976; Shen et al. 1986a), is by expression of an energy-dependent drug-efflux pump (Dano 1973; Elder et al. 1988; Gottesman and Pastan 1988; Horio et al. 1988). The human gene that encodes this multidrug transporter is called *MDR*1. This gene, as well as the mouse and hamster *mdr* genes, has been cloned and shown to encode a 4.5-kb mRNA that is present in multidrug-resistant cell lines (Riordan et al. 1985; Scotto et al. 1986; Shen et al. 1986b; Van der Bliek et al. 1986), in the large and small intestine, in the kidney, liver, and adrenal glands (Fojo et al. 1987a), and in many tumors (Fojo et al. 1987a; Fojo et al. 1987b; Goldstein et al. 1989). Expression vectors carrying cloned human *MDR*1 or mouse *mdr* cDNAs confer multidrug resistance on mouse and human cells after transfection or infection with retroviral vectors (Gros et al. 1986a; Guild et al. 1988; Pastan et al. 1988; Ueda et al. 1987a).

The protein product of the *MDR*1 gene is a 170-kDa membrane glycoprotein, termed P-glycoprotein (Ueda et al. 1986). Use of specific antisera has permitted identification of P-glycoprotein in various multidrug-resistant cell lines (Beck et al. 1979; Juliano and Ling 1976; Riordan et al. 1985; Scotto et al. 1986; Shen et al. 1986a and 1986b; Van der Bliek et al. 1986). P-glycoprotein is found in the plasma membrane of resistant cells (Willingham et al. 1987) and binds both cytotoxic drugs (Cornwell et al. 1986) and ATP-binding subunits of transport proteins (Cornwell et al. 1987). Sequence analysis has shown homology to ATP binding subunits of transport proteins (Chen et al. 1986; Gros et al. 1986b).

The effectiveness of cancer chemotherapy is limited by the drug-resistance of tumor cells and by toxicity of drugs to normal cells, such as those in bone marrow. In the case of many drugs, myelosuppression is dose-limiting and more chemotherapy could be used for treatment if the bone marrow could be protected. We reasoned that expression of a human *MDR*1 cDNA sequence under the control of an appropriate heterologous promoter might allow generation of transgenic mice that express the *MDR*1 gene in bone marrow and in other tissues in which it is not normally expressed. We chose the β-actin promoter for our original studies since β-actin is abundantly expressed in a wide range of eucaryotic cells and is evolutionarily conserved (Elder et al. 1988; Ng et al. 1985; Pollard and Cooper 1986). In this chapter, we demonstrate that an *MDR*1 cDNA, under the control of a chicken promoter for β-actin, confers drug resistance when stably expressed in human and mouse cells. In transgenic mice, the *MDR*1 transgene is expressed in cells of the bone marrow, spleen, skeletal muscle, and ovary. As a result of this expression, transgenic mice become resistant to reduced peripheral white blood counts (leukopenia) induced by the cytotoxic drug daunomycin.

10.1 CONSTRUCTION AND TRANSFECTION OF β-ACTIN PROMOTER-*MDR*1 PLASMIDS

To produce a fragment for microinjection into mouse embryos, a 330-bp fragment of the promoter of the β-actin gene from chicken was first inserted 5' to *MDR*1 cDNA in a unique *Sal*I site of a human *MDR*1 cDNA clone (pHG1, Figure 10–1A). To remove all potentially deleterious vector sequences from the microinjected fragment, an internal *Xho*I β-actin promoter site was utilized which, upon digestion with *Xho*I, left a 270-bp promoter

10.1 Construction and Transfection of β-Actin Promoter-MDR1 Plasmids

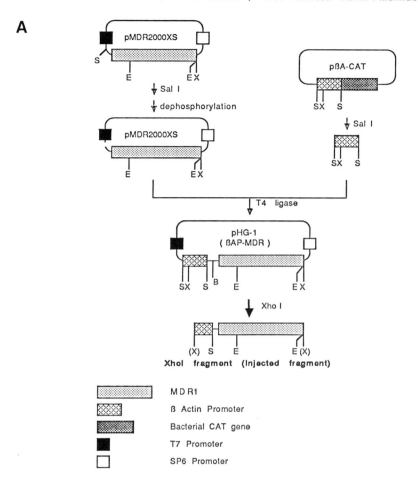

FIGURE 10-1 Scheme for the construction of expression vectors pHG1 and pHG2. (A) Construction of pHG1 and isolation of the pronuclear injected fragment. Letters indicate restriction enzymes used for cloning and orientation characterization: B, *BamH*I; E, *Eco*RI; N, *Nde*I; S, *Sal*I; X, *Xho*I. Restriction enzyme sites within parentheses indicate those sites destroyed after ligation (continued next page).

fragment adjacent to *MDR*1 that contained the TATA and CAAT boxes (Figure 10-2A). To determine whether this truncated promoter was still active, a second β-actin promoter-*MDR*1 plasmid was constructed, which contains the most 3' 270-bp promoter sequences but is identical to pHG1 in all other respects (pHG2, Figure 10-1B). Both the original (pHG1) and the truncated (pHG2) β-actin gene promoter-*MDR*1 constructions were then tested by transfection into drug-sensitive KB-3-1 cells. pHaMDR, a retroviral expression vector containing Ha-MSV LTRS (Ueda et al. 1987a), was

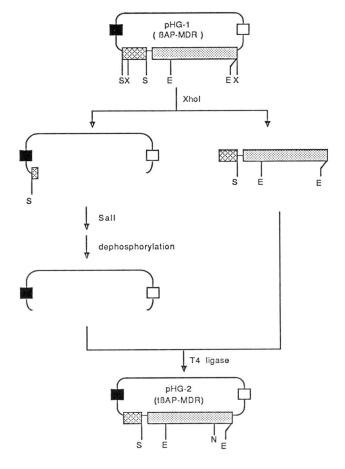

FIGURE 10-1 continued Scheme for the construction of expression vectors pHG1 and pHG2. (B) Construction of pHG2. Letters indicate restriction enzymes used for cloning and orientation characterization: B, *BamH*I; E, *EcoR*I; N, *Nde*I; S, *Sal*I; X, *Xho*I. Restriction enzyme sites within parentheses indicate those sites destroyed after ligation.

used as a positive control. For a negative control, cells were treated by the same transfection protocol in the absence of DNA or with a vector DNA without *MDR*1 sequences.

10.1 Construction and Transfection of β-Actin Promoter-MDR1 Plasmids

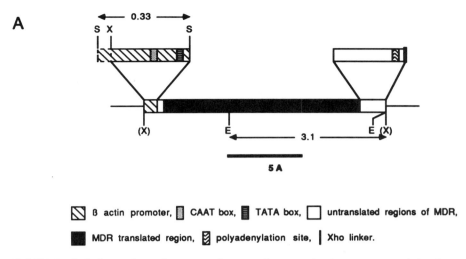

FIGURE 10-2 Screening of mouse tail genomic DNA by Southern hybridization. (A) Diagrammatic representation of the β-actin promoter-*MDR*1 fusion transgene integrated in the mouse genome. A 3.1-kb labeled fragment is expected after *Eco*RI digestion of a mouse genomic DNA and hybridization with the 5A probe, which is derived from the middle part of *MDR*1 cDNA, if the transgene is integrated. Symbols and abbreviations of restriction sites are described in the legend to Figure 10-1 and in that figure. This figure is reproduced with permission from Galski et al. (1989) (continued next page).

10.1.1 Transfection of the Transgene to Human Cells

KB-3-1 cells were transfected with 10 μg of either pHG1, pHG2, or pHaMDR DNA and selected with colchicine at 5–8 ng/ml, a concentration that is three to five times as high as the LD_{50} of colchicine for the parental KB-3-1 cells. The experiments summarized in Table 10-1 show that pβAP-MDR (pHG-1) and pHaMDR produced about the same number of resistant colonies at 5 ng/ml of colchicine. The average size of colonies was, however, smaller for the pβAP-MDR transfectants. This difference in size, as well as the differences in the number of colonies at higher concentrations of colchicine, suggests that the HaMSV promoter is more active than the β-actin promoter in KB cells. Expression of *MDR*1 under control of the truncated β-actin promoter (tβAP-MDR) was similar to or higher than the expression of *MDR*1 under the larger β-actin promoter.

10.1.2 Transfection of the Transgene to Mouse Cells

To confirm that the β-actin *MDR*1 expression vectors function in mouse cells, we also introduced these vectors into NIH 3T3 cells and measured the colony-forming ability of parental NIH 3T3 cells and transfectants at

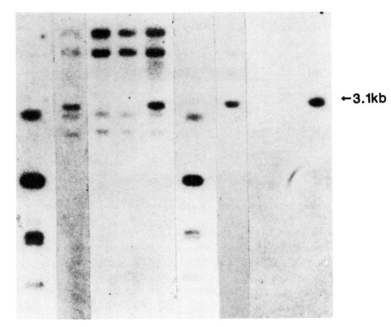

FIGURE 10-2 continued Screening of mouse tail genomic DNA by Southern hybridization. (B) Southern blot analysis of genomic DNA (20 μg) digested with *EcoRI* and hybridized with the 5A probe under conditions of low stringency (lanes 1–5), and under conditions of high stringency (lanes 6–10) (Galski et al. 1989). Genomic DNA isolated from: KB-3-1 cells (lanes 1 and 6); normal mouse cells mixed with 10 pg of injected DNA fragment (lanes 2 and 7); normal mouse (lanes 3 and 8); negative mouse from a litter produced after pronuclear injection (lanes 4 and 9); transgenic mouse from the same litter (lanes 5 and 10). This figure is reproduced with permission from Galski et al. (1989).

60 ng/ml colchicine. In this experiment (data not shown), a similar, relative colony-forming ability in the presence of colchicine was achieved for NIH 3T3 transfectants with either the longer or truncated actin-promoter construction. Therefore, it appeared that the truncated β-actin promoter was a reasonable promoter for expression of the *MDR1* cDNA in transgenic mice.

10.2 PRODUCTION OF MICE THAT CARRY THE βAP-*MDR1* TRANSGENE

The 4.7-kb fragment obtained from an *XhoI* digest of βAP-*MDR1* (see Figure 10–1A) was microinjected into fertilized eggs to generate *MDR1* transgenic mice. The mice were screened by Southern blot analysis of tail genomic

10.3 Transmission of MDR1 to Progeny

TABLE 10-1 Frequency of Drug-Resistant Colonies[a]

Plasmid	Colonies per Dish at Given Concentrations of Colchicine (ng/ml)		
	5	6	8
pHAMDR	620	130	25
pHG1 (pβAP-MDR)	680	32	1
pHG2 (ptβAP-MDR)	890	46	7
No DNA	0	0	0

[a]KB-3-1 (1×10^6 cells/10-cm dish) were transfected with 10 μg of DNA. Forty-eight hours after transfection, cells were split into four dishes and cultured in the presence of colchicine at the indicated concentration. On day 12, cells were stained and colonies counted. Reproduced with permission from Galski et al. (1989).

DNA digested with *Eco*RI. Blots were hybridized with the 5A probe (see Figure 10-2A). Southern blot analysis of mouse tail DNA (see Figure 10-2B) showed the expected unique 3.1-kb internal fragment from the human *MDR*1 cDNA as well as the mouse endogenous fragments, which could be detected under hybridization conditions of low stringency (lanes 1–5) but not under high stringency (lanes 6–10). For each blot, the 4.7-kb injected fragment was mixed with negative mouse genomic DNA, digested with *Eco*RI, and applied to the gel as an internal control for complete *Eco*RI digestion of the genomic DNAs. This sample was also used as a standard to indicate the size and to estimate the number of copies of *MDR*1 in each transgenic mouse (see Figure 10-2B, lanes 2 and 7). An *Eco*RI digest of human KB-3-1 genomic DNA was also used to confirm the estimated number of copies of *MDR*1 (see Figure 10-2B, Lanes 1,6). The DNA samples from the *MDR*1-positive mice were also analyzed by slot-blot analysis to confirm the copy number in each transgenic mouse (Figure 10-3A).

Five founder mice containing integrated human *MDR*1 sequences were generated (see Figure 10-3): Mouse 39 (female) had a low copy number (1 to 3 copies); mouse 132 (female) contained 100 to 200 copies of *MDR*1 DNA; mouse 168 (male) carried approximately 50 copies; and mouse 93 (male) had a high copy number and an additional *MDR*1 fragment detected by a Southern analysis, which probably represented rearranged DNA sequences or junction fragments with host DNA (data not shown). This mouse died soon after weaning. Mouse 104 (female), which had around 10 copies, killed her first litter and died during her second pregnancy.

10.3 TRANSMISSION OF MDR1 TO PROGENY

Each of the three remaining transgenic founder mice (M39, M132, and M168) was mated with the parental C57Bl/6XSJL (F_1) mouse and tail DNA from their progeny was analyzed. Two of the founder mice (M39 and M132)

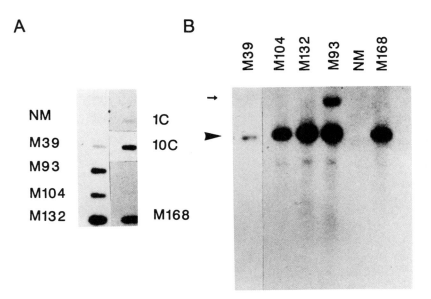

FIGURE 10-3 Blot-hybridization analyses of DNA from founder mice that carry the *MDR*1 transgene. (A) DNA slot-blot analysis of 10 μg of tail DNA, denatured and hybridized with the 5A probe under high-stringency conditions (Galski et al. 1989). NM, normal mouse; M39-M168, transgenic founder mice; 1C and 10C indicate numbers of copies of the transgene, as explained in the legend to Figure 10–2B for lanes 2 and 7. (B) Southern blot analysis of tail DNA from founder mice described in the legend to Figure 10–3A. Hybridization was performed with the 5A probe under conditions of high stringency. ▶, 3.1-kb fragment; →, fragment of a rearranged product of the transgene.

transmitted the integrated *MDR*1 sequence through the germline to about 50% of the progeny. No major variations in the number or structure of the acquired DNA sequences could be detected in F_1 and F_2 generations and their descendants (Figure 10-4, and data not shown). The DNA analyses of the inheritance of the human *MDR*1 cDNA in the third founder (M168) revealed that out of the 32 first-generation offspring, only three carried the introduced *MDR*1 gene. This third founder is probably a mosaic.

10.4 EXPRESSION STUDIES IN *MDR*1 TRANSGENIC MICE

Founder mouse M39 and its positive progeny were mated with parental C57Bl/6XSJL mice to generate a *MDR*1 heterozygous mouse line, termed line MDR-39. Total RNA was prepared from 18 tissues (brain, liver, kidney, spleen, heart, lung, stomach, small intestine, colon, skin, tail, bone, bone marrow, skeletal muscle, ovary, uterus, oviduct, and testes) of seven F_2

10.4 Expression Studies in *MDR*1 Transgenic Mice

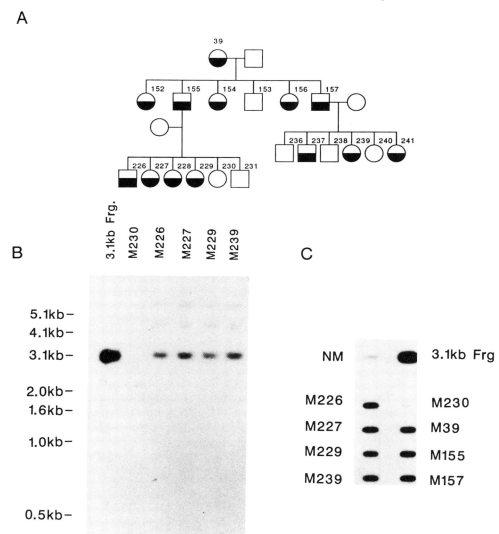

FIGURE 10–4 Inheritance of the *MDR*1 transgene in mice generated from founder M39 (see Figure 10–3). Panel A. Pedigree of M39 and descendant mice. Squares, males; circles, females; half-filled symbols, carrier mice; open symbols, noncarriers. (B) Southern blot analysis of tail DNA from F_2 generation mice. Mouse numbers correspond to numbers shown in the pedigree above. Molecular weight standards, using a 1-kb ladder, are indicated. (C) Slot-blot analysis of tail DNA from mouse 39 and from its F_1 and F_2 generation progeny. Probe and hybridization conditions as described in the legend to Figure 10–3.

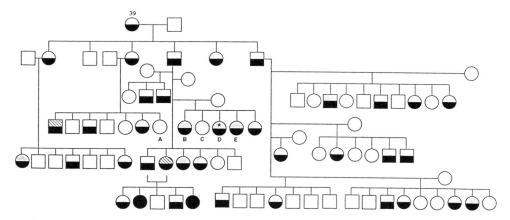

FIGURE 10-5 Pedigree expression analyses of the MDR-39 mouse line. Squares, males; circles, females; open symbols, noncarrier mice; half-filled symbols at bottom, heterozygotes; completely filled symbols, homozygotes; half-stippled symbols at top, mice expressing the transgene as detected by RNA studies; half-hatched symbols at top, mice expressing the transgene as detected by protein immunofluorescence localization. *, Mice expressing the transgene in spleen but not in bone marrow. Mice marked by letters are those for which RNA analysis is shown in Figure 10-6.

transgenic mice (for pedigree see Figure 10-5) as well as from 10 negative sibling mice. These RNA samples were analyzed by slot-blot hybridization with *MDR*1 probes (Figures 10-6 and 10-7). The blots were subsequently rehybridized with a γ-actin probe to control for accurate loading and filter transfer of RNA (see Figure 10-6, and data not shown). In each experiment, total RNA samples from multidrug-resistant KB cell lines, known to contain elevated levels of *MDR*1 mRNA, were included so that the unknown samples could be directly compared to samples with known levels of *MDR*1 RNA and known multidrug resistance. Relative to drug-sensitive KB-3-1, the multidrug-resistant subline KB-8-5 has a 40-fold increase in MDR1 and KB-V-1 (Vbl) has a greater than 500-fold increase (Fojo et al. 1987a).

The experiments summarized in Figure 10-6 reveal that the human *MDR*1 RNA is expressed at significant levels in bone marrow and spleen. Lower expression was detected in skeletal muscle and ovary (see Figure 10-7B). The same blots were cross-hybridized with three different contiguous *MDR*1 probes (10, 5A, and PVUII #6) which, under these high-stringency conditions, define approximately 85% of the *MDR*1 cDNA (see Figure 10-

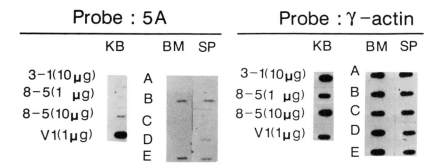

FIGURE 10-6 Analysis of expression of human *MDR*1 RNA in normal and transgenic mice. A slot-blot of total RNA samples (10 μg) extracted from bone marrow and spleen of mice A through E (as indicated in Figure 10-5) was hybridized with the *MDR*1 5A probe (left) and human γ-actin probe (right) under conditions of high stringency and under conditions of low stringency, respectively (Galski et al. 1989). 3-1, Parental drug-sensitive KB cell line; 8-5 and V-1, multidrug-resistant sublines; BM, bone marrow; SP, spleen. Reproduced with permission from Galski et al. (1989).

7A). All positive tissues showed similar results with all three probes (see Figure 10-7B), with the exception of skeletal muscle, which hybridized to a significantly greater extent to the 5' probe (probe 10) than to the others. These results suggest that in all of the tissues which express human *MDR*1 RNA, with the exception of skeletal muscle, the whole *MDR*1 mRNA is expressed. Northern blot analysis (data not shown) showed a diffuse band of RNA, which ranged from 4.5 (the expected full-length *MDR*1 message) to around 11 kb. These results suggest that the endogenous polyadenylation signal at the 3' end of the *MDR*1 cDNA was only weakly effective in the transgenic mice.

Of seven MDR-39 mice tested, six showed relatively high levels of expression in bone marrow, with somewhat lower expression in spleen. However, one mouse showed expression in spleen but not in bone marrow. One male mouse showed expression in kidney and liver in addition to skeletal muscle, as well as a high level of expression in bone marrow and spleen. Some expression was detected in the ovary, but none in the uterus or oviduct of transgenic females.

The human *MDR*1 RNA was expressed in bone marrow and spleen at levels as least as high as those in KB-8-5 cells (see Figure 10-6). Slot-blot analysis of RNA isolated from tissues of transgenic and control sibling mice

114 Expression of MDR1 under the Control of a β-Actin Promoter

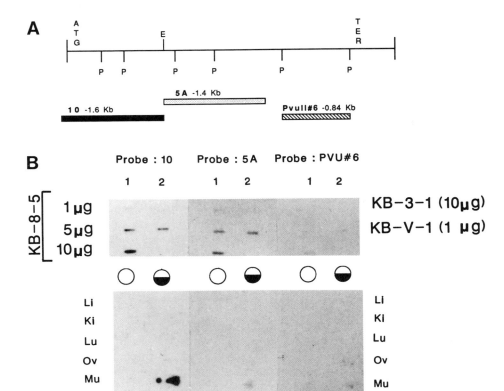

FIGURE 10-7 Tissue-specificity of expression of the transgene. (A) Simplified restriction map of *MDR*1 cDNA and probes used in this study. E, *Eco*RI; P, PVUII. (B) Slot-blot hybridization of total RNA (10 µg) extracted from: upper, 1, KB-8-5 drug-resistant cell lines; lower, 1, *MDR*1 noncarrier mouse; 2, *MDR*1 carrier mouse. Li, liver; Ki, kidney; Lu, lung; Ov, ovary; Mu, skeletal muscle; BM, bone marrow; Sp, spleen; He, heart; Br, brain. The same blot was hybridized with different probes, as indicated.

under low-stringency hybridization conditions showed that the levels of the human *MDR*1 RNA in bone marrow were at least 20-fold higher than those of the *mdr* (Galski et al. 1989).

To confirm the results obtained by RNA-blot hybridization and to determine whether the human *MDR*1 cDNA is expressed under the control of the β-actin promoter, primer-extension studies of RNA extracted from tissues that expressed the transgene were performed with a human-specific, 35-base synthetic oligonucleotide (Figure 10–8, and data not shown). This primer does not cross-hybridize to mouse *mdr* sequences under the conditions of high stringency used in these studies (Ueda et al. 1987b). The primer-extension analysis of RNA samples from the transgenic mice resulted in a unique and specific extension product of around 192 bases, as expected if the *MDR1* cDNA is expressed under the control of the β-actin promoter. These results suggest that there is a single, major site of initiation of transcription of the transgene under control of the β-actin promoter rather than under control of a mouse endogenous promoter adjacent to the integration site.

10.5 EXPRESSION OF P-GLYCOPROTEIN DETECTED BY IMMUNOFLUORESCENCE

To confirm that the RNA was expressed as protein, immunofluorescence studies were performed with a mouse monoclonal antibody directed against the human P-glycoprotein (MRK-16) (Hamada and Tsuruo, 1986). Surface expression of human P-glycoprotein was detected in most of the bone marrow cells of two transgenic mice examined (Figure 10–9), showing that the *MDR1* RNA is translated. The type of bone-marrow cells expressing the *MDR1* gene and their relative levels of expression of P-glycoprotein have not been determined.

10.6 STUDIES IN VIVO

To determine whether expression of human *MDR1* confers drug resistance in the bone marrow of transgenic mice and, therefore, prevents bone-marrow suppression, groups of transgenic and nontransgenic sibling mice were injected with a single dose of the cytotoxic drug daunomycin. The drug was injected intraperitoneally at a dose that is twofold higher than the LD_{50}. This dose is highly toxic to bone marrow of mice and humans, with its greatest deleterious effect on the population of white blood cells (WBC), which results in a reduced number of leukocytes (leukopenia) measured in bone marrow and in peripheral blood (Gabizon et al. 1986; Maral and Jouanne 1981).

The number of WBC in peripheral blood of the daunomycin-injected mice was measured both before and after injection of the drug (Figure 10–10). Eight days after treatment, the WBC counts of the *MDR1*-negative mice dropped to a level which was, on average, threefold lower than the normal counts. The WBC counts of the *MDR1*-positive mice were, however, significantly higher than those measured in the *MDR1*-negative mice, and on

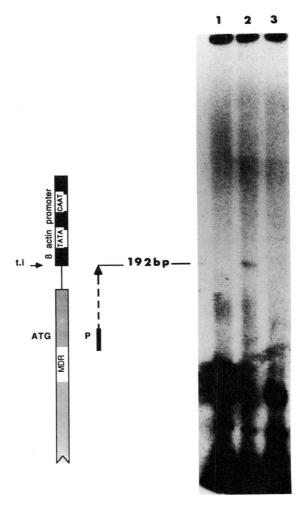

FIGURE 10-8 Primer-extension analysis. 30 μg of RNA extracted from spleen of *MDR*1 non-carrier (1) and carrier (2) mice or NIH 3T3 cells (3) were hybridized to a 5' end-labeled, synthetic oligonucleotide of 35 bases, as indicated. Primer-extended products were analyzed in a 5% polyacrylamide gel that contained 7 M urea. The major expected product is indicated by the arrow.

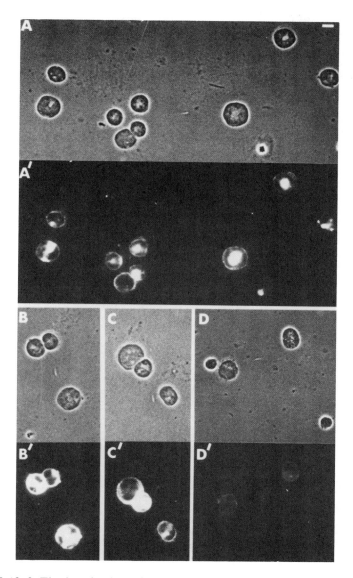

FIGURE 10–9 The localization of human P-glycoprotein by immunofluorescence in bone-marrow cells from an *MDR*1 transgenic mouse. Bone-marrow samples from either an *MDR*1 transgenic mouse (A,B,C) or a normal sibling (D) were smeared on glass slides, air-dried and then fixed in formaldehyde. The smears were labeled using monoclonal antibody MRK16 (antihuman P170) and indirectly labeled with rhodamine. Equal-time exposures show a high level of expression of P-glycoprotein in all cells from the MDR mouse (A'–C'), but not in cells from the control mouse (D'). Panels (A) through (D) represent phase-contrast images of the cells shown in (A') through (D'). (Magnification, ×630; bar, 6.5 μm). Reproduced with permission from Galski et al. (1989).

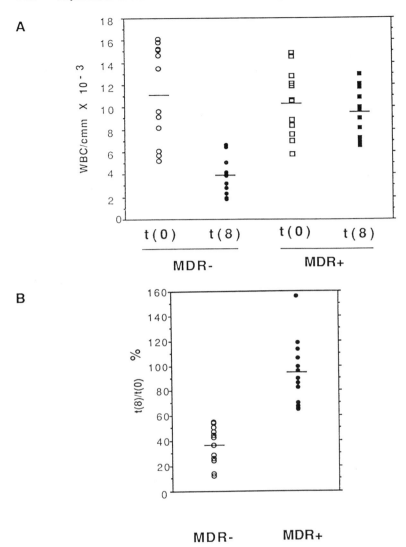

FIGURE 10-10 The effects of daunomycin on white blood cells (WBC) from normal and *MDR*1 transgenic mice in peripheral blood. Groups of transgenic (n = 13) and nontransgenic (n = 13) sibling mice (six-month-old males, F_2 generation) were injected intraperitoneally with 8.5 mg/kg daunomycin. The WBC in peripheral blood were counted just before treatment (to) and then eight days after injection (t8). (A) Scattergram of total WBC counts of the individual mice; (B) Scattergram of the WBC counts eight days after injection, as percentages of the initial counts. The bars indicate group means. Reproduced with permission from Galski et al. (1989).

average were similar to the normal counts. However, it should be noted that in some of the treated, $MDR1$-negative mice, the WBC level was elevated after the drug treatment. This elevation was not detected in five saline-injected, control mice (data not shown) and it might represent a selection in vivo and an over-proliferation of bone marrow cells that are highly resistant to the drug. The analysis of bone marrow from mice sacrificed eight days after daunomycin treatment and a study of individual populations of WBC are in progress.

The hematocrits of the $MDR1$ and the control mice were unchanged after treatment (data not shown). The numbers of platelets were not determined but none of the mice exhibited bleeding during the experiments.

10.7 EXPRESSION OF THE TRANSGENE: NO OBVIOUS PHENOTYPIC EFFECTS

Histopathological studies of tissues that express the human $MDR1$, as well as those tissues in which expression was not detected, failed to reveal any morphological abnormalities (data not shown). To date, neither obvious phenotypic changes nor functional and behavioral disturbances have been detected in heterozygous and homozygous animals that carry and express the transgene. These data, as well as the fact that the transgene was stably expressed by several generations of mice, suggest that the expression de novo of the product of the $MDR1$ gene in mouse tissue does not affect the normal function of these tissues.

10.8 DISCUSSION

10.8.1 General Aspects

In this chapter, we have described the production of transgenic mice that carry and express the human $MDR1$ cDNA under the transcriptional control of a β-actin promoter. The $MDR1$ transgene is expressed primarily in hemopoietic tissues, such as bone marrow and spleen, and to a lesser extent in skeletal muscle and ovary. Immunofluorescence studies localized the protein product of $MDR1$, namely P-glycoprotein, to the surface of most of the bone marrow cells of the transgenic animals. Expression on the cell surface is required for P-glycoprotein to function as an efflux pump to remove chemotherapeutic drugs from cells (Willingham et al. 1987) and to prevent bone-marrow suppression caused by chemotherapy. The marrow of the transgenic mice was also resistant to leukopenia induced by injection of the anticancer drug daunomycin, a result that strongly suggests that the human P-glycoprotein is a functional protein in the mouse bone marrow cells. It has previously been shown that cloned human $MDR1$ cDNA can confer multidrug-resistance on mouse and human drug-sensitive cells after transfection or infection with retroviral vectors (Pastan et al. 1988; Ueda et al. 1987a). The present study demonstrates that acquired drug resistance can result from expression of the human *MDR*1 gene in culture and in vivo, as was previously suggested by the correlation of drug resistance and expression of *MDR*1 in tumors (Goldstein et al. 1989). These results therefore support the model (Gottesman and Pastan 1988) that P-glycoprotein acts as a multidrug transporter, where its presence in normal or tumor cell results in drug-resistance.

10.8.2 Characterization of the Integration and the Inheritance of the Transgene in Line MDR-39

From the *MDR*1 founder mice and their progeny, we have chosen one line (MDR-39) for expression studies. Analysis of restriction fragment patterns of genomic DNAs from this founder and from over 500 descendant mice (see Figure 10-5, and data not shown) indicate that one to three copies of the injected fragments were integrated into the mouse DNA. Internal rearrangements, deletions, and duplications were not detected. Pedigree analyses indicate that the transgene is: (a) transmitted through the germ line; (b) integrated into a single chromosome, since the transgene is stably inherited by about 50% of the progeny of matings of heterozygous and normal mice; and (c) inherited in an autosomal fashion. It is neither X- nor Y-linked, since male-to-male and female-to-female transmission of the transgene occurred.

It is known that the insertion of foreign DNA sequences into the cellular genome can cause mutational changes by disrupting the function of endogenous genes or of control elements. Most insertional mutations in transgenic mice are recessive but, in most of the matings between heterozygous mice, their embryonic lethal phenotype is demonstrated by a significant reduction in litter size and by the inability to produce homozygous mice (Jaenisch 1988). Since heterozygous matings of the progeny of MDR-9 resulted in litters of normal size and, since phenotypically normal homozygotes could be generated (data not shown), it seems likely that no insertional mutation of endogenous housekeeping or control genes occurred during the integration of the transgene in this mouse line.

10.8.3 Utilizing the Chicken β-actin Promoter in Culture and in the Mouse

10.8.3.1 Expression under Control of the β-Actin Promoter in Culture

The β-actin promoter was used as a heterologous promoter for expression of the human *MDR*1 cDNA because β-actin is abundant in all animal tissues and the promoter should therefore be active in most cell types. The activity of the promoter of the β-actin gene can be inhibited by a region 3' to the promoter sequences and by a small region 5' to the polyadenylation signal (DePonti-Zilli et al. 1988; Elder et al. 1988). These inhibitory regions were not part of our constructions. A comparative study of DNA-mediated transfer of expression vectors that contain a cloned, bacterial gene for chloramphenicol acetyltransferase (CAT) into a wide range of cell types has shown that the β-actin promoter activity is at least as strong as that of the SV40 early promoter (Gunning et al. 1987). When the *MDR*1 cDNA was placed under the control of the β-actin gene promoter and transfected into either human KB cells or mouse 3T3 cells, the cells acquired a drug-resistant phenotype (Galski et al. 1989).

10.8.3.2 The Effect of Deleted Sequences in the Promoter The presence of procaryotic vector sequences can be highly inhibitory to expression of certain genes in transgenic mice. Moreover, in mice carrying a transgene adjacent to procaryotic vector sequences, aberrations such as local instability at the site of integration and insertional mutagenesis has been observed (Jaenisch 1988). To eliminate vector sequences in the injected DNA, we isolated an XhoI fragment of pHG1. This strategy also deleted 60 bp in the 5' region of the β-actin gene promoter. In DNA-mediated transfer experiments, the expression ability of this truncated β-actin promoter was found to be similar to that of a promoter that included the upstream 60 bp of the β-actin promoter. Therefore, we reasoned that the XhoI β-actin promoter-*MDR*1 fragment is adequate for introduction of the human *MDR*1 cDNA into mice.

10.8.3.3 Expression of the Transgene Under the β-Actin Promoter in the Mouse Studies of RNA from the MDR-39 transgenic mice showed that the transgene is expressed in bone marrow, spleen, skeletal muscle, and ovary. Although the *MDR*1 transgene is expressed in several tissues, there is no detectable expression in many other tissues, as might be expected when using a universal promoter such as the promoter of the β-actin gene. Moreover, since we found that DNA-mediated transfer of β-actin promoter-*MDR*1 cDNA to NIH 3T3 fibroblasts confers drug-resistance, we expected to detect expression in tissues rich in fibroblasts. Our failure to find more widespread expression is probably explained by different requirements for expression in transfected somatic cells as compared to those for developmentally regulated expression in transgenic mice, which is demonstrated in the case of some other transgenes and promoters when they are introduced into mice rather than into cultured cells (Brinster et al. 1988; Palmiter and Brinster 1986). Differences in transcriptional activity of the introduced β-actin promoter-*MDR*1 transgene could be accounted for by one or by a combination of the following: the absence of 60 bp of the 5' sequences in the truncated promoter; the presence of mouse endogenous control elements adjacent to the site of integration of the transgene; tissue-specific transcriptional inactivation caused by methylation de novo of the transgene in certain tissues (Jaenisch 1988; Razin 1984); the existence of intrinsic control elements in the *MDR*1 cDNA itself; and preferential mRNA instability or splicing in mouse tissues (Stiles et al. 1976). The last phenomenon may be exemplified by our observation of predominant expression of the 5' portion of the *MDR*1 cDNA in skeletal muscle (see Figure 10–7). It should be noted in this regard that the *MDR*1 transcripts in tissues that express the transgene are heterogeneous in size, probably indicating partial activity of the endogenous polyadenylation signal of the transgene or predominant activity of a mouse endogenous polyadenylation signal that is 3' to the insertion site of the transgene. This phenomenon was recently observed by us in DNA-mediated transfer of an *MDR*1 expression vector that contains long terminal repeats (LTRs) from Moloney murine leukemia virus (MLV) (pGMDR). In that study, competition between polyadenylation signals of the LTR and the *MDR*1 endogenous polyadenylation signal resulted in two distinct, major sizes of message (Ueda et al. 1987a).

10.8.3.4 Variability in Expression of RNA from the Transgene The levels of expression of *MDR*1 RNA in bone marrow of MDR-39 transgenic mice are relatively high. The level of expression is at least as high as that of *MDR*1 mRNA in KB-8-5 drug-resistant cells. These levels (which are 40- to 120-fold higher than the basal level of expression in drug-sensitive cells) are equal to or greater than the levels of *MDR*1 RNA in multidrug-resistant human tumors (Fojo et al. 1987a and 1987b; Goldstein et al. 1989). The variability in levels of expression of RNA in individual transgenic mice can be explained by genetic variations in mice that were generated from crosses of two distinct inbred lines. Moreover, the level of expression might be age-, stress-, or sex-dependent. Sex-dependence is suggested by the observation of expression of *MDR*1 RNA in kidney and liver of transgenic males, but not in females. The analysis of more transgenic mice, including *MDR*1 homozygous and inbred mice, is needed to confirm this observation.

10.8.4 Applications for the MDR Mice
The MDR-39 transgenic mouse line will serve as a model for studies of *MDR*1-dependent multidrug-resistance in vivo. It is likely that these mice may also be useful for testing high-dose chemotherapy regimens and for the development of novel chemotherapeutic agents and treatment protocols aimed at overcoming the obstacle of multidrug-resistance. MDR-39 mice can also be used to help elucidate the normal function of P-glycoprotein in vivo. Because the major site of expression of *MDR*1 in the MDR-39 transgenic mouse line is bone marrow, within a wide range of cell types these transgenic mice should aid in the evaluation of somatic gene therapy in

which expression vectors that contain the *MDR*1 gene and a nonselectable gene are introduced into bone marrow and then selected in vivo. These *MDR*1 expression vectors could also be used to render human bone-marrow cells resistant to multiple drugs, so that cancers can be treated more aggressively.

REFERENCES

Akiyama, S.-I., Fojo, A., Hanover, J.A., Pastan, I., and Gottesman, M.M. (1985) *Somat. Cell Mol. Genet.* 11, 117–126.

Beck, W.T., Mueller, T.J., and Tanzer, L.R. (1979) *Cancer Res.* 39, 2070–2076.

Biedler, J.L., and Riehm, H. (1970) *Cancer Res.* 30, 1174–1184.

Brinster, R. L., Allen, J.M., Behringer, R.R., Gelinas, R.E., and Palmiter, R.D. (1988) *Proc. Natl. Acad. Sci. USA*, 85, 836–840.

Chen, C-J., Chin, J., Ueda, K., et al. (1986) *Cell* 47, 381–389.

Cornwell, M.M., Safa, A.R., Felsted, R.L., Gottesman, M.M., and Pastan, L. (1986) *Proc. Natl. Acad. Sci. USA* 83, 3847–3850.

Cornwell, M.M., Tsuruo, T., Gottesman, M.M., and Pastan, I. (1987) *FASEB J.* 1, 51–54.

Dano, K. (1973) *Biochim. Biophys. Acta.* 323, 466–483.

DePonti-Zilli, L., Seiler-Tuyns, A., and Paterson, B.M. (1988) *Proc. Natl. Acad. Sci. USA* 85, 1389–1393.

Elder, P.K., French, C.L., Subramaniam, M., Schmidt, L.J., and Getz, M.J. (1988) *Mol. Cell Biol.* 8, 480–485.

Fojo, A., Ueda, K., Slamon, D., et al. (1987a) *Proc. Natl. Acad. Sci. USA* 84, 265–269.

Fojo, A.T., Shen, D.-W., Mickley, L.A., Pastan, I., and Gottesman, M.M. (1987b) *J. Clin. Oncol.* 5, 1922–1927.

Gabizon, A., Meshorer, A., and Barenholz, Y. (1986) *J. Natl. Cancer Inst.* 77, 459–469.

Galski, H., Sullivan, M., Willingham, M.C., et al. (1989) *Mol. Cell Biol.* 9, 4357–4363.

Goldstein, L.J., Galski, H., Fojo, A., et al. (1989) *J. Natl. Cancer Inst.* 81, 116–124.

Gottesman, M.M., and Pastan, I. (1988) *J. Biol. Chem.* 263, 12163–12166.

Gros, P., Ben-Neriah, Y., Croop, J., and Housman, D.E. (1986a) *Nature* 323, 728–731.

Gros, P., Croop, J., and Housman, D. (1986b) *Cell* 47, 371–380.

Guild, B.C., Mulligan, R.C., Gros, P., and Housman, D. (1988) *Proc. Natl. Acad. Sci. USA* 85, 1595–1599.

Gunning, P., Leavitt, J., Muscat, G., Ng, S.-Y, and Kedes, L. (1987) *Proc. Natl. Acad. Sci. USA* 84, 4831–4835.

Hamada, H., and Tsuruo, T. (1986) *Proc. Natl. Acad. Sci.* USA 83, 7785–7789.

Horio, M., Gottesman, M.M., and Pastan, I. (1988) *Proc. Natl. Acad. Sci. USA* 85, 3580–3584.

Jaenisch, R. (1988) *Science* 240, 1468–1473.

Juliano, R.L. and Ling, V. (1976) *Biochim. Biophys. Acta* 455, 152–162.

Maral, R.J., and Jouanne, M. (1981) *Cancer Treat. Rep.* 65 (Suppl. 4), 9–18.

Ng, S.-Y, Gunning, P., Eddy, R., et al. (1985) *Mol. Cell Biol.* 5, 2720-2732.
Palmiter, R.D., and Brinster, R.L. (1986) *Annu. Rev. Gen.* 20, 465-499.
Pastan, I., Gottesman, M.M., Ueda, K., et al. (1988) *Proc. Natl. Acad. Sci. USA* 85, 4486-4490.
Pollard, T.D., and Cooper, J.A. (1986) *Annu. Rev. Biochem.* 55, 987-1035.
Razin, A. (1984) in *DNA Methylation: Biochemistry and Biological Significance* (Razin, A., Cidar, H., and Rigs, A., eds.), pp. 127-147, Springer-Verlag, New York.
Riordan, J.R., Deuchars, K., Kartner, N., et al. (1985) *Nature* 316, 817-819.
Scotto, K.W., Biedler, J.L., and Melera, P.W. (1986) *Science* 232, 751-755.
Shen, D.-W., Cardarelli, C., Hwang, J., et al. (1986a) *J. Biol. Chem.* 261, 7762-7770.
Shen, D.-W., Fojo, A., Chin, J.E., et al. (1986b) *Science* 232, 643-645.
Stiles, C.D., Lee, K.L., and Kenney, F.T. (1976) *Proc. Natl. Acad. Sci. USA* 73, 2634-2638.
Ueda, K., Cornwell, M.M., Gottesman, M.M., et al. (1986) *Biochem. Biophys. Res. Commun.* 141, 956-962.
Ueda, K., Cardarelli, C., Gottesman, M.M., and Pastan, I. (1987a) *Proc. Natl. Acad. Sci. USA* 84, 3004-3008.
Ueda, K., Clark, D.P., Chen, C.-J., et al. (1987b) *J. Biol. Chem.* 262, 505-508.
Van der Blick, A.M., Van der Belde-Koerts, Ling. V., and Borst, P. (1986) *Mol. Cell. Biol.* 6, 1671-1678.
Willingham, M.C., Richert, N.D., Cornwell, M.M., et al. (1987) *J. Histochem. Cytochem.* 35, 1451-1456.

CHAPTER 11

Insertion of a Disease Resistance Gene into the Chicken Germline

Donald W. Salter
Lyman B. Crittenden

Avian leukosis virus (ALV) infects chicken cells by first attaching its envelope glycoprotein to cell-membrane receptors. Subsequent transport of virion contents to the cytoplasm and reverse transcription results in the integration of the ALV genome into the chicken chromosomes. The integration near the c-*myc* proto-oncogene in bursal cell DNA can lead to its enhanced expression, which results in the induction of lymphomas and other neoplasms (Fadly 1986; Varmus 1988).

ALV can be classified into different subgroups by the specific interaction of the viral envelope glycoprotein and its cellular receptor. Infection of chicken cells with a particular ALV subgroup will prevent subsequent infection with viruses of the same subgroup by a phenomenon called interference (Vogt and Ishizaki 1966). Based on a report demonstrating that chicken embryo fibroblasts (CEF) expressing subgroup E envelope glyco-

---

This chapter was prepared on official government time and reports research paid for by the American taxpayer. The chapter and the research information, are therefore in the public domain and are not copyrighted.

proteins from the endogenous proviral genes, *ev*3 and *ev*6, were resistant to subgroup E Rous sarcoma virus (RSV) (Robinson et al. 1981), we and others proposed a method for providing resistance to infection by subgroup A ALV (Crittenden and Salter 1985 and 1986; Freeman and Bumstead 1987; Payne 1985). Thus, the insertion of the subgroup A ALV envelope gene into the germline of chickens and its subsequent expression could provide resistance to infection by ALV through interference with binding or penetration of virus. A number of transgenic chickens containing ALV proviral inserts have been produced by injecting replication-competent ALV near the developing embryo of fertile line-0 eggs on the first day of incubation and testing succeeding generations for genetic transmission of proviral DNA (Salter et al. 1986 and 1987; Crittenden and Salter 1988). One proviral insert, *alv*6 (Salter and Crittenden 1989; Crittenden et al. 1989), was originally detected as a dot-blot positive female progeny of line-0 females and a viremic line-0 male made tolerant to a recombinant ALV, RAV-0-A(1) (Wright and Bennett 1986). Unlike the remainder of the transgenic chickens, the blood of this progeny was negative for infectious ALV and p27 antigen (one of the products of the *gag* gene). Since the *env* gene product is translated from a spliced mRNA different from the *gag* or *pol* gene products (Coffin 1985), transcription and translation of *env* may still occur. This brief report describes the characteristics of transgenic chickens carrying the *alv*6 insert.

11.1 MATERIALS AND METHODS

Details on materials and methods have been published in detail elsewhere (Salter et al. 1986 and 1987; Crittenden and Salter 1988; Salter and Crittenden 1989; Hughes et al. 1986).

11.2 RESULTS AND DISCUSSION

11.2.1 In Vitro Envelope Expression

The expression of ALV envelope genes can be measured in a chicken-helper-factor assay (Crittenden et al. 1980). Two sires hemizygous for the *alv*6 proviral insert were mated to line-0 females and CEF prepared from individual 11-day-old embryos. CEF were first co-cultivated with 16Q quail cells, a cell line transformed by the envelope-defective Bryan high-titer (BH) RSV, which produces particles that lack the envelope glycoprotein (Murphy 1977). The CEF spontaneously fuse with 16Q cells and, if the CEF are expressing envelope glycoprotein, they yield focus-forming pseudotype virus that can infect susceptible CEF and form foci. Table 11–1 shows that only CEF containing the *alv*6 insert produced pseudotype virus that transformed CEF resistant to subgroup E ALV. Thus, the defective *alv*6 proviral insert expresses the subgroup A ALV envelope glycoprotein.

TABLE 11–1 Presence of Chicken Helper Factor (Envelope Glycoprotein) in Chicken Embryo Fibroblasts (CEF) That Carry *alv6*

Sire	*alv6* CEF[a]	Number of Foci on C/E CEF[b]	
857	−	0	1
	+	500	386
	+	420	512
	−	0	0
	−	0	1
	−	0	0
858	−	0	0
	+	342	404
	+	416	412
	−	0	0
	+	384	532
	−	0	0
Line 0	Control	0	0

[a] The presence or absence of the *alv6* insert was determined by restriction enzyme analysis of embryo DNA (Crittenden et al. 1989).
[b] CEF, Chicken embryo fibroblasts resistant to subgroup E ALV.

11.2.2 Interference In Vitro

Since subgroup A envelope glycoprotein is expressed, then specific inhibition of focus formation should occur when CEF that contain the *alv6* insert are infected with subgroup A RSV. A portion of the same CEF as described above was infected with tenfold dilutions of subgroup A [BH-RSV (RAV-1)] or subgroup B [BH-RSV (RAV-2)] virus and foci counted after seven days. Table 11–2 shows that only CEF carrying the *alv6* defective proviral insert are highly resistant to focus formation after subgroup A RSV infection. The same cells remain fully susceptible to subgroup B RSV. Based on the differences in the number of foci with subgroup A RSV, CEF containing the *alv6* insert are more than 3,000-fold more resistant to infection by subgroup A sarcoma virus than are CEF lacking *alv6*. This degree of resistance is similar to that reported for CEF containing the *ev6* locus, which expresses subgroup E envelope glycoprotein and blocks subgroup E RSV infection (Robinson et al. 1981).

11.2.3 Interference In Vivo

Since there is significant expression of subgroup A envelope glycoprotein in CEF, chickens containing the *alv6* insert may be protected from infection by pathogenic subgroup A retroviruses. One-week-old progeny chicks from

TABLE 11-2 Interference with Infection by Subgroup A Sarcoma Virus in CEF That Carry *alv6*

Sire	alv6 CEF[a]	Number of Foci on C/E[b] CEF					
		BH-RSV(RAV-1)[c]			BH-RSV(RAV-2)[d]		
		Und	10^{-1}	10^{-2e}	Und	10^{-1}	10^{-2}
857	−	C[f]	C	846	C	C	344
	+	28	2	0	C	C	572
	+	20	8	0	C	C	402
	−	C	C	818	C	C	526
	−	C	C	756	C	C	286
	−	C	C	1024	C	C	446
858	−	C	C	768	C	C	292
	+	4	0	0	C	C	364
	+	20	4	0	C	C	486
	−	C	C	666	C	C	330
	+	28	2	0	C	C	350
	−	C	C	1100	C	C	290
Line 0	Control	C	C	900	C	C	380

[a]The presence or absence of the *alv6* insert was determined by restriction enzyme analysis of embryo DNA (Crittenden et al. 1989).
[b]CEF, Chicken embryo fibroblasts resistant to subgroup E ALV.
[c]Subgroup A RSV.
[d]Subgroup B RSV.
[e]Dilutions of undiluted (Und) virus stock.
[f]Confluent.

TABLE 11-3 Interference of Sarcoma Induction in Chickens That Contain the *alv6* Insert

Progeny[a]	Number Positive for Sarcomas[b]
alv6 +	0/6
alv6 −	14/17

[a]Presence or absence of *alv6* insert determined by dot-blot analysis (Crittenden et al. 1989).
[b]Five hundred focus-forming units of BH-RSV(RAV-1) injected into wing-web at day 7 and sarcomas assayed at two weeks.

matings described above were injected into the wing-web with subgroup A sarcoma virus. None of the six chicks containing the *alv6* proviral insert developed sarcomas whereas 14 of 17 chicks lacking the *alv6* proviral insert had palpable tumors two weeks after injection (Table 11-3).

A similar experiment measured the long-term response of transgenic chickens to infection by a pathogenic field strain of subgroup A ALV. Four *alv*6 hemizygous males were mated with line-0 females. Progeny chicks were injected intra-abdominally with RPL-42 (a field strain of subgroup A ALV; Fadly and Okazaki 1982) at hatching and then reared intermingled in two separate isolators. Sera were collected at the indicated times from random samplings and the presence of subgroup A ALV and antibody to subgroup A ALV determined by standard procedures. Pathogenicity due to ALV was recorded from two through 40 weeks of age. Table 11-4 shows that the *alv*6 chickens showed no evidence of infection to 40 weeks of age as measured by sensitive assays for virus and antibody and none developed lymphoid leukosis. Infection did not occur even though the *alv*6 chickens were constantly exposed to virus shed by their infected hatch-mates. All nontransgenic chickens became viremic, some produced antibody, and many had tumors characteristic of lymphoid leukosis.

11.2.4 Tolerance to Other Pathogenic ALV Subgroups

A possible complication of the expression of the ALV envelope gene throughout the life of the chicken is the induction of immunological tolerance to glycoproteins shared by subgroups of ALV (Crittenden et al. 1987). Progeny chicks from a mating similar to that described in the previous section were injected intra-abdominally with RAV-2 (subgroup B ALV) on day 7 after hatching and then reared intermingled in two separate isolators. Sera were collected at the indicated times from random samplings and the presence of subgroup B ALV and antibody to subgroup B ALV were determined by standard procedures. Pathogenicity due to ALV was recorded from two through 30 weeks of age. Table 11-5 shows that both transgenic and nontransgenic chickens responded similarly to RAV-2. Both populations of chickens developed antibody to RAV-2 and similar numbers had tumors characteristic of lymphoid leukosis. Therefore, there is little induction of tolerance to another subgroup of ALV in *alv*6 chickens expressing significant levels of subgroup A envelope glycoproteins.

TABLE 11-4 Interference with Infection by Subgroup A ALV and Oncogenicity in Transgenic Chickens That Carry *alv*6

Progeny[a,b]	Viremia				Antibody			LL[e]
	2[c]	7	16	40	7	16	40	Total
*alv*6 +	0/36[d]	0/24	0/27	0/27	0/24	0/27	0/27	0/36
*alv*6 −	39/39	20/23	23/25	0/1	1/23	3/25	1/1	22/39

[a]Presence or absence of *alv*6 insert determined by dot-blot analysis (Crittenden et al. 1989).
[b]Chicks were injected intra-abdominally with 10^4 infectious units of field strain subgroup A ALV RPL-42 at hatching.
[c]Weeks of age.
[d]Positive per number of chickens assayed.
[e]Lymphoid leukosis.

TABLE 11-5 Lack of Tolerance to Infection by Subgroup B ALV in Transgenic Chickens That Carry *alv*6

Progeny[a,b]	Viremia		Antibody		Lymphoid Leukosis
	7[c]	16	7	16	
*alv*6 +	3/16[d]	4/28	16/16	26/28	3/30
*alv*6 −	0/31	1/35	29/31	34/35	4/39

[a]Presence or absence of *alv*6 insert determined by dot-blot analysis (Crittenden et al. 1989)
[b]Chicks were injected intra-abdominally with 10^5 infectious units of subgroup B ALV RAV-2 on day 7.
[c]Weeks of age.
[d]Positive per number of chickens assayed.

11.2.5 Other Observations on the *alv*6 Insert

In order to monitor the stability of the defective insert, chickens from similar matings were monitored for 40 weeks. All remained free of subgroup A ALV and subgroup A ALV antibody. Thus, the defective insert appears to be stable for an extended period. The defect in the proviral DNA responsible for this phenotype has not been determined. Both 5' and 3' long terminal repeats and normal-size, proviral internal fragments were present in DNA restricted with *Bam*HI and *Eco*RI. However, one of the two *Sac*I restriction enzyme sites was missing, but its absence may be due to a heterogenous mixture of recombinant retroviruses in the virus stock used to produce the transgenic chickens (Wright and Bennett 1986).

Transgenic chickens homozygous for *alv*6 have been sucessfully produced. Homozygous males and females produce semen and eggs of reasonable fertility. Thus, the *alv*6 proviral insert has not disrupted endogenous genes necessary for development or reproduction. Large-scale productivity trials are planned for the future to determine whether any changes in genes involved in egg productivity or growth traits have been induced.

11.3 CONCLUSIONS

This potentially useful transgenic chicken line was detected in our lines of transgenic chickens that had been generated by infecting germ cells with replication-competent, recombinant ALV. Thus, the successful insertion of a beneficial gene into chicken germ cells and its subsequent expression demonstrate that ALV retroviral vectors can be used to insert foreign genes into the chicken germline.

The remarkable resistance to ALV infection in transgenic chickens carrying *alv*6 suggests that germline insertion of viral envelope genes may be a general approach for producing animals resistant to at least some classes of viruses.

REFERENCES

Coffin J. (1985) in *RNA Tumor Viruses: Molecular Biology of Tumor Viruses* (Weiss, R., Teich, N., Varmus, H., and Coffin, J., eds.), 2nd edition, pp. 17–74, Cold Spring Harbor Laboratory, Cold Spring Harbor, NY.
Crittenden, L.B., Gulvas, G.A., and Eagen, D.A. (1980) *Virology* 103, 400–406.
Crittenden, L.B., McMahon, S., Halpern, M.S., and Fadly, A.M. (1987) *J. Virol.* 61, 722–725.
Crittenden, L.B. and Salter, D.W. (1985) *Can. J. Animal Sci.* 65: 553–562.
Crittenden, L.B. and Salter, D.W. (1986) *Avian Dis.* 30: 43–46.
Crittenden, L.B. and Salter, D.W. (1988) in *Proc. Second Intl. Conf. Quantitative Genet.* (Weir, B.S., Eisen, E.J., Goodman, M.M., and Namkoong, G., eds.), pp. 207–214, Sinauer Associates, Sunderland, MA.
Crittenden, L.B., Salter, D.W., and Federspiel, M. (1989) *Theoret. Appl. Genet.* 77:505–515.
Fadly, A.M. (1986) in *Avian Leukosis* (deBoer, G.F., ed.), pp. 197–211, Martinus Nijhoff Publishing, Boston.
Fadly, A.M. and Okazaki, W. (1982) *Poultry Sci.* 61, 1055–1060.
Freeman, B.M. and Bumstead, N. (1987) *World's Poultry Sci. J.* 43, 180–189.
Hughes, S.H., Kosik, E., Fadly, A.M., Salter, D.W., and Crittenden, L.B. (1986) *Poultry Sci.* 65, 1459–1462.
Murphy, H.M. (1977) *Virology* 77, 705–721.
Payne, L.N. (1985) in *Poultry Genetics and Breeding* (Hill, W.G., Manson, J.M. and Hewitt, D.), pp. 1–16, British Poultry Science Ltd., Longman Group, Harlow, England.
Robinson, H.L., Astrin, S.M., Senior, A.M., and Salazar, F.H. (1981) *J. Virology* 40, 745–751.
Salter, D.W., Smith, E.J., Hughes, S.H. et al. (1986) *Poultry Sci.* 65, 1445–1458.
Salter, D.W., Smith, E.J., Hughes, S.H., Wright, S.E., and Crittenden, L.B. (1987) *Virology*, 157, 236–240.
Salter, D.W. and Crittenden, L.B. (1989) *Theoret. Applied Genet.* 77:457–461.
Varmus, H. (1988) *Science*, 240, 1427–1435.
Vogt, P.K. and Ishizaki, R. (1966) *Virology*, 30, 368–374.
Wright, S.E. and Bennett, D.D. (1986) *Virus Res.* 6, 173–180.

PART III

Use of Transgenics in Understanding Biology

CHAPTER 12

Analysis of Regulatory Genes Using the Transgenic Mouse System

Guerard W. Byrne, Claudia Kappen
Klaus Schughart, Manuel Utset
Leonard Bogarad, Frank H. Ruddle

The mouse is an ideal organism for studies of mammalian embryogenesis and for the development of models for human diseases. Over its long history as an experimental animal, the mouse has accumulated a very large number of developmental, immunological, and biochemical mutations. These mutations arose either spontaneously in inbred lines or as a result of chemical or radiation-induced mutagenesis. Although many of these mutations have been informative, they have been exceedingly challenging with respect to studies at the molecular level because of difficulties in identifying and isolating the mutant genes. Moreover, attempts to identify genes that are important in the regulation of development by randomly mutating a genome the size of the mouse's would be prohibitively laborious and expensive. Thus, these technical problems have greatly limited the degree to which classical genetic techniques can be applied to studies of mouse development.

We are fortunate that recent developments have for the first time permitted us to begin a detailed study of mammalian development at the molecular level and provided broadly applicable tools for engineering defined mutations in the mouse for future study. These developments, which we will discuss in this chapter, include the isolation of mammalian homeobox genes and their presumptive role in controlling vertebrate development and a new technique for regulating genes in transgenic mice, which has been designed to produce gain-of-function mutations.

12.1 DISCOVERY OF THE HOMEOBOX

The homeobox was first identified in *Drosophila* (McGinnis et al. 1984; Scott and Weiner 1984) where it was found as part of the coding region of many of the polarity, segmentation (Nusslein-Volhard and Wieschaus 1980), and homeotic genes (Lewis 1978). These *Drosophila* loci control early embryogenesis by defining the embryonic axis, the number of body segments, and the characteristic differentiation of each body segment (Gehring 1987). While the genes that control polarity and segmentation map to unlinked locations in the genome, the *Drosophila* homeotic genes are arrayed in two major clusters, the *Antennapedia* and bithorax complexes. An interesting aspect of these gene clusters is that the distribution of the genes in the complex is collinear with the region of the fly that is altered when the genes are mutated (Gehring 1987). When the first homeotic genes from *Drosophila* were cloned, it was found that many of them contained a highly conserved sequence of 180–183 bp, which we now call the homeobox. It was also observed that the homeobox sequences were present in many other organisms, including the mouse and in humans. The use of homeobox probes derived from the *Antennapedia* complex has permitted the identification of approximately 50 cross-reactive sequences in the mouse and in humans. This surprising conservation made it possible for researchers to jump from the homeotic genes of *Drosophila* to the homeobox genes of vertebrates, the hope being that the vertebrate homeobox genes might have a developmental function similar to those observed in *Drosophila*. Although there are no known mutations in any of the Antennapedia-type (A-type) vertebrate homeobox genes, there are a number of striking similarities to *Drosophila* in the organization and patterns of homeobox expression that support a potential function in the regulation of development (Ruddle 1989).

The A-type homeobox genes of the mouse are arrayed in four major complexes in a manner similar to the organization of the *Drosophila* genes. In the mouse, the homeobox complexes designated *Hox*-1, 2, 3, and 4 are located on chromosomes 6, 11, 15, and 2, respectively. A similar pattern of clusters is also present in humans, and the homologous complexes have been mapped to chromosomes 2, 7, 17, and 12. Within a complex, all of the *Hox* genes have the same transcriptional orientation. The gene transcripts are typically about 1.5–3.5 kb long with two exons that encode a protein of 25–30 kDa. The homeobox sequence is always in the 5' portion of the second exon.

There is strong evidence that the products of both the *Drosophila* and mammalian homeobox genes act as transcription factors by binding to DNA. The homeodomain encodes a helix-turn-helix motif that exhibits partial homology to yeast and bacterial transcription factors (Laughon and Scott 1984; Shepherd et al. 1984). Analysis in vitro of homeodomain fusions with β-galactosidase or with bacterially produced homeobox gene products have demonstrated that the homeobox sequence provides a specific DNA-binding function (Desplan et al. 1985 and 1988; Fainsod et al. 1986; Hoey and Levine 1988). Several of the *Drosophila* homeobox genes have been demonstrated to be either transcriptional activators or antagonists in transfection studies in yeast, *Drosophila*, and mammalian systems (Fitzpatrick and Ingles 1989; Jaynes and O'Farrell 1988; Thali et al. 1988). Furthermore, the mammalian transcription factors Oct-1 and Oct-2 contain homeodomains (Sturm et al. 1988; Ko et al. 1988; Ingraham et al. 1988). Although the homeobox sequences of Oct-1 and Oct-2 are significantly diverged from the A-type sequence, these transcription factors still retain the helix-turn-helix structural motif. Disruption of this protein structure, at least in the case of Oct-2, abolishes its ability to bind DNA (Ko et al. 1988). It is interesting that the products of homeobox genes often bind to similar *cis*-DNA sequences, albeit with different affinities and with quantitatively different degrees of transcriptional modulation. Thus, it appears that the homeobox proteins function as transcription factors, forming a network of interacting gene products that control both transcription of homeobox genes and the transcription of additional target genes.

12.2 EVOLUTION OF HOMEOBOX GENES

We have catalogued all the available sequence data for the homeobox genes in a variety of vertebrate species and made comparisons within and between species. This analysis has provided insight into the evolution of the homeobox system and underscores the highly conserved nature of these loci (Kappen et al. 1989; Schughart et al. 1989). Figure 12–1 illustrates the organization of the A-type clusters of homeobox genes. There are 23 individual genes. However, the boundaries for these complexes has still not yet been unequivocally defined. For each linkage group, the genes are aligned into nine groups that reflect their degree of sequence similarity (shaded areas). Not explicitly shown in Figure 12–1 is the fact the spacing of the individual genes has been conserved between the various linkage groups. Thus, for example, in *Hox*-1 and *Hox*-2, both the linear order of homologous genes is the same and the spacing between genes along the chromosomes is quite similar (Schughart et al. 1988b). Although data for the other linkage groups is less complete, this conservation of structure appears to be present in all four clusters. Similar conservation of structure is observed in the human.

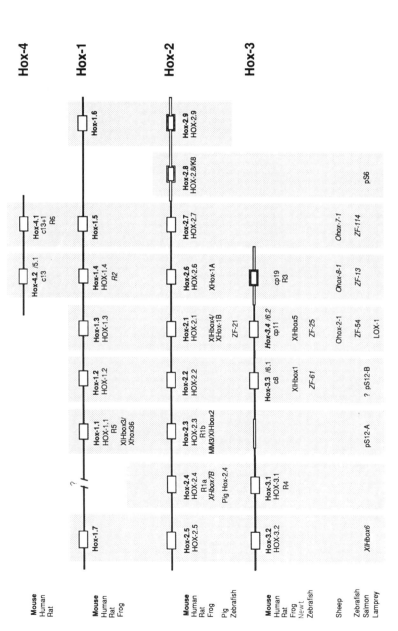

FIGURE 12-1 Classification of nucleotide sequences of homeobox genes. The chromosomal arrangement of the four major mouse homeobox gene clusters is shown schematically. Open lines and boxes represent additional information from the human system. Shaded areas delineate cognate groups of related homeoboxes. Sequences at the bottom of the figure could not be unambiguously assigned to the particular cognate group. Assignment of pS12-B to the *Hox-2*-B to the *Hox-2.2* group is tentative since it is equally similar to the *Hox-2.3* group. Where sequences are believed to represent the same gene, we have included all of the currently used names. Italics indicate incomplete or missing sequences. No attempt has been made to represent the actual physical distances between the genes.

The A-type homeobox sequences used in our evolutionary analysis have at least 55% homology to the Antennapedia homeobox sequence from *Drosophila*. When homologous human and mouse homeobox sequences are compared, there are on average 10.7 bp substitutions with only 6% of these changes producing replacements that alter the amino acid sequences. This high degree of conservation indicates that each such cognate pair shares an ancestral origin and that the formation of the four linkage groups occurred prior to the divergence of mouse and human species. A similar degree of conservation is observed when the mouse homeobox sequences in one linkage group are compared to their cognate sequences in the other linkage groups. In this case, we find an average of 29 bp changes of which only 13% produce amino acid substitutions. This result implies that the four linkage groups arose from an ancestral complex and that, once formed, the homeobox sequences were strongly conserved.

The highest degree of sequence divergence is observed when the individual homeobox sequences within a linkage group are compared to one another. Our analysis of the *Hox*-2 complex indicates that within this linkage group there are on average 51 bp differences between the individual homeobox sequences and that 52% of these differences lead to amino acid substitutions. A similar result is obtained for *Hox*-1, with an average of 56 bp changes. The high proportion of replacement substitutions observed within a linkage group suggests that divergence of homeobox sequences was enhanced during the formation of the ancestral complex through the linear expansion of a primordial homeobox sequence.

We conclude from these studies that the vertebrate homeobox system evolved in two discrete steps. First, during the Precambrian period, prior to the separation of the arthropod and chordate lineages, an ancestral homeobox gene underwent a series of linear duplications to produce a complex. During this expansion phase, the homeobox sequences diverged. After the initial expansion, additional loci were formed in the chordate lineage, apparently through chromosomal or genomic duplications. After the duplication phase, the homeobox sequences were conserved. It is tempting to speculate that the initial expansion and divergence of homeobox genes may have been in part responsible for the diverse proliferation of body forms that are observed in the fossil record from the Precambrian era. Further evolutionary comparisons and experiments that define the developmental function of the homeobox genes will be needed to support this hypothesis. The conservation and absence of rearrangements in the *Hox* complexes may further suggest that the integrity of the *Hox* loci is important for appropriate function. As we discuss below, the transgenic mouse system is now providing the first experimental evidence that this may be the case.

12.3 EXPRESSION OF HOMEOBOX GENES

The expression of many murine homeobox genes has been analyzed by a variety of techniques. It is not our intention to review this work in detail; instead, we will illustrate the more general properties of the homeobox regulatory network by describing the patterns of expression of *Hox*-3.1, 2.5, 2.2, and 2.1, in particular in the developing central nervous system (CNS). Several recent reviews provide a more thorough summary of homeobox expression (Holland and Hogan 1988; Fienberg et al. 1987). The earliest documented expression of a murine homeobox gene occurs late in gastrulation on embryonic day 7.5 to 7.75. In these presomitic embryos, *Hox*-1.5 is detected in the posterior neuro-ectoderm, but not in more anterior domains. Expression of *Hox*-3.1 is detected in the allantois (Gaunt et al. 1986; Guant 1988). Transcription of *Hox*-2.1 and *Hox*-1.3 can also be detected at this time by an RNase protection assay but the level of expression is apparently just beneath the level of sensitivity for in situ hybridization (Jackson et al. 1985; Odenwald et al. 1987).

Later in development, expression of homeobox genes is typically observed in the developing spinal cord, spinal ganglia, ectoderm, endoderm, and lateral and somitic mesoderm. Each of the *Hox* genes has a unique and often developmentally dynamic spatiotemporal pattern of expression in a subset of these tissues. For example, on embryonic day 9.5, the embryo is well formed within 13 to 20 somites and is in the early to intermediate stages of organogenesis. At this time *Hox*-2.5 is expressed in the neural tube, with an anterior boundary at the third cervical somite, and in the lateral and somitic mesoderm up to the level of the fifth cervical somite. Expression of *Hox*-2.5 is also detected in the mesonephric mesoderm that surrounds the hindgut (Bogarad et al. 1989). This pattern of expression, which is essentially the same on day 8.5, changes dramatically from days 10.5 to 13.5. In the 11.5-day embryo, expression of *Hox*-2.5 in the CNS is shifted rostrally so that the anterior boundary of spinal cord expression is now at the first cervical somite. Additionally, expression in the spinal cord, which had previously exhibited no dorsal-ventral localization, is now dorsalizing such that the cells along the outer edge of the ventricular zone have the greatest abundance of *Hox*-2.5 transcripts. The polarization of expression of *Hox*-2.5 continues, so that by day 13.5 only the cells in the developing dorsal horn are labeled (Figure 12-2). During this dorsalizing phase, mesodermal expression of *Hox*-2.5 decreases dramatically. Dynamic changes similar to those described for *Hox*-2.5 are also observed for some other, but not all, homeobox genes. Prior to day 10.5, expression of *Hox*-3.1 is observed in the posterior third of the embryo in all three germ layers. After day 10.5, expression of *Hox*-3.1 is predominantly observed in the developing spinal cord with an anterior limit at the third cervical vertebra. In contrast, expression at this time is greatly reduced or absent in both lateral and somitic mesoderm, in endoderm, and in the spinal ganglia (Awgulewitsch et al. 1986; Le Mouellic et al. 1988).

The developing CNS is a generic site of expression of A-type homeobox genes. Within the developing spinal cord, each *Hox* gene has a well-defined anterior limit at which expression is high. Posterior to this site, expression of a particular *Hox* gene is often lower and more diffuse but generally con-

12.3 Expression of Homeobox Genes 141

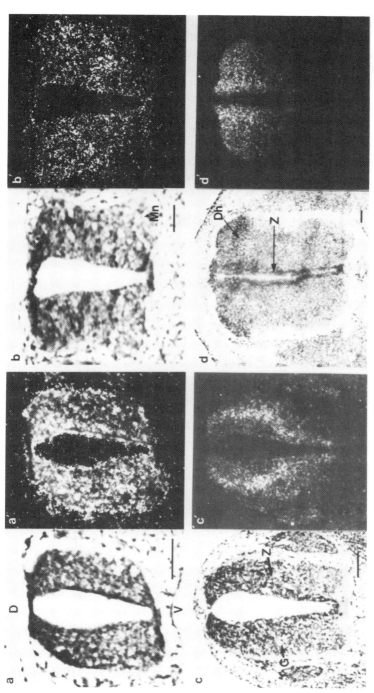

FIGURE 12-2 Expression of *Hox-2.5* along the dorsoventral axis of the developing neural tube. (a, a'–d, d') Cross sections through the upper thoracic spinal cord of 9.5-, 10.5-, 11.5-, and 13.5-day embryos. (a, a') Initially, expression of *Hox-2.5* occurs throughout the dorsoventral axis of the spinal cord. (b, b') Dorsalization of expression of *Hox-2.5* is first detected. No expression of *Hox-2.5* is found in the differentiating motor neurons of the ventral horn. (c, c') At 11.5 days, the outer edge of the ventricular zone is strongly labeled. In more anterior sections, the lateral edge of the dorsal horn is significantly labeled (data not shown). (d, d') Expression of *Hox-2.5* is confined to the dorsal horn by 13.5 days. D, dorsal; Dh, dorsal horn; G, grey zone; Mn, motor neuron; V, ventral; Z, ventricular zone. Scale bars equal 100 μm.

tinues throughout the spinal cord. As described above, the anterior most limit of expression of *Hox*-2.5 on embryonic day 10.5 is found at the first cervical somite. When the expression of additional genes from the *Hox*-2 complex is examined, we observe a series of overlapping domains with each gene having a unique anterior boundary of expression (Figure 12–3). For example, *Hox*-2.1 and 2.2, which are transcriptionally downstream of *Hox*-2.5, have more anterior limits of expression. Both *Hox*-2.2 and 2.1 are expressed in the posterior hindbrain but, in parallel sections, the boundary of expression of *Hox*-2.1 extends more rostrally than that of *Hox*-2.2 (Schughart et al. 1988a and 1988b). It appears, therefore, that the anterior boundary of expression of particular *Hox* genes parallels the chromosomal organization of the genes (Utset et al. 1987). Recently, Wilkinson et al. (1989)

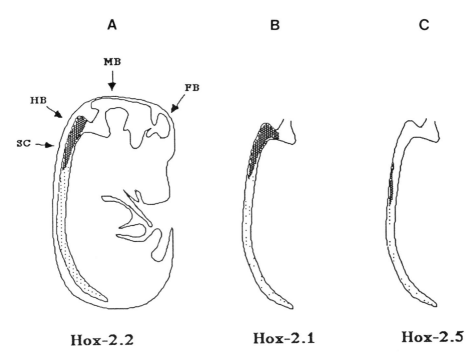

FIGURE 12–3 Schematic representation of expression of *Hox*-2.2, 2.1, and 2.5 in the developing spinal cord of 13.5-day embryos. (A) Expression of *Hox*-2.2 is detected within the medulla and the spinal cord. (B) and (C) Schematic representation of developing spinal cords, indicating the patterns of expression of *Hox*-2.1 and 2.5, respectively. (B) Expression of *Hox*-2.1 is found within the medulla and spinal cord. In parallel sections, the anterior boundary of expression of *Hox*-2.1 is clearly more rostral than that of *Hox*-2.2. (C) Transcripts of *Hox*-2.5 accumulate within the dorsal half of the spinal cord (see Figure 12–2), posterior to the first cervical vertebra. FB, forebrain; MB, midbrain; HB, hindbrain; SC, spinal cord.

extended the analysis of the *Hox*-2 complex by examining the expression of *Hox*-2.6, 2.7, 2.8, and 2.9. They demonstrated that these genes are expressed in progressively more anterior positions of the developing hindbrain. Specifically, these genes are expressed in the rhombdomeres of the 9.5-day embryo with the *Hox*-2.6 gene expressed in r7, *Hox*2.7 in r5, and *Hox*-2.8 in r3. Unexpectedly, the *Hox*-2.9 gene is expressed in r4, a position posterior to the site of expression of *Hox*-2.8. Similar posterior-to-anterior polarity of the *Hox* gene clusters has also been observed in the *Hox*1 complex (Gaunt et al. 1988).

The significance of the anterior boundaries of the expression of *Hox* genes in the unsegmented spinal cord is not clear. In the developing hindbrain, some of the *Hox* genes, together with other transcription factors such as *Krox*-1 (Wilkinson et al. 1989), may be involved in segmentation of the hindbrain into rhombdomeres. In the spinal cord, the overlapping domains of the expression of *Hox* genes may provide positional information that defines the axial location. Although the spinal cord has no apparent segmentation, it does produce a repetitive axial pattern of sensory and motor neurons that integrate in the appropriate fashion with the highly patterned body. It may be, therefore, that the expression of *Hox* genes in the ectoderm, mesoderm, and endoderm functions to define the spatial patterning and differentiation of these tissues and that expression in the spinal cord provides the axial landmarks that are required for integration of the CNS with the body.

12.4 REGULATION OF HOMEOBOX GENES

The spatial and temporal patterns of expression of *Hox* genes are complex and many laboratories are now using the transgenic mouse system to map their *cis*-regulatory elements. The basic approach is to link a reporter gene, frequently β-galactosidase (*lacZ*), to various 5' and 3' regulatory domains. The reporter construct is microinjected into mouse embryos, which are then used to establish transgenic lines. Alternatively, the embryos can be analyzed directly for β-galactosidase activity at some later stage in development. The advantage of using the *lacZ* gene is that β-galactosidase can be detected with exquisite sensitivity by a histochemical stain in both whole mounts and sections of embryos.

To a first approximation, each *Hox* gene has its own 5' regulatory sequences. However, in many cases, the spacing of the *Hox* genes within a linkage group can be very close. For instance, the homeobox-to-homeobox distance between *Hox*-2.2 and *Hox*-2.1 is approximately 4.5 kb. This space must encompass the 3' exon and translation signal sequences for *Hox*-2.2, as well as the 5' regulatory domain, first exon, and intron of *Hox*-2.1. Similar intragenic distances are observed between *Hox*1.2 and 1.3, *Hox*-3.3 and 3.4, *Hox*-2.4 and 2.3, and between *Hox*-5.3 and 5.2. Although the other *Hox* genes are more widely separated, in general the amount of space for promoter sequences that control these complex patterns is surprisingly limited. This structural feature of the *Hox* loci suggests to us that some domains that are important for the regulation of transcription of these genes may be dispersed throughout the complex or located at distant sites on either side of the complexes.

Partial reproduction of the patterns of expression of *Hox* genes has been detected in transgenic mice using the *lacZ* reporter system. Zakany et al. (1988) have observed spinal cord-specific expression of a *Hox*-1.3 regulated *lacZ* transgene. However, in the case of these constructs, the normal pattern of expression of *Hox*-1.3 in tissues outside the CNS is not present. We have also begun to analyze the regulatory domains for *Hox*-3.1 using the *lacZ* reporter system (Ruddle, in preparation). Using 5 kb of 5' genomic DNA from *Hox*-3.1, we observe *Hox*-3.1-like expression of *lacZ* in the posterior embryonic tissues. In our experiments, it is possible that some important regulatory elements might reside further in the 5' direction or in the intron and 3' untranslated sequences of *Hox*-3.1, since these were not included in our construct; this possibility does not appear to be likely in the case of *Hox*-1.3. In the experiments of Zakany and colleagues, the presence of intronic or 3' untranslated sequences did not affect the pattern of expression of *lacZ*. It also appears that, in the case of *Hox*-1.3, inclusion of additional 5' sequences may start to impinge on the 3' untranslated portions of the preceding gene (Fibi et al. 1988).

We interpret these early results as being consistent with the proposal that additional regulatory elements may either be dispersed throughout the complex or may be organized as distant enhancer elements flanking the complex. Thorough promoter-mapping studies of this type on additional *Hox* genes will be required to test this interpretation. The presence of flanking enhancer elements might, however, exert selective pressure for the conservation of homeobox linkage groups and, thus, be consistent with the highly conserved nature of the *Hox* complexes. Precedent for this interpretation is provided by the globin complex, where distant enhancer elements are required for quantitative and temporal control of globin expression (Grosveld et al. 1987).

12.5 TRANSGENIC MICE AND THE FUNCTION OF HOMEOBOX GENES

The available evidence strongly suggests that the homeobox genes play a major role in the regulation of vertebrate development. Increasing efforts are underway to determine the developmental functions of the homeobox genes using transgenic mice (Gordon et al. 1980). Two basic approaches are being used in this effort. The first approach utilizes embryonic stem cells and homologous recombination to mutate specific homeobox genes. This approach, which is summarized in Chapter 9 of this book, is directed at producing loss-of-function mutations. The second approach, which we will review here, is to produce gain-of-function mutations by engineering mice with inappropriate patterns of expression of homeobox genes.

The initial attempts at generating gain-of-function mutations involved the introduction of additional homeobox genes or homeobox genes that were controlled by heterologous promoter elements. Wolgemuth et al. (1989) introduced an additional *Hox*-1.4 gene. The *Hox*-1.4 transgene exhibited an elevated level of expression in the embryonic gut, a normal site for expression of *Hox*-1.4. This elevated level of transcription was apparently responsible for a megacolon-like phenotype that was frequently lethal within the first few weeks after birth. However, occasional animals reached maturity and were able to breed to establish a transgenic line. Using a heterologous promoter, Balling et al. (1989) introduced a *Hox*-1.1 gene controlled by a chick β-actin promoter. This construct produced a broad pattern of expression of *Hox*-1.1 and induced a lethal phenotype characterized by cleft palate and multiple cranio-facial abnormalities. In this case, no transgenic line was established since all of the animals that expressed *Hox*-1.1 died within two weeks of birth. Using an alternative approach, Harvey and Melton (1988) investigated the role of *Xhox*-1A (the cognate of mouse *Hox*-2.6) in the development of *Xenopus* by micro-injecting large amounts of mRNA into fertilized eggs. These experiments produced a dramatic disruption of somitogenesis.

From the results of the abovementioned experiments, it is difficult to deduce a general description of the developmental function of any of the genes tested. It is clear, however, that inappropriate expression of *Hox* genes produces dominant phenotypic perturbations. Unlike the results of the transgenic studies, the results of Harvey and Melton (1988) revealed a rather dramatic phenotypic disruption that supported a role for homeobox function during the formation of somites. Compared to the transgenic studies, the strong effect of *Xhox*-1A may reflect a greater sensitivity to ectopic *Xhox*–1A expression during the Xenopus development, or differences in the techniques used, or it may indicate that the results observed in the mouse studies were weak phenotypic effects. This last explanation seems the most likely if during development the microinjected embryos with high levels of expression and, thus, strong phenotypic effects were selected against and resorbed. If this type of selection is occurring, then only weak "alleles" would be isolated.

12.6 MULTIPLEX GENE REGULATION

We have recently developed a two-tiered system of gene regulation, which we believe offers several advantages for the production of dominant embryonic gain-of-function mutations in transgenic mice (Byrne and Ruddle,

1989). The multiplex gene regulatory (MGR) system is based on the observed transactivation of the immediate early (IE) genes of *Herpes simplex* virus (HSV) by the transactivator virion polypeptide VP16 (Mackem and Roizman 1982; Kristie and Roizman 1987; O'Hare and Hayward 1987). The MGR system consists of two transgenic lines and is illustrated in Figure 12–4. One transgenic line, the transactivator, expressed VP16. The pattern of expression of VP16 can be regulated by either inducible or tissue-specific promoter elements. The second transgenic line, the transreponder, contains the gene of interest regulated by one of the HSV-1 immediate early promoter elements (ICP4). In the absence of VP16, there is little or no expression of the gene of interest in the transresponder line. When the two transgenic lines are mated, the gene of interest is induced by VP16 with a pattern of expression that parallels the pattern of VP16 activity.

For our initial test of the MGR system, we produced a series of transgenic mice that contained a gene for chloramphenicol acetyltransferase (CAT) controlled by the 360-bp ICP4 promoter (IE-CAT) and a transactivator mouse that expressed VP16 under the regulation of the murine neurofilament L promoter (NFT). Using Southern blot hybridization, we found that the transactivator mouse contained two separate sites of integration, only one of which produced an active VP16 product (Figure 12–5A). When this NFT transactivator was crossed to an IE-CAT transresponder, CAT activity was detected only in those offspring with an IE-CAT and an active VP16 transgene. The pattern of expression of CAT was specific for the spinal cord and brain, as would be expected from the neurofilament promoter (see Figure 12–5).

Three important characteristics of the MGR system are necessary if it is to be useful for the production of gain-of-function mutations in transgenic mice. First, the IE promoter element must be associated with little or no expression in the absence of the VP16 transactivator. This characteristic is required if ectopic expression of the gene being studied has a dominant developmental effect. Using our IE-CAT lines, we have been unable to detect significant activity of CAT in transgenic embryos or in a variety of tissues from transgenic adults. More recently, we have successfully used this system to introduce IE-regulated *Hox*-3.1 and 2.2 genes into mice. Together, these results strongly indicate that the ICP4 promoter has a very low basal level of activity in the absence of VP16. Secondly, expression of the VP16 transactivator cannot itself have oncogenic or developmental effects. Although VP16 is a very potent transcriptional activator, we have successfully derived transgenic lines of mice that express VP16 using both the neurofilament promoter and the 5-kb *Hox*-3.1 promoter element described above. None of these animals exhibit any oncogenic or developmental effects as a result of the expression of VP16. While this observation does not preclude the possibility of subtle developmental or neurological changes, it does indicate that VP16 expression is well tolerated in a variety of tissues. Finally, if the MGR system is to be generally useful, it is important that VP16 induction

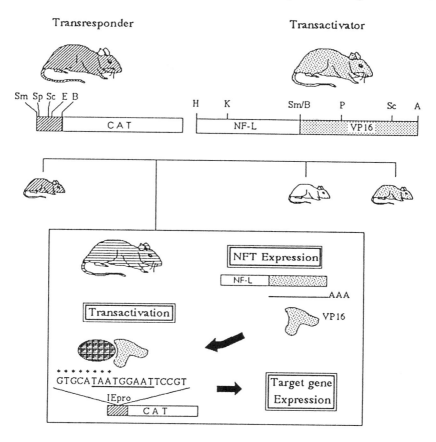

FIGURE 12-4 A diagrammatic representation of the two-tiered multiplex regulatory system. Two transgenic mouse lines are represented. One transgenic line, the transresponder, contains a target gene (CAT) regulated by the HSV-1 immediate early promoter element (cross-hatched area). In the transresponder line, there is no expression of the target gene. The second transgenic line, the transactivator (right), contains the HSV-1 transactivator VP16. In this example, VP16 is regulated by a neuro-specific promoter element from the murine neurofilament L gene. The two lines are crossed. The offspring inherit the transresponder (left), transactivator (far right), neither (small white animal), or both transgenes (boxed region). In the double-transgenic offspring, VP16 is expressed in a neuro-specific pattern. In the cells that express VP16, VP16 forms a protein-protein complex with the ubiquitous octamer binding protein (oval). This protein complex specifically activates the immediate early promoter, which results in neuro-specific expression of the target gene. One of the TAATGARAT domains (underlined) in the 360-bp ICP4 promoter is shown. Immediately 5' of the TAATGARAT sequence (*) is a potential octamer-binding site. Abbreviations for restriction enzymes: B, *Bam*HI; E, *Eco*RI; K, *Kpn*I; P, *Pvu*II; Sp, *Sph*I; Sc, *Sac*II; and Sm, *Sma*I.

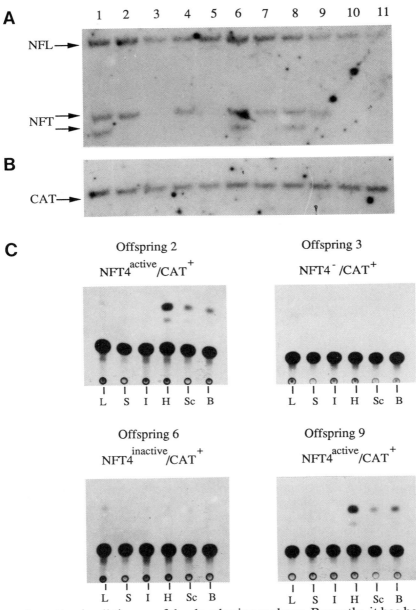

be active in all tissues of the developing embryo. Recently, it has been shown that VP16 induction depends on the presence of the ubiquitous cellular transcription factor, Oct-1 (O'Hare and Goding 1988; O'Hare et al. 1988). While octamer binding sites have been detected in a variety of genes, the presence of an octamer site is not by itself sufficient for VP16 induction. Since the ubiquitous Oct-1 protein is a co-transcription factor with VP16,

FIGURE 12–5 Induction of CAT by mating with an NFT mouse. A female homozygous for the IE-CAT8 transgene was mated to the NFT4 founder male. (A) Southern blot hybridization analysis of the newborn offspring hybridized with a 370-bp *Pst*I fragment of the NF-L promoter. Two different restriction patterns are evident. Offspring 3, 5, 10, and 11 did not inherit the VP16 transgene. Offspring 1, 6, and 8 exhibit two transgene-specific restriction fragments (NFT arrows) of 2.8 and 3.2 kb. Offspring 2, 4, 7, and 9 exhibit a single 3.2-kb transgene restriction fragment. (B) The same filter probed with the 280-bp *Bam*HI-*Eco*RI fragment of CAT. All offspring inherited the IE-CAT8 transgene. All the offspring in this litter were examined for expression of CAT in a variety of tissues. (C) Expression of CAT in representative offspring. Samples of liver (L), spleen (S), intestine (I), and heart (H), in addition to the spinal cord (SC), and brain (B), were assayed for CAT. A 50-fold excess of protein was used to assay the non-neuronal tissues relative to the samples of brain and spinal cord. Offspring with an active site of integration of VP16 (offspring 2, 4, 7, and 9) exhibit CAT activity in the brain, spinal cord, and heart. Animals that did not inherit the VP16 gene (offspring 3, 5, 10, and 11) or inherited the inactive site of integration (offspring 1, 6, and 8) exhibited no signficant CAT activity in any of the tested tissues.

it seems likely that VP16 induction should occur in most or all tissues. Our results with the induction of CAT have revealed VP16-specific activation in the CNS, most probably in neuronal tissue, in the heart, and in adult eyes. In other experiments, using a *Hox*-3.1-regulated VP16 transactivator, we have detected VP16 induction in 10.5-day embryos, apparently in all three germ layers of the tail and posterior embryo. We are convinced, therefore, that induction by VP16 will operate in most embryonic tissues.

The MGR system is particularly useful for engineering gain-of-function mutations because, once the appropriate animals are isolated, the mutations can be generated at will and studied in a reproducible manner. This system, therefore, eliminates the necessity for continuous micro-injection of dominant genes for the reproduction of a particular phenotype and minimizes the phenotypic variability caused by integration-site-specific modulation of the transgene. Although the MGR system initially requires the production of two transgenic lines, once a transactivator line is made, it can be saved and used to induce a variety of transresponders. As different lines of transactivator mice are accumulated, therefore, it will become relatively easy to test a transresponder line rapidly under a variety of different patterns of induction. This potential represents an important feature of the system since it seems likely that no single gain-of-function mutation will completely illuminate the developmental function of a gene. Instead, we feel that it is important to test a gene in a variety of tissues and at different stages of development to discern any commonality between the phenotypes. Finally, the simple MGR system can be expanded into a network by including multiple transresponder transgenes or multiple transactivators. This possibility could prove very useful for studies of the interactions between two different regulatory genes.

12.7 FUTURE PROSPECTS

The identification of the vertebrate homeobox genes has provided a molecular foothold from which to begin a protracted study of mammalian development. All the available data indicate that these genes play a critical role in controlling vertebrate development and their function as transcription factor suggests that these genes may produce an informational network. For example, there are several instances in which homeodomain proteins bind to cis-flanking sites in their own regulatory regions. Thus, single homeobox genes can potentially act as genetic switches. If this is the case, our evolutionary analysis suggests that during the initial expansion of the homeobox system a complex set of switching patterns was formed, since each gene would presumably have not only the potential for self-regulation but also the ability to cross-regulate the other genes in the complex. During the divergence of the homeobox sequences, the specificity of the individual genes would change to produce a spatially complex pattern of expression of homeobox genes. Furthermore, this newly evolved specificity of the individual homeobox genes must have become integrated with the rest of the genome to control the developmental expression of a large number of target genes. Further expansion of the primordial homeobox complex could have increased the informational capacity of such a system in an exponential manner. Thus, it is perhaps appropriate to consider the homeobox genes as an information processing network that interprets spatial developmental cues and translates these cues into a pattern of differentiation by affecting the transcription of a much larger set of target genes.

The transgenic mouse system is a rapidly maturing technique which has the potential to circumvent many of the problems associated with a classical genetic approach to mouse development. The use of embryonic stem cells, together with an ability to recognize homologous recombination events, suggests that it will prove possible to mutate any mammalian gene specifically and to incorporate this gene into the mouse germline. Likewise, the multiplex regulatory system, although a less established technique, has the potential for producing conditional dominant mutations in any gene of interest. Although these techniques are laborious and complex, together they furnish the means for a detailed molecular genetic analysis of gene function in the mouse and in many instances they will be useful for developing model systems for human disease.

There are many genes that have been isolated which are already known to have important regulatory functions in the control of cellular growth or development. This list would include the oncogenes, the genes for growth factors and hormone receptors, numerous genes involved in cellular adhesion, and many of the genes that encode the components of the extracellular matrix. While these genes are obvious candidates for analysis, the identification of new mammalian genes that control development is still problematic. It is encouraging, however, to find so many instances of mutations in lower organisms that are later discovered to occur in genes that are at least partially homologous to known mammalian genes. These observations and the conservation of the homeobox system suggest that many of the fundamental genes that regulate multicellular development may have ancient evolutionary origins and should, therefore, be present in most experimental organisms. Furthermore, the evolution of the homeobox system suggests that increases in developmental complexity may have been due to duplications and divergence of ancestral genes, such that many of the genes that regulate development may also be members of extended families of genes. Thus, by taking advantage of these cross-species homologies and the recent advances in transgenic technology, we should be able to begin a detailed analysis of mammalian development and, at the same time, to produce numerous models for human disease.

REFERENCES

Awgulewitsch A., Utset M.F., Hart C.P., McGinnis W., and Ruddle F.H. (1986) *Nature* 320, 328–335.
Balling, R., Mujtter, G., Gruss, P., and Kessel, M. (1989) *Cell* 58, 337–347.
Bogarad, L.D., Utset, M.F., Awgulewitsch, A., et al. (1989) *Dev. Biol.* 133, 537–549.
Byrne, G.W., and Ruddle, F.H. (1989) *Proc. Natl. Acad. Sci. USA* 86, 5473–5477.
Desplan, C., Theis, J., and O'Farrell, P.H. (1985) *Nature* 318, 630–635.
Desplan, C., Theis, J., and O'Farrell, P.H. (1988) *Cell* 54, 1081–1090.
Fainsod, A., Bogarad, L.D., Ruusala, T., et al. (1986) *Proc. Natl. Acad. Sci. USA* 83, 9532–9536.
Fibi, M., Zink, B., Kessel, M., et al. (1988) *Development* 102, 349–359.
Fienberg, A.A., Utset, M.F., Bogarad, L.D., et al. (1987) *Current Topics in Developmental Biology*, Academic Press, San Diego, CA.
Fitzpatrick, V.D., and Ingles, C.J. (1989) *Nature* 337, 666–668.
Gaunt, S.J. (1988) *Development* 103, 135–144.
Gaunt, S.J., Miller, J.R., Powell, D.J., and Deboule, D. (1986) *Nature* 324, 662–664.
Gaunt, S.J., Sharpe, P.T., and Deboule, D. (1988) *Development* 104, Supplement, 169–179.
Gehring, W.J. (1987) *Science* 236, 1245–1252.
Gordon, J.W., Scangos, G.A., Plotkin, D.J., Barbosa, J.A., and Ruddle, F.H. (1980) *Proc. Natl. Acad. Sci. USA* 77, 7380–7384.
Grosveld, F., van Assendelft, G.B., Greaves, D.R., and Kollias, G. (1987) *Cell* 51, 975–985.
Harvey, R.P., and Melton, D.A. (1988) *Cell* 53, 687–697.
Hoey, T., and Levine, M. (1988) *Nature* 332, 858–861.
Holland, P.W.H., and Hogan, B.L.M. (1988) *Genes Dev.* 2, 773–782.
Ingraham, H.A., Chen, R.P., Mangalam, H.J., et al. (1988) *Cell* 55, 519–529.
Jackson, I.J., Schofield, P., and Hogan, B.L.M. (1985) *Nature* 317, 745–748.
Jaynes, J.B., and O'Farrell, P.H. (1988) *Nature* 336, 744–749.
Kappen, C., Schughart, K., and Ruddle, F.H. (1989) *Proc. Natl. Acad. Sci. USA* 86, 5459–5463.

Ko, H.S., Fast, P., McBride, W., and Staudt, L.M. (1988) *Cell* 55, 135–144.
Kristie, T.M., and Roizman, B. (1987) *Proc. Natl. Acad. Sci. USA* 84, 71–75.
Laughon, A., and Scott, M.P. (1984) *Nature* 310, 25–31.
Lewis, E.B. (1978) *Nature* 276, 565–570.
Mackem, S., and Roizman, B. (1982) *J. Virol.* 44, 939–949.
McGinnis, W., Levine, M.S., Haften, E., Kuroiwa, A., and Gehring, W. J. (1984) *Nature* 308, 428–433.
Le Mouellic, H., Condamine, H., and Brulet, P. (1988) *Genes Dev.* 2, 125–135.
Nusslein-Volhard, C., and Wieschaus, E. (1980) *Nature* 287, 795–801.
Odenwald, W.F., Taylor, C.F., Palmer-Hill, F.J., et al. (1987) *Genes Dev.* 1, 482–496.
O'Hare, P., and Goding, C.R. (1988) *Cell* 52, 435–445.
O'Hare, P., Goding, C.R., and Haigh, A. (1988) *EMBO J.* 7, 4231–4238.
O'Hare, P., and Hayward, G.S. (1987) *J. Virol.* 61, 190–199.
Ruddle, F.H. (1989) in *The Physiology of Growth* (Tanner, J.M., and Priest, M.A., eds.), pp. 47–66 Cambridge University Press, Cambridge, England.
Schughart, K., Kappen, C., and Ruddle, F.H. (1988b) *Br. J. Cancer* 58, 9–13.
Schughart, K., Kappen, C., and Ruddle, F.H. (1989) *Proc. Natl. Acad. Sci. USA* 86, 7076–7071.
Schughart, K., Utset, M.F., Awgulewitsch, A., and Ruddle F.H. (1988a) *Proc. Natl. Acad. Sci. USA* 85, 5582–5586.
Scot, M.P., and Weiner, A.J. (1984) *Proc. Natl. Acad. Sci. USA* 81, 4115–4119.
Sheperd, J.C.W., McGinnis, W., Carrasco, A.E., DeRobertis, E.M., and Gehring, W.J. (1984) *Nature* 310, 70–71.
Staudt, L.M., Clerc, R.G., Singh, H., et al. (1988) *Science* 241, 577–580.
Sturm, R.A., Das, G., and Herr, W. (1988) *Genes Dev.* 2, 1582–1599.
Thali, M., Muller, M.M., DeLorenzi, M., Matthias, P., and Bienz, M. (1988) *Nature* 336, 598–601.
Utset, M.F., Awgulewitsch, A., Ruddle, F.H., and McGinnis, W. (1987) *Science* 235, 1379–1382.
Wilkinson, D.G., Bhatt, S., Cook, M., Boncinelli, E., and Krumlauf, R. (1989) *Nature* 337 405–409.
Wolgemuth, D.J., Behringer, R.R., Mostoller, M.P., Brinster, R.L., and Palmiter, R.D. (1989) *Nature* 337, 464–467.
Zakany, J., Tuggle, C.K., Patel, M.D., and Nguyen-Huu, M.C. (1988) *Neuron* 1, 679–691.

CHAPTER 13

Regulation of Expression of a Class I MHC Transgene

Dinah S. Singer
William I. Frels
Rachel Ehrlich

The class I major histocompatibility complex (MHC) family of genes encodes a series of structurally related cell-surface glycoproteins (Klein 1979). The class I antigens can be divided into subfamilies, based on their function and tissue distribution. One subfamily consists of the transplantation antigens. Although originally identified through their ability to mediate graft rejection, the major function of the transplantation antigens is to serve as receptors for peptide antigens, inducing a response by T cells (Zinkernagel and Doherty 1979). The transplantation antigens are ubiquitously expressed, as assessed by cell-surface staining (Klein 1975). Two or three transplantation antigens, depending on the species, are expressed on individual cells. Other subfamilies of class I antigens contain differentiation antigens, which are expressed in a tissue-restricted fashion. Among these are the Qa and TL antigens in the mouse (Flaherty 1980). The function(s) of these differentiation antigens is unknown. The molecular mechanisms that regulate the differential expression of these subfamilies are also largely unknown.

In all vertebrate species, the class I MHC genes are linked on a single chromosome and define the major histocompatibility complex (Klein 1979).

The size of the family of class I MHC genes varies among species. The largest families are found in rodents: mice contain between 35 to 40 class I homologous sequences in their genomes (Steinmetz et al. 1982; Weiss et al. 1984). Smaller families are found in other species, for example, in humans and rabbits (Cohen et al. 1983; Orr et al. 1982). The smallest identified family of class I genes occurs in miniature swine and contains only seven class I MHC genes (Singer et al. 1982). Because of its small size, the class I family of the miniature swine is particularly amenable to intensive analysis of its organization, patterns of expression, and regulation.

Genetic analysis of the class I MHC locus of the miniature swine demonstrated that these genes are linked (Leight et al. 1977). In situ hybridization localized the MHC to the short arm of chromosome 7 (Rabin et al. 1985). Pulsed-field analysis of the class I genes indicated that all the class I homologous sequences are contained on a DNA segment of approximately 500 kb (D. Singer, unpublished observation). The structures of four class I genes have been determined; these genes have been shown to have an organization typical of class I genes (Ehrlich et al. 1987; Satz et al. 1985). That is, they consist of eight exons that encode a leader, three extracytoplasmic domains, a transmembrane domain, and intracytoplasmic domains. From homologies in DNA sequence, three subfamilies can be defined among the seven sequences (Singer et al. 1988). One subfamily consists of three members, two of which encode the major transplantation antigens in the miniature swine. One of these, PD1, has been extensively studied in an attempt to elucidate the regulatory mechanisms that determine its patterns of expression in vivo (Singer et al. 1988; Ehrlich et al. 1988; Singer and Ehrlich 1988).

In vivo, PD1 is expressed in a variety of tissues (Table 13–1). The highest levels of expression are observed in the spleen and lymph nodes with varying lower levels elsewhere. Even within the lymphoid compart-

TABLE 13–1 Expression of the PD1 Gene In Situ and in Transgenic Mice

Tissue	In Situ	Transgenic
Kidney	1.0	1.0
Heart	3.4	0.1
Testis	1.3	4.0
Lymph nodes	7.6	13.2
Spleen	4.5	11.0
T cells	1.0	1.0
B cells	1.6	1.8

Relative amounts of PD1 RNA were calculated by densitometric analysis of autoradiograms either after analysis by cleavage with S1 nuclease of total RNA from various tissues from miniature swine or after direct Northern blot analysis of RNA from transgenic B10.PD1 tissues. The densitometric values were first normalized for the amount of RNA and then normalized to levels in kidney in group A or to levels in T cells in group B.

ment, PD1 is differentially expressed. Approximately twofold higher steady-state levels of PD1 RNA are found in B cells, relative to levels in T cells (see Table 13-1; Ehrlich et al. 1988a). In order to determine whether the pattern of expression of PD1 observed in swine tissue is inherent to the PD1 gene itself or is conferred upon it by positional or other extragenic factors, a segment of DNA containing the entire PD1 coding sequence, as well as 1.1 kb of upstream DNA sequences, was introduced into the germline of a mouse (Frels et al. 1985). The resulting transgenic line was then tested for its ability to express the PD1 gene. As shown in Table 13-1, PD1 was expressed in the transgenic mouse; its pattern of expression paralleled that observed in situ in the miniature swine. That is, the highest levels were observed in the lymphoid tissues with varying lower levels elsewhere. In a further parallel with expression in situ, expression of PD1 in transgenic B cells is approximately twofold higher than it is in T cells. From these studies, it is concluded that the regulatory signals necessary for the normal expression of PD1 are contained within the 9-kb segment of DNA introduced into the transgenic mouse. In addition, the cellular *trans*-acting factors that interact with the DNA regulatory elements have been conserved to such an extent during evolution that the murine factors still recognize swine DNA.

In addition to control of basal levels of expression, transcription of the PD1 gene is also known to be responsive to a variety of immunomodulators (Satz and Singer 1984; Parent et al. 1987). Among these immunodulators, the best characterized is interferon. Murine L cells stably transfected with PD1, that is, L(PD1) cells, express the PD1 MHC antigen on their cell surface. Treatment of L(PD1) cells with α,β-interferon results in an approximate twofold increase in expression of PD1 antigen on the cell surface, which is the result of increased transcription (Satz and Singer 1984). Similar increases occur in the expression of the endogenous murine H-2 antigens. Despite extensive studies on the response to interferon of class I genes from humans, mice, and swine, little is known about the nature of this response in vivo. The reason for this lack of knowledge is primarily that the complexity of the family and the high degree of homology among its members confound the analysis. Thus, it is very difficult to distinguish the effects of interferon on specific genes from the broad effects of interferon on the whole family of genes. The PD1 transgenic mouse has provided us with the opportunity to examine the response to interferon of a single class I gene.

Following treatment of mice with α,β-interferon, expression of PD1 antigen on the cell surface was assessed in splenocytes and thymocytes (Table 13-2). Both types of cell showed a small but significant and reproducible increase in the expression of PD1 on the cell surface. This 1.5-fold increase was quantitatively similar to that observed in L(PD1) cells. Using monoclonal antibodies putatively specific for a murine transplantation antigen, H-2 K^b, and for a differentiation antigen, Qa-2, we observed net increases in the response to interferon. However, the response of the *Qa*-2 gene in the spleen was restricted to the B cell population (data not shown). Analysis

TABLE 13–2 Increase in Expression of PD1 Antigen and Endogenous Antigens on the Cell Surface Following Treatment of B10.PD1 Transgenic Mice with α,β-Interferon In Vivo

Cell-Surface Antigen	Relative Level of Expression After Treatment with Interferon	
	Thymus	Spleen
PD1	1.7	1.5
H-2Kb	1.1	1.4
Qa-2[a]	1.2	1.2
Thy-1	1.0	1.1
CD8	1.0	—
CD4	1.0	—
Ly1	1.0	—
IL2 receptor	1.0	—

Mice were treated with a total of 1.8×10^6 units of α,- interferon, administered in six doses. Sex- and age-matched control mice were injected, in parallel, with buffer alone. Suspensions of thymocytes and splenocytes were prepared and stained as described elsewhere (Ehrlich et al., 1988a), and subjected to analysis in a fluorescence-activated cell sorter. Background staining values (as detected with a nonspecific antibody) were subtracted from experimental values. The ratio between net fluorescence intensity of cells from interferon-treated and control mice was determined.
[a]The staining profile of Qa-2 was bimodal with selective changes induced by interferon. The calculated value is derived from the mean fluorescence of the total population.

of the thymocyte population showed no elevation in levels of other cell-surface markers (see Table 13–2). Analysis of RNA from these tissues revealed a parallel increase in steady-state levels of RNA (data not shown).

Analysis of the 5' flanking region of the PD1 gene has revealed a variety of DNA regulatory sequences (Singer et al. 1988; Ehrlich et al. 1988; Singer and Ehrlich 1988). In addition to the canonical promoter sequences, TCTAA (a variant of the TATA box) and CCAAT, both positive and negative regulatory sequence elements, and an interferon-response element have been identified. These elements have been functionally mapped by transfection of various cell lines with 5' end-deletion mutants of PD1. The finding that the PD1 transgenic mouse regulates expression of PD1 in a normal fashion affords the possibility of an analysis of the chromatin structure and the nature of murine DNA-binding factors associated with PD1. Sites hypersensitive to DNAse I in chromatin have been identified as regions of binding of trans-acting factors. In the PD1 transgenic mouse, a single constitutive site of hypersensitivity to DNAse I, mapping immediately 5' to the interferon-response element (IRE), is found in both spleen and thymus (Figure 13–1). Treatment with interferon does not alter this site, nor does it induce a new site (data not shown). Thus, from these data, we can conclude that alterations in the patterns of expression of the PD1 gene are not reflected

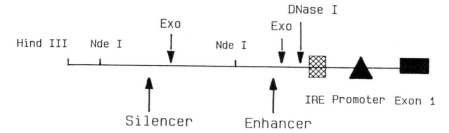

FIGURE 13-1 Physical and functional map of the 5' flanking region of PD1. The relative positions of various regulatory elements are noted: exon 1 (■), promoter (▲), interferon-response element (IRE, ▨). The positions of relevant restriction sites are indicated. Arrows above the line indicate the sites of DNAse I hypersensitivity and protection from exonucleolytic digestion. The arrows below the line indicate the general regions of functional activity and are not meant to imply specific locations. These functional elements may or may not coincide precisely with the binding sites defined by exonucleolytic digestion.

in changes in the structure of its chromatin, as assessed by hypersensitivity to DNAse I.

For a more detailed identification of PD1-associated sequences that bind *trans*-acting factors, an exonuclease protection assay was used (Gimble et al. 1987). A segment of PD1 DNA containing the 1.1 kb of 5' flanking region present in the transgenic mouse was incubated with nuclear extracts, to allow binding of *trans*-acting factors, and then treated with exonuclease. The presence of a binding factor was indicated by the appearance of a new band in the analytic gel (Figure 13-2). To provide internal controls, the DNA fragment was ligated to the bacterial *lac* operon and also incubated with the lac repressor, which binds the operon with high affinity. Exonucleolytic activity is blocked by the presence of this repressor, and cleavage products of known length are formed, even in the absence of cell extracts (see Figure 13-2). As shown in Figure 13-2, two extract-dependent stop sites, at which exonucleolytic cleavage is interrupted, are generated on the PD1 DNA fragment by the extract. These sites map to regions in which regulatory elements have been identified (see Figure 13-1).

In summary, the combined approaches of molecular analysis and transgenic technology provide a powerful means for dissecting the molecular mechanisms involved in regulation of gene expression. Using such techniques, we have demonstrated that a 9-kb fragment of swine DNA contains all of the necessary regulatory signals and coding information for the proper expression of a class I antigen. In addition, the responses of this transgene to the immunomodulator, α,β-interferon have been assessed in a variety of tissues. Finally, it has been demonstrated that changes in the structure of the chromatin of the PD1 segment of DNA are not prerequisites for changes

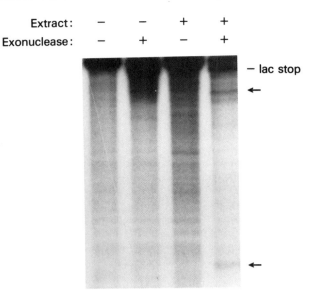

FIGURE 13-2 Exonuclease protection assay of the 5′ flanking region of PD1 reveals the presence of two binding sites. A 1.1-kb 5′ noncoding segment of PD1 DNA ligated to the bacterial *lac* operon was incubated with lac repressor and subjected to digestion with T7 exonuclease in either the presence or absence of Cos 7 nuclear extract. The DNA was then resolved by electrophoresis on an acrylamide gel. Stop sites, corresponding to the binding of the lac repressor to its operon, are indicated. Arrows point to the locations of specific stop sites generated by the nuclear extract.

in its patterns of expression. Indeed, exonuclease protection studies indicate that at least two major DNA-binding factors that interact with PD1 DNA do not contribute to the pattern of sensitivity to DNAse I. Studies are currently underway to define more accurately the relationships between specific DNA-regulatory elements, such as the enhancer and silencer, their cognate binding factors, and patterns of expression in vivo. To this end, a series of transgenic animals carrying various segments of the regulatory region of the PD1 gene have been constructed. Analysis of these animals in conjunction with molecular analyses should shed further light on the regulatory mechanisms that govern expression of this family of genes.

REFERENCES

Cohen, D., Paul, P., Font, M., et al. (1983) *Proc. Natl. Acad. Science, USA* 80, 6289–6292.

Ehrlich, R., Lifshitz, R., Pescovitz, M., Rudikoff, S., Singer, D.S. (1987) *J. Immunol.* 139, 593–602.

Ehrlich, R., Maguire, J., and Singer, D.S. (1988a) *Immunogenetics* 30, 18–26.
Ehrlich, R., Maguire, J., and Singer, D.S. (1988b) *Mol. Cell. Biol.* 8, 695–703.
Flaherty, L. (1980) in *Biological Basis for Immunodeficiency* (Gelfand, E.W. amd Dosch, H.M., eds.), p. 99, Raven Press, New York.
Frels, W., Bluestone, J., Hodes, T., Capecchi, M., and Singer, D. (1985) *Science* 228, 577.
Gimble, J., Levens, D., and Max, E. (1987) *Mol. Cell. Biol.* 7, 1815–1822.
Klein, J. (1975) *Biology of the Mouse Histocompatibility Complex*, Springer Verlag, New York.
Klein, J. (1979) *Science* 203, 516–518.
Leight, G.S., Sachs, D.H., and Rosenberg, S.A. (1977) *Transplantation* 23(3), 271–276.
Orr, H., Bach, R., Ploegh, H., et al. (1982) *Nature* 296, 454–456.
Parent, L.J., Ehrlich, R., Matis, L., and Singer, D.S. (1987) *FASEB J.* 1, 469–473.
Rabin, M., Fries, R., Singer, D.S., and Ruddle, F. (1985) *Cytogenet. Cell. Genet.* 39, 206–209.
Satz, M., and Singer, D.S. (1984) *J. Immunol.* 132, 496–501.
Satz, M., Wang, L.C., Singer, D.S., and Rudikoff, S. (1985) *J. Immunol.* 135, 2167–2175.
Singer, D.S., Camerini-Otero, M., Satz, M., et al. (1982) *Proc. Natl. Acad. Sci., USA* 79, 1403–1407.
Singer, D.S., and Ehrlich, R. (1988) *Cur. Topics Microbiol. Immunol.* 137, 148–154.
Singer, D.S., Ehrlich, R., Golding, H., et al. (1988) in *Molecular Biology of the Major Histocompatibility Complex of Domestic Animal Species* (Warner, C., Rothschild, M., and Lamont, S., eds.), Iowa State University Press, Ames.
Steinmetz, M., Winoto, A., Minard, K., and Hood, L. (1982) *Cell* 28, 489–498.
Weiss, E.H., Golden, L., Fahrner, K., et al. (1984) *Nature* 310, 650–655.
Zinkernagel, R.M., and Doherty, P.C. (1979) *Adv. Immunol.* 27, 51–177.

CHAPTER 14

Generation of Transgenic Mice with Major Histocompatibility Class II Genes

Javier Martin
Bing-Yuan Wei
Roger Little
Suresh Savarirayan
Chella S. David

The major histocompatibility complex (MHC) of the mouse is located on chromosome 17 and encodes surface glycoproteins intimately involved in immune function (Klein 1975). The I region encodes the class II molecules, immune-response-associated (Ia) antigens, which play a role in the induction and regulation of the immune response. T helper (Th) cells recognize foreign antigens only when they are associated with Ia antigens on the surface of antigen-presenting cells. The activation of Th cells is required for both cell-mediated and most B-cell antibody responses (Nagy et al. 1981; Corley et al. 1985).

The murine Ia molecules, designated A (A_α, A_β) and E (E_α, E_β) are transmembrane glycoproteins, each consisting of a 33-kd α chain noncovalently associated with a 29-kd β chain. Both the α and the β chains contain two extracellular domains, designated $\alpha1$ and $\alpha2$, and $\beta1$ and $\beta2$, respectively (Kaufman and Strominger 1983). Sequence analysis of allelic α and β chains revealed that most polymorphic residues are located in the NH_2-terminal of α1 and β1 domains (Cook et al. 1981; Choi et al. 1983; Benoist et al. 1983; Estess et al. 1986). The major types of murine cells expressing Ia antigens are B lymphocytes, macrophages, dendritic cells, and thymic epithelium. The expression of Ia on these different types of cell can be modulated by exposure to various agents. Interferon-*gamma* has been shown to induce both class I and class II antigens on macrophages and a limited number of other cells of nonhematopoietic origin (Steeg et al. 1982; Pober et al. 1983), while IL-4 has been shown to be active on B cells and a variety of macrophages (Noelle et al. 1984; Stuart et al. 1988). Bacterial lipopolysaccharide (LPS) has also been shown to induce Ia on B cells (Cambier et al. 1987). Furthermore, the antigen-presenting ability of cells has been demonstrated to be augmented as a result of treatment with these lymphokines (Zlotnik et al. 1983 and 1987).

In the past few years, accumulated data on the nucleotide sequences of MHC genes has allowed us to understand more about their structure, function, and regulation (Steinmetz et al. 1982; Devlin et al. 1984). Furthermore, DNA transfection of cloned human and mouse class II genes into cultured cells has been successfully carried out as part of an attempt to study a number of features of the structure and function of the MHC (Germain and Malissen 1986). However, a full understanding of the roles of Ia antigens in the immune response requires the evaluation of effects of the gene products in a whole organism. The recent development of the technology for introducing cloned DNA into the germline of mice provides additional opportunities for studying the regulation and function of specific genes in a whole organism (Palmiter and Brinster 1986; Scangos and Bieberich 1987). To study the regulation of expression of class II MHC genes and the effects of specific class II genes on the development of responsiveness of the immune system, we have generated a number of lines of transgenic mice that carry a wild-type A_α^k gene and a mutant A_β^k MB gene. The two genes were injected either alone or in combination. Mutant A_β^k DNA (A_β^k MB) generates amino acid substitutions characteristic of the A_β^d polypeptide at positions 63 and 65 through 67 in the β1 domain of the A_β^k molecule. By in vitro transfection, it was demonstrated that these mutations result in the loss of the binding to A_β^k-reactive monoclonal antibodies (mABs) but in the gain of reactivity with most A_β^d-binding mABs (Buerstedde et al. 1988). Initial analysis by fluorescence-activated cell sorter (FACS) of spleen and peripheral blood cells showed no expression of A_α^k or A_β^k MB, but when these cells were stimulated with LPS in vitro prior to FACS analysis, expression of A_α^k and A_β^k MB was detected. The results of skin graft and MLR experiments suggest that the product of the transgene is expressed on the cell surface prior to stimulation by LPS, but at levels below the limit of detection of our immunofluorescence assay.

14.1 MATERIALS AND METHODS

14.1.1 Microinjection of DNA into Mouse Embryos

A_α^k genomic DNA obtained from the *Hind*III fragment of cosH-2^k 8.1 (>9 kb) and mutant A_β^k DNA (A_β^k MB), an A_β^k-puc12 DNA linearized with *Hind*III used for microinjection (Figures 14–1 and 14–2), were excised from agarose gels, purified by "gene-clean" procedures, and dialyzed against 10 mM Tris-HCl, 0.25 ml EDTA. DNA fragments were diluted to 1–2 µg/ml and injected into the male pronucleus of fertilized embryos from (B6 × SJL)F_2, (SJL × SWR)F_2, or (B10.M × CBA)F_2 mice. Embryos surviving microinjection were reimplanted into the oviducts of pseudopregnant foster mothers.

14.1.2 Identification of Transgenic Mice

Tail-skin DNA was prepared from three-week-old mice. Tails were minced in small pieces in a tube that contained 500 µl NETS buffer (100 mM NaCl, 50 mM EDTA, 50 mM Tris-Cl, pH 7.6, 1% SDS). The preparation was digested with proteinase K and RNase A, extracted with phenol:chloroform, and extensively dialyzed against TE buffer (10 mM Tris, pH 7.6, 1 mM EDTA) for at least 24 hours at 4°C. 15 µg of DNA was digested with *Pst*^{r}I or *Eco*RI, subjected to electrophoresis in 0.7% agarose and transferred to nylon membrane. Hybridization of DNA blots was carried out at 65°C in 6× SSC/5× Denhart/0.5% SDS for 24 hours with A_α^k cDNA and A_β^k cDNA probes labelled with an oligolabelling kit (Pharmacia).

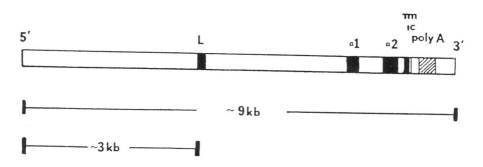

FIGURE 14–1 Diagram of the A_α^k gene construct used for microinjection. A_α^k genomic DNA obtained from *cos*H-2^k 8.1 *Hind*III fragment (>9 kb), including 5' (approximately 3 kb) and 3' flanking sequences, which contained promoter and polyadenylation sites. Black boxes represent the A_α^k exons as follows: L, leader peptide; α1 and α2, extracellular domains; TM, transmembrane segment; Ic intracytoplasmic tail; and poly A, polyadenylation site.

164 Mice with Major Histocompatibility Class II Genes

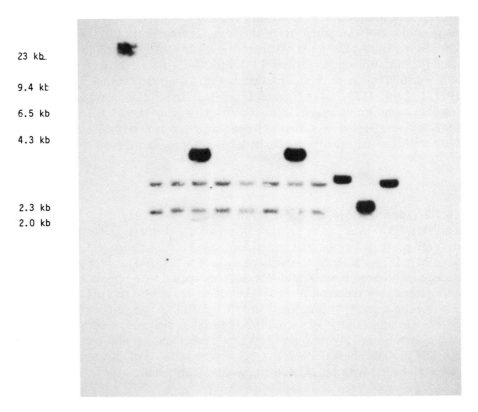

FIGURE 14-2 Integration of the A_α^k gene into the genome of transgenic mice. Southern blot analysis of tail DNA digested with *Pst*I and hybridized with an A_α^k cDNA probe. Mice numbers 3 and 7 are transgenic.

14.1.3 Monoclonal Antibodies and Fluorescence Cytometry

Expression on the cell surface of products of the A_α^k and A_β^k MB genes was detected with biotinylated mAbs 39J (anti-A_α^k) and 25-9-17 (anti-A_β^d) (courtesy of Dr. McKean), respectively. Spleen and lymph nodes were incubated with LPS (80 μg/ml) for 72 hours in medium (RPMI, 5% horse serum, 5 × 10⁻⁵ M β-mercaptoethanol plus 100 μg/ml penicillin, 10% streptomycin, 2 uM glutamine). Lymphocytes from peripheral blood and cultured spleen and lymph nodes cells were separated on a Ficoll-Hypaque gradient (Pharmacia) and washed twice with 1% bovine serum albumin in PBS (pH 7.2) that contained 0.1% sodium azide. The lymphocytes were incubated with mouse anti-mouse biotinylated mAbs 39J and 25-9-17 for 30 min at 4°C, washed twice and then further with fluorescein-streptoavidin (Becton Dickinson Monoclonal Center, Mountain View, CA) for 30 min at room temperature. After washing, the cells were examined in a FACS-IV flow cytometer (Becton Dickinson, Mountain View, CA).

14.1.4 Skin Grafting

Skin grafts were performed by an adaptation of the method of Billingham and Medawar (1951). Briefly, tail samples of skin (approximately 1 cm²) were taken from either a founder transgene-positive mouse or a transgene-negative (SJL × SWR)F_1 mouse and transplanted onto the backs of sex-matched (SJL × SWR)F_1 mice. Bandages were removed after seven days and grafts were monitored daily until rejection (13–16 days postgraft) of the transgene-positive skin.

14.2 RESULTS

14.2.1 A_α^k Transgenic Mice

A_α^k genomic DNA obtained from the HindIII fragment of *cos* H-2^k 8.1 (>9 kb), including 5′ (approximately 3 kb) and 3′ flanking sequences, which contained promoter and polyadenylation sites (see Figure 14–1), was injected into the male pronucleus of F_2 hybrids derived from (C57BL/6 × SJL) mice. Of 1,136 treated embryos, 696 were implanted in the oviducts of pseudopregnant mothers and resulted in 105 live offspring (Table 14–1). Integration of the transgenes into the genome of the recipient mice was confirmed by hybridization of *Pst*I-digested tail DNA, with an A_α^k cDNA probe that contained all the coding sequences of A_α^k. Fifteen mice were found to have integrated the foreign genes and six transgenic lines were obtained. The results of Southern blot analysis of two A_α^k transgenic lines are illustrated in Figure 14–2.

The number of integrated copies of the transgene ranged from one to four per cell, as determined by densitometric comparisons of the intensities of the bands that represent the transgene with those of bands derived from endogenous genes. The pattern of inheritance of the transgenes in the off-

TABLE 14–1 Production of I-A Transgenic Mice

Gene	Embryos		Number of Ova		Founder Mice
	Strain	MHC	Injected	Implanted	
A_α^k	(B6 × SJL)F_2	H-$2^{b,s}$	1,136	696	12
$A_\alpha^k + A_\beta^k$ (MB)	(SJK x SWR)F_2	H-$2^{s,q}$	858	423	6
A_β^k (MB)	(B10.M × CBA)F_2	H-$2^{f,k}$	1,080	633	4

spring of all transgenic mouse lines was consistent with normal Mendelian transmission except for the case of one line that appeared to be a germline mosaic since its rate of transmission of the transgene was only about 10%.

Expression of the product of the A_α^k transgene on the cell surface was analyzed in peripheral blood and spleen cells from different A_α^k transgenic lines using an A_α^k-specific mAb (39J). No expression of A_α^k was detected on the cell surface by FACS analysis. Northern blot analysis of total RNA from different organs with an oligonucleotide probe specific for A_α^k failed to detect transcription of A_α^k at the mRNA level. However, with the same blot and with cDNA specific for A_α^k as the probe, it was possible to demonstrate that the splenic RNA contained intact endogenous A_α^k mRNA message. The liver and kidney RNA did not hybridize with the A_α^k cDNA, indicating that there was no aberrant expression of the A_α^k transgene.

We were surprised to find that when spleen and lymph node cells from these mice were stimulated with LPS in vitro for 72 hours prior to FACS analysis, expression of A_α^k was readily detected with the A_α^k-specific mAB 39J (Figure 14–3). However, expression and the percentage of A_α^k-positive cells varied from line to line. There seemed to be no correlation between the level of expression and the number of copies of the transgene.

Transcomplementation studies have indicated that the transgene A_α^k can associate with A_β^b regardless of the mouse strain that serves as background (B10, A.BY, C3H.SW) but it cannot associate with A_β^s.

14.2.2 A_α^k + A_β^k MB Transgenic Mice

The mutant A_β^k DNA (A_β^k MB) used for injections was A_β^k-puc12 DNA linearized with HindIII, with mutations at positions 63 and 65 through 67 of the $\beta 1$ domain characteristic of the A_β^d amino acid sequence (Figure 14–4). Previous transfection studies in vitro revealed that these A_β^d substitutions result in the loss of binding of all the A_β^k-reactive mAbs but also result in a gain of the capacity for binding of most A_β^d-reactive mAbs (Buerstedde et al. 1988). Meanwhile, the pairing between A_α^k and A_β^k is not impaired. The A_β^k MB gene was either injected alone (B10.M × CBA/n)F$_2$ or in combination with the wild-type A_α^k gene into the pronuclei of (SJL × SWR)F$_2$ fertilized eggs (see Table 14–1). Southern blot analysis showed that integration of both the wild-type A_α^k and the A_β^k MB genes had occurred in six cases. The pattern of segregation suggested that the two genes were integrated at the same sites of chromosomes. In addition to these six mice, four A_β^k MB transgenic mice were also obtained. Figure 14–5 shows that mouse number 3 integrated both the A_α^k and the A_β^k MB genes.

The expression of these transgenes on the surface of spleen cells and peripheral blood cells was not detected by the initial FACS analysis with either an A_α^k-specific mAb (39J) or an A_β^d-specific mAb (25-9-17). However, after incubation in vitro with LPS for 72 hours, these cells become positive

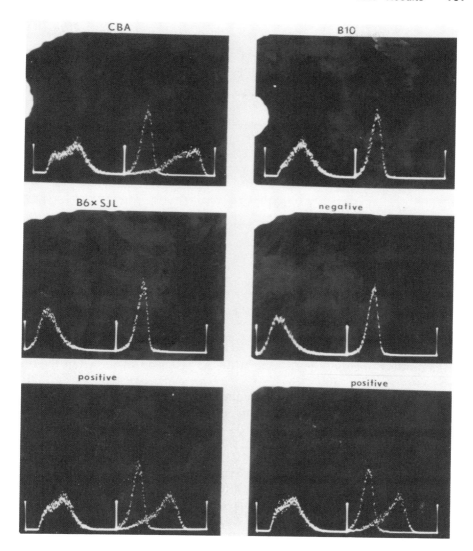

FIGURE 14-3 Expression of the A_α^k antigens on spleen cells, as detected by fluorescence-activated cell sorting. Spleen cells, acitvated with lipopolysaccharide, were stained with biotinylated mAb 39J (anti-A_α^k) and incubated with fluorescein-streptoavidin. CBA (H-2^k) was used as a positive control. B10 (H-2^b) and B6 $\times$ SJL (H-2^b $\times$ H-2^s) were used as a negative control. "Positive" and "negative" denote transgenic mice and their negative littermates, respectively.

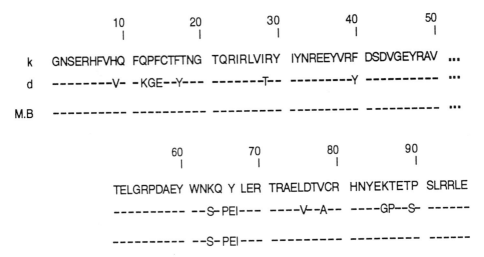

FIGURE 14-4 Sequences of the $\beta 1$ domain of the A_β^k and A_β^d polypeptides and of the mutant A_β^k MB polypeptide. Mutant A_β^k MB DNA contains amino acid substitutions characteristic of the A_β^d polypeptide at positions 63 and 65 through 67 in the $\beta 1$ domain of the A_β^k molecule.

for the product of either the A_α^k gene or the A_β^k MB gene, as shown in Figure 14-6.

To exclude the possibility that either immunofluorescence or immunoprecipitation was insufficiently sensitive to reveal the existence of low levels of the products of the A_α^k and A_β^k transgenes on the cell surface, we performed skin graft experiments. Tail skin from $A_\alpha^k + A_\beta^k$ MB founder mice numbers 20 and 29 was engrafted to (SJL × SWR)F$_1$ recipient mice. These grafts were both rejected with a survival time of approximately 15 days (Figure 14-7). However, the control grafts were not rejected. Thus, the tail skin from $A_\beta^k + A_\beta^k$ MB transgenic mice functioned as an allograft in (SJL × SWR)F$_1$ recipients. In addition, mixed lymphocyte reactions (MLR) were performed with the abovementioned mice. Spleen cells of the $A_\alpha^k + A_\beta^k$ MB founder transgenic mice were shown to induce significant proliferation of lymph node cells from the engrafted (SJL × SWR)F$_1$ animals, indicating that the transgene antigens can be successfully recognized as alloantigens by the primed T-lymphocytes from (SJL × SWR)F$_1$ mice. The results of MLR suggest that the transgene products are expressed on uninduced spleen cells when primed T-lymphocytes are used as an indicator. However, we were unable to detect the products of the transgenes in a primary MLR. Preliminary data indicate that this proliferation can be inhibited by A_α^k-reactive mAbs, while most A_β^k-reactive mAbs showed no effect (data not shown). These experimental results need to be confirmed by more extensive studies with inbred strains of these transgenic mice.

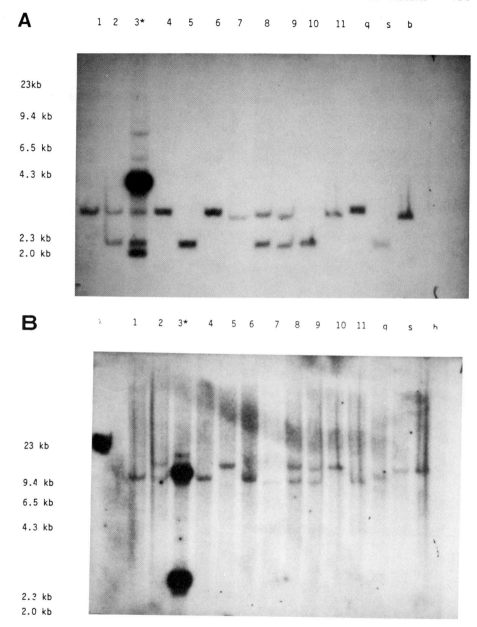

FIGURE 14-5 Integration of A_α^k and A_β^k MB into the genome of transgenic mice (SJL × SWR)F_2. (A) Southern blot analysis of tail DNA digested with *Pst*I hybridized with an A_α^k cDNA probe. (B) Same blot as in (A) stripped of the A_α^k probe and hybridized with A_β^k cDNA probe. Mouse number 3 shows evidence of integration of both the A_α^k and A_β^k genes.

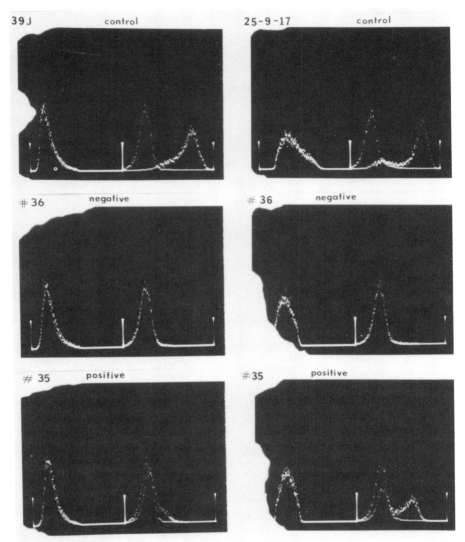

FIGURE 14–6 Expression of the products of the A_α^k and A_β^k MB genes on spleen cells as detected by fluorescence-activated cell sorting. Spleen cells, activated with lipopolysaccharide, were stained with biotinylated mAB 39J (anti-A_α^k) and 25-9-17 (anti-A_β^d), then incubated with fluorescein-streptavidin. CBA (H-2^k) and B10.M (H-2^f) were used as positive and negative controls, respectively, in the left panel. In the right panel, DBA/2J (H-2^d) and B10.M (H-2b) were used as positive and negative controls, respectively. Mouse number 35 is an $A_\alpha^k + A_\beta^k$ MB transgenic mice and mouse number 36 is its negative littermate.

FIGURE 14-7 (A) Rejection of a graft of tail skin from founder number 20 (female) by a (SJL × SWR)F_1 female, 13 days after the graft. (B) A negative control for skin grafting shows the absence of any evidence of rejection after 13 days of tail skin from a female donor (SJL × SWR)F_1 by a female (SWR × SJL)F_1 recipient.

14.3 DISCUSSION

Many of the genes within the MHC have recently been cloned and sequenced. Subsequently, several have been reintroduced into cells in culture so that structure/function relationships can be examined. Gene transfer into somatic cells is, however, of limited use. Questions concerning the expression and regulation of MHC genes during development in a whole organism, the influence of particular MHC genes on the induction of self-tolerance, the composition of the T-cell repertoire, distinctions between responders and nonresponders, and the maturation of the immune response are best approached in transgenic mice by microinjection of genes into the pronuclei of fertilized eggs, so that this DNA can be incorporated into the germline (Billingham and Medawar 1951).

The results presented in this paper indicate that A_α^k genomic DNA, i.e., the H-2 cos8.1 HindIII fragment (>9 kb), which contains approximately 3 kb of 5' and 3' region can be expressed in transgenic mice. Previous transfection studies in vitro showed that the same A_α^k gene is functional and can be expressed in B lymphoma cells (Buerstedde et al. 1988). Recently, nucleotide sequence analysis has identified highly conserved sequences of DNA in the 5' flanking region of all MHC class II genes, namely, TATA box, Y-box, X-box, Z-box and the IFN consensus sequence. These conserved upstream promoter elements govern expression of MHC class II genes (Sullivan et al. 1987). Early data from transgenic mice indicated that the introduction of the entire I-E_α gene, including 2 kb of 5' flanking noncoding I-E_α DNA, into the germ line leads to tissue-specific expression of the transgene (LeMeur et al. 1985). The deletion of 600 nucleotides of the 5' I-E_α

flanking sequence (to position −1400 bp) abolishes expression of the I-E gene in B cells while expression continues in cells in the thymus, spleen, and peritoneal exudate (Yamamura et al. 1985). This result suggests that elements within a region of approximately 2 kb of the 5′ flanking sequence are important in the regulation of expression of the I-E$_\alpha$ gene and that multiple upstream elements control the tissue-specific expression of MHC class I genes in transgenic animals. Since we injected A$_\alpha^k$ genomic DNA with approximately 3 kb of 5′ flanking region, we expected to observe expression of the transgene in a tissue-specific fashion.

Initial FACS analysis of spleen and peripheral blood cells from A$_\alpha^k$ transgenic mice, using A$_\alpha^k$-specific mAbs, revealed no expression of the A$_\alpha^k$ gene. L-cell transfectants revealed that *cis*-chromosomal α:β pairs (for example, A$_\beta^b$A$_\alpha^b$) were always associated with higher levels of expression than were *trans*-pairs (for example, A$_\beta^b$A$_\alpha^k$) (Germain et al. 1985). We introduced the A$_\alpha^k$ gene into H-2^b mice. In such a case, there is competition between the A$_\alpha^k$ transgene and the wild-type A$_\alpha^b$ for the A$_\beta^b$ protein, so the level of expression in an A$_\alpha^k$:A$_\beta^b$ pair should be reduced. However, when spleen cells from such mice were stimulated with LPS prior to FACS analysis, expression of A$_\alpha^k$ was detected. It is hypothesized that the majority of A$_\alpha^k$ and/or A$_\beta^b$ chains produced after stimulation with LPS will pair with each other even in the presence of competition by A$_\alpha^b$. LPS has been shown to induce expression of Ia on B cells (Monroe and Cambier 1983). The mechanism by which LPS increases expression of the MHC is probably related to its ability to induce the release of lymphokines such as IFN-*gamma*, IFNα/β, and TNF (Jephthah-Ochola et al. 1988). However, our preliminary results indicate that INF-*gamma* does not significantly enhance the expression of transgenes. The results above suggest that the extent of A$_\alpha^k$A$_\beta^b$ chain-pairing is very low and that stimulation by LPS enhances the expression of the transgenes such that they are detectable by immunofluorescence. L-cell transfection analyses have shown that α:β heterodimers are required for surface expression of Ia (Malissen et al. 1983 and 1984; Rabourdin-Combe and Mach 1983; Norcross et al. 1984). Ia-positive cells from heterozygous individuals might contain a mixture of Ia molecules derived from the free assortment of allelic α and β chains of a single isotype in all possible combinations. Thus, in (H-2^b × H-2^k)F$_1$ mice, one would find A$_\beta^b$A$_\alpha^b$, A$_\beta^b$A$_\alpha^k$, A$_\beta^k$A$_\alpha^b$, and A$_\beta^k$A$_\alpha^k$ heterodimers (Fathman and Kimoto 1981; Silver et al. 1980). However, data from transfection of L-cells revealed a restriction on cross-isotype α:β assembly. A$_\alpha^k$ can pair with A$_\beta^k$ and A$_\beta^b$ but not with A$_\beta^d$ (Germain et al. 1985). To address the question of α:β pairing between different H-2 haplotypes, transcomplementation studies with A$_\alpha^k$ transgenic mice are being performed. Our initial results indicate that A$_\alpha^k$ can associate with A$_\beta^b$ regardless of the strain background, but there is no association between A$_\alpha^k$ and A$_\beta^s$. Further analysis using different H-2 haplotypes is in progress.

A$_\alpha^k$ + A$_\beta^k$ MB transgenic mice showed the integration of both genes at the same sites on chromosomes. These mice (H-2^s, H-2^f) expressed the products of their transgenes on spleen cells stimulated with LPS, as detected by FACS analysis using the A_α^k-specific mAb (39J) and the A_β^d-specific mAb (25-9-17). Since the A_α^k molecule cannot pair with either A_β^s or possibly A_β^f, and since A_β^k has been shown to be able to associate with A_α^k by analysis in vitro, the above results suggest that the A_α^k:A_β^k MB complex may be the heterodimer expressed on the cell surface. Immunoprecipitation with an A_α^k-specific mAb and two-dimensional gel electrophoretic may confirm this possibility.

The A_β^k MB DNA used for microinjection contains sequences from *puc*12. Studies in transgenic mice have shown that the presence of plasmid vector sequences attached to the injected gene influences the expression of some genes (Townes et al. 1985; Widera et al. 1987). The puc sequences attached to the A_β^k MB gene seemed not to interfere significantly with the expression of the transgene in spleen and lymph node cells.

Skin grafts performed with $A_\alpha^k + A_\beta^k$ MB transgenic mice indicate that the transgenes can be expressed in the tail skin and that the products can induce immunologic rejection in (SJL × SWR)F$_1$ recipient mice. It also appears that skin grafting provided a more sensitive method for the detection of products of the MHC transgene on the cell surface than did immunofluorescence. Studies on the presentation of antigen and susceptibility to disease are currently underway using Ia transgenic mice.

REFERENCES

Benoist, C., Gerlinger, P., LeMeur, M., and Mathis, D. (1986) *Immunol. Today* 7, 138–141.
Benoist, C.O., Mathis, D.J., Kanter, M.R., Williams, V.E. II, and McDevitt, H.O. (1983) *Cell* 34, 169–177.
Billingham, R.E., and Medawar, P.B. (1951) *J. Exp. Med.* 28, 235–239.
Buerstede, J.M., Pease, L.R., Bell, M.P., et al. (1988) *J. Exp. Med.* 167, 473–487.
Cambier, J.C., Justement, L.B., Newell, M.K., et al. (1987) *Immunol. Rev.* 95, 37–57.
Choi, E., McIntyre, K., Germain, R.N., and Seidman, J.G. (1983) *Science* 221, 283–286.
Cook, R., Capra, J.D., Uhr, J.W., and Vitetta, E.S. (1981) in *Structural Studies of the Murine Ia Alloantigens. Current Trends in Histocompatibility* (Reisfeld, I.R.A. and Ferrone, S., eds.), pp. 249–289, Plenum Press, New York.
Corley, R.B., LoCascio, Y., Ounic, H., and Haughton, G. (1985). *Proc. Natl. Acad. Sci. USA* 82, 516–520.
Devlin, J.J, Wake, C.T., Allen, H., et al. (1984) in *Regulation of the Immune System* (Cantor, H., Chess, L., and Sercarz, E., eds.), pp. 57–66, Alan R. Liss, New York.
Estess, P., Begovich, A.B., Koo, M., Jones, P.P., and McDevitt, H.O. (1986) *Proc. Natl. Acad. Sci. USA* 83, 3594–3598.
Fathman, C.G., and Kimoto, M. (1981) *Immunol. Rev.* 43, 57–79.
Germain, R., and Malissen, B. (1986) *Ann. Rev. Immunol.* 4, 281–315.
Germain, R.N., Bentley, D.M., and Quill, H. (1985) *Cell* 43, 233–242.

Jephthah-Ochola, J., Urwson, J., Farkas, S., and Halloran, P.F. (1988) *J. Immunol.* 141:792–800.

Kaufman, J.F., and Strominger, J.L. (1983) *J. Immunol.* 130, 808–817.

Klein, J. (1975) *Biology of the Mouse Histocompatibility-2 Complex*, pp. 541–602, Springer-Verlag, Berlin.

LeMeur, M., Gerlinger, P., Benoist, C., and Mathis, D. (1985) *Nature* 316, 38–42.

Malissen, B., Peele-Price, M., Goverman, J.M., et al. (1984) *Cell* 36, 319–327.

Malissen, B., Steinmetz, M., McMillan, M., Pierres, M., and Hood, L. (1983) *Nature* 305, 440–443.

Monroe, J.G., and Cambier, J.C. (1983) *J. Immunol.* 130, 626–631.

Nagy, Z.A., Baxevanis, C.N., Ishii, N., and Klein, J. (1981) *Immunol. Rev.* 60:59–83.

Noelle, R., Krammer, P.H., Ohara, J., Uhr, J.W., and Vitetta, E.S. (1984) *Proc. Natl. Acad. Sci. USA* 81, 6149–6153.

Norcross, M.A., Bentley, S.M., Margulies, D.H., and Germain, R.N. (1984) *J. Exp. Med.* 160, 1316–1337.

Palmiter, R.D., and Brinster, R.L. (1986) *Ann. Rev. Genet.* 20, 465–499.

Pober, J.S., Collins, T., Bimgrone, Jr., M.A., et al. (1983) *Nature* 305, 726–729.

Rabourdin-Combe, C., and Mach, B. (1983) *Nature* 303, 670–674.

Scangos, G., and Bieberich, C. (1987) *Adv. Genet.* 24, 285–322.

Silver, J., Swain, S.L., and Hubert, J.J. (1980) *Nature* 286, 272–274.

Steeg, P.S., Moore, R.N., Johnson, H.M., and Oppenheim, J.J. (1982) *J. Exp. Med.* 156, 1780–1793.

Steinmetz, M., Minard, K., Horvath, S., et al. (1982) *Nature* 300, 35–42.

Stuart, P.M., Zlotnik, A., and Woodward, J.G. (1988) *J. Immunol.* 140, 1542–1547.

Sullivan, K.E., Claman, A.F., Nakanishi, M., et al. (1987) *Immunol. Today* 8, 289–293.

Townes, T.M., Lingrel, J.B., Chen, H.Y., Brinster, R.L., and Palmiter, R.D. (1985) *EMBO J.* 4, 1715–1723.

Widera, G., Buekly, L.C., Pinkert, C.A., et al. (1987) *Cell* 51, 175–187.

Yamamura, K., Hikutani, H., Folsom, V., et al. (1985) *Nature* 216, 67–69.

Zlotnik, A., Fischer, M., Roehm, N., and Zipon, D. (1987) *J. Immunol.* 138, 4275–4279.

Zlotnik, A., Shimonkevitz, R.P., Gefter, M.L., Kappler, J., and Marrack, P. (1983) *J. Immunol.* 131, 2814–2820.

CHAPTER 15

Mice Transgenic for a Gene That Encodes a Soluble, Polymorphic Class I MHC Antigen

Rosemarie Hunziker
David H. Margulies

The central function of the immune system is to maintain a battery of cells and molecules capable of potent and specific reaction to challenges by microbial, viral, or malignant invaders and, simultaneously, to discriminate between these foreign structures and a large number of endogenous molecules. These requirements are elegantly satisfied by mechanisms that first generate a diverse repertoire of T-cell and B-cell receptors and subsequently inactivate those cells that carry receptors that happen to be reactive with self antigens. The remaining receptors are poised to deal with foreign antigens. Surface receptors serve as the first arm in the cellular response by which helper and cytolytic T cells (T_H and T_C, respectively) are activated

---

Rosemarie Hunziker is a postdoctoral fellow of the Leukemia Society. We are grateful to our colleagues, Drs. S. Kozlowski, R. Ribaudo, and J. Schneck, for their comments on the manuscript and to Ms. L. Boyd for assistance in various aspects of the work. The anti-Q10[b] antipeptide antiserum was a generous gift from Drs. W.L. Maloy and J.E. Coligan of the Laboratory of Immunogenetics, NIAID, National Institutes of Health, Bethesda, MD. The research for this chapter was carried out on behalf of the United States Government and is, therefore, in the public domain and cannot be copyrighted.

to produce lymphokines or to kill cells that carry altered self molecules. In addition, the B-cell receptor, cell surface membrane immunoglobulin (mIg), serves as an initial element in the stimulation of B cells to differentiate into cells capable of secreting soluble antibody at high rates. The phenomenon of the organism's nonresponsiveness to its own molecules ("self") is known as immunological tolerance, and it is a major goal of current immunological research to understand its molecular basis in detail.

Cell-surface proteins encoded by the major histocompatibility complex (MHC) associate with peptide antigens, and this MHC molecule/antigen complex serves as the ligand for the stimulation of T-lymphocytes through their T-cell receptor (Schwartz 1986). Class I and class II MHC molecules are structurally related, but for the purposes of this discussion we will describe class I molecules only. Class I proteins consist of an integral membrane glycoprotein, a heavy chain of 45 kd, encoded within the MHC (known as H-2K, D, or L in the mouse; and HLA-A, B, or C in the human), which is noncovalently associated with a light chain, β2-microglobulin (12 kd). Nucleotide sequence analysis of genes for class I heavy chains has lead to the general conclusion that distinct regions or domains of the encoded protein are derived from separate genomic coding blocks (Maloy 1987). Thus, there are unique exons for the signal (leader) peptide, the extracellular protein domains (α1, α2, and α3), the transmembrane domains, and the cytoplasmic domains.

The membrane-distal α1 and α2 domains of the MHC class I molecules are the most critical ones in the binding of antigen and in the interaction with T-cell receptors. This conclusion was reached in studies using in vitro derived, recombinant, class I molecules generated from different allelic and nonallelic members of this multigene family (Margulies and McCluskey 1985). Further support for this conclusion has been provided by the determination of the three-dimensional crystallographic structure of the human class I HLA-A2 molecule. These studies show that the α1 and α2 domains form a unitary structure that defines a cleft for the binding of antigenic peptides (Bjorkman et al. 1987a and 1987b). The ligand for a given T-cell receptor, then, is a combinatorial structure comprised of the distal domains of an MHC molecule and an antigenic peptide fragment. How does the immune system select and expand those T-cell clones that are capable of reacting with foreign antigen bound to self MHC molecules and how, at the same time, does it eliminate those clones that react with self-antigens bound to MHC molecules? T-cell precursors migrate from the bone marrow to the thymus where they are exposed to soluble and membrane-bound molecules and, in particular, to cell-surface MHC class I and class II molecules. It is thought that those T cells that express receptors with moderate affinity, sufficient to bind to MHC molecules, are selected and expanded (positive selection). Subsequently, T cells with particularly high affinity binding to endogenous MHC molecules or to MHC/peptide complexes are either deleted or rendered unresponsive (negative selection). Thus, the highest affinity "self-reactive" cells are eliminated, reducing the probability of the maturation of autoreactive T cells. Mature T cells that leave the thymus bear receptors with a moderate affinity for self MHC and are prepared for high-affinity stimulation by an appropriate self MHC/antigenic peptide complex. The positive selection of moderate-affinity T cells is termed "thymic education," while the negative selection against high-affinity T cells generates "thymic tolerance." While many of the details of this scenario of T-cell ontogeny are hypothetical, there is evidence to support aspects of both positive and negative selection (Kappler et al. 1987 and 1988; Kisielow et al. 1988; MacDonald et al. 1988).

Our current view of the processes that lead to T-cell education and tolerance in the thymus is based on our understanding of the requirements for activation of mature, peripheral T cells. Evidence suggests that T cells can be stimulated by multivalent or insoluble arrays of MHC or MHC/antigen complexes, but not by soluble, monovalent complexes (Herrmann and Mescher 1986; McCluskey et al. 1988). A question of critical importance to our understanding of T-cell maturation, then, is whether the signals that convey the developmental messages, such as tolerance and education, also depend upon multivalent interactions of the T-cell receptor with MHC or MHC/peptide complexes. Transgenic mouse technology offers a unique opportunity whereby we can ask whether a soluble analog of an MHC class I molecule can provide the molecular signals necessary for appropriate positive and negative selection.

In this chapter we describe the preliminary characterization of a C57BL/6 mouse strain that expresses an obligately secreted counterpart of the classical murine class I MHC molecule, $H-2D^d$. Subsequent immunological studies will allow definitive tests of the hypothesis that expression of multivalent cell-surface molecules is a requirement for the development of immunological tolerance to an MHC class I molecule.

15.1 EXPERIMENTAL PROCEDURES

15.1.1 Construction of the $H-2D^d/Q10^b$ Gene

The plasmid containing the chimeric gene used in this study has been described elsewhere (Margulies et al. 1986).

15.1.2 Generation of the $H-2D^d/Q10^b$ Transgenic Mouse Line

Procedures for the production of transgenic mice have been described (Hogan et al. 1986).

15.1.3 Western Blotting

Western blots of samples of murine serum were performed after electrophoresis in nonreducing, sodium dodecyl sulfate (SDS) polyacrylamide gels. After electrophoretic transfer to nitrocellulose, the blots were developed with an antibody raised in rabbit against mouse Q10 carboxyl-terminal peptide, with subsequent treatment with ^{125}I-protein A, essentially as described by Lew et al. (1986).

15.1.4 Immunoprecipitations

Immunoprecipitations were performed as described elsewhere (McCluskey et al. 1985), except that subsequent detection was performed by Western blotting as described above.

15.2 RESULTS AND DISCUSSION

15.2.1 Construction of the Gene and Characteristics of the Soluble Class I Protein

Several possible strategies for the production of soluble class I molecules all rely on disruption of the "stop transfer" signal, the hydrophobic membrane-spanning peptide that arrests the nascent protein chain's translocation across the rough endoplasmic reticulum (reviewed in Margulies et al. 1987). Studies of the structure of the class I-like Q10^b protein show that this monomorphic molecule is secreted, rather than bound to membrane, as the result of a deletion within the transmembrane exon that leads to a frameshift mutation and premature termination of translation within the transmembrane region (Kress et al. 1983). Since the α1 and α2 domains of the class I molecule are most critical to its function, it was expected that a molecule that consisted of the polymorphic amino-terminal α1 and α2 domains of H-2D^d and the membrane-proximal α3 domain of the nonpolymorphic, class I-like gene, Q10^b, would preserve the major H-2D^d functional structure and have the secretory phenotype of Q10^b. We constructed a gene that contained the genomic DNA for the 5' flanking region, the leader peptide, and the α1 and α2 exons of H-2D^d covalently coupled to the α3 exon and 3' flanking sequences of Q10^b (Figure 15-1). Upon transfection into mouse L cells, this gene directs the synthesis of a biologically active, serologically intact, chemically purifiable, secreted chimeric protein (Margulies et al. 1986). When displayed in a multivalent array, whether covalently coupled to agarose beads, covalently coupled to high-molecular-weight soluble dextran, or noncovalently adsorbed to polystyrene plates, this molecule stimulates an H-2D^d-reactive T-cell hybridoma to secrete interleukin 2 (IL-2) (McCluskey et al. 1988).

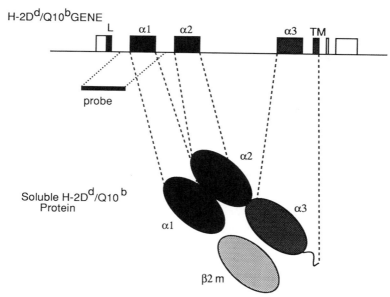

FIGURE 15-1 The H-2D^d/Q10^b gene and its encoded protein. A schematic representation is depicted in which open boxes indicate 5' and 3' untranslated regions, solid black boxes indicate translated coding blocks derived from the H-2D^d gene, and dark stippling indicates coding regions derived from Q10^b. The origin of the 5'H-2D^d α1 exon probe employed in the experiment whose results are shown in Figure 15-2 is indicated. As described in detail elsewhere, approximately 500 bp of the intron between α2 and α3 were deleted in the construction of the chimeric gene (Margulies et al. 1986).

15.2.2 Generation of a Line of Mice Transgenic for H-2D^d/Q10^b.

The recombinant H-2D^d/Q10^b gene was released from the plasmid by digestion with appropriate restriction endonucleases and, after purification, the DNA was introduced into the most accessible pronucleus of fertilized C57BL/6 mice ova by microinjection. The eggs surviving this procedure (about 50–75%) were cultured overnight in M16 medium (Hogan et al. 1986), and those embryos that had advanced to the two-cell stage (75–90%) were transferred to the right uterine horn of CB6/F$_1$ pseudopregnant foster mothers.

A total of 43 pups were born, of which 41 survived to weaning. DNA was extracted from the distal centimeter of the tail of each weanling, digested with an appropriate restriction endonuclease, and analyzed by electrophoresis and Southern blotting (Figure 15-2). In the construction of the chimeric gene, approximately 500 bp of DNA from the third intron was deleted (see legend to Figure 15-1). Thus, hybridization analysis would be expected to

180 A Gene That Encodes a Soluble, Polymorphic Class I MHC Antigen

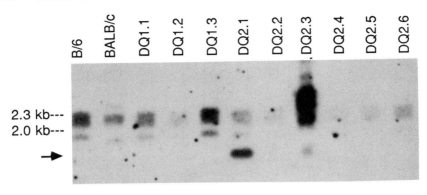

FIGURE 15-2 Southern blot analysis of DNA from mice injected as zygotes with the $D^d/Q10^b$ gene. Ten micrograms of DNA from the tail was digested to completion with Bam H1, electrophoresed through 0.8% agarose, denatured, blotted to nitrocellulose, and probed with the α1-derived fragment shown in Figure 15-1. The arrow indicates the location of the 1.6-kb fragment of DNA unique to the chimeric gene.

reveal a restriction fragment unique to the chimeric gene. One mouse of the 41 tested contained the diagnostic restriction fragment length polymorphism (see Figure 15-2).

The founder female was bred to a B6 male and 60% of the subsequent litter contained the transgene. Further breeding of this mouse and her offspring revealed an autosomal, Mendelian pattern of inheritance (Figure 15-3). Animals homozygous for the transgene do not show any abnormal behavior or health problems and are fertile.

15.2.3 Expression and Quantitation of the H-2D^d/Q10^b Protein

The presence of the H-2D^d/Q10^b protein in serum was assessed by the Western blot technique. The peptide-specific antiserum used to develop the Western blot binds to only one band (Q10^b) in control serum (Figure 15-4). However, in addition to the Q10^b band, a second, more slowly migrating, diffuse band is seen in the serum of the transgenic mouse DQ2.1 (see Figure 15-4). The H-2D^d/Q10^b protein contains one more glycosylation site than does the Q10^b protein, so the chimeric molecule, which contains an additional carbohydrate moiety, migrates more slowly in the gel during electrophoresis. Also, the slower band co-migrates with the H-2D^d/Q10^b protein purified from the supernatants of L cells transfected with the H-2D^d/Q10^b gene (data not shown).

Another experiment confirmed the presence of H-2D^d epitopes on the slowly migrating band (Figure 15-5). Serum was reacted with either 34.5.8 (a monoclonal antibody that recognizes an epitope in the amino-terminal

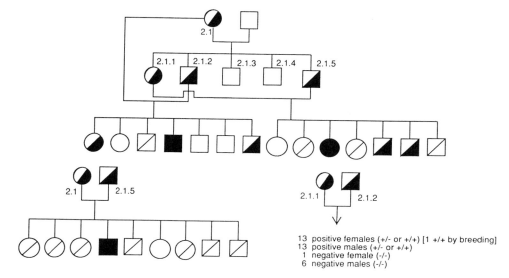

FIGURE 15-3 Pedigree of the B6.tD^d/Q10^b line. The founder mouse, DQ2.1, and its progeny were tested for the presence of the transgene by Southern blot analysis and for the presence of the chimeric protein in the serum by Western blotting analysis. Males (squares) and females (circles) are indicated, and those definitively shown to be heterozygous (half-filled) or homozygous (completely filled) are shown. Because of ambiguity in the discrimination of homozygotes from heterozygotes in the biochemical assays, only homozygotes that have been proven by backcross and progeny testing are shown as filled symbols. Those mice that are either heterozygotes or homozygotes are indicated by a single line through the symbol.

portion of the H-2D^d molecule [Evans et al 1982]), 34.2.12 (a monoclonal antibody that recognizes an α3 H-2D^d determinant), or phosphate-buffered saline (PBS) alone. Protein A-Sepharose was added to the mixtures to precipitate any antigen-antibody complexes. The bound and supernatant fractions were separately recovered, mixed with SDS-loading buffer, subjected to electrophoresis, and transferred to nitrocellulose. This Western blot was developed with the antiserum raised against Q10^b peptide. Neither 34.2.12 nor the PBS precipitated any bands, but the 34.5.8 antibody precipitated only the slowly migrating band, leaving the nonpolymorphic Q10^b band unprecipitated in the supernatant (see Figure 15-5). This preclearing experiment indicates that the more slowly migrating, heterogeneous molecule is seen by the H2D^d-specific monoclonal antibody 34.5.8 as well as by the Q10^b-specific antiserum.

The concentration of H-2D^d/Q10^b in the serum of the transgenic mice was 100–200 μg/ml, as estimated by reference to a standard curve of protein purified from the supernatant of H-2D^d/Q10^b transfected L cells grown in serum-free medium (Figure 15-6).

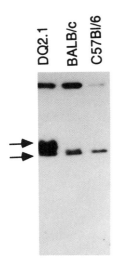

FIGURE 15–4 Western blot analysis of mice injected as zygotes with the $D^d/Q10^b$ gene. One milliliter of serum was subjected to electrophoresis through a non-reducing 10% polyacrylamide gel that contained SDS, and electroblotted onto nitrocellulose. The blot was developed with the specific antibody (raised in rabbit) against $Q10^b$ carboxyl-terminal peptide antibody, as described in the Experimental Procedures, and exposed to X-ray film. Lane 1: transgenic founder mouse, DQ2.1; lane 2: BALB/c ($H-2^d$ control); lane 3: non-transgenic littermate of founder mouse.

15.3 CONCLUSIONS

We have been studying molecularly engineered soluble analogs of the murine MHC antigens in order to investigate the role played by histocompatibility antigens in the generation and maintenance of immunological tolerance and education. We have previously described a recombinant gene comprised of 5′ sequences derived from the polymorphic murine MHC class I gene H-$2D^d$, and 3′ sequences derived from the nonpolymorphic, but obligately secreted class I-like gene $Q10^b$. Upon transfection into mouse L cells, the chimeric gene (H-$2D^d/Q10^b$) directs the synthesis of a biologically active, serologically intact, and chemically purifiable soluble molecule. To ask whether the function of MHC molecules in tolerance and education is related to expression of multivalent antigens on the cell surface, we have produced a strain of transgenic mice that express this soluble class I protein. The transgene is inherited in an autosomal, Mendelian fashion and homozygous mice are healthy and fertile. All mice bearing the injected transgene expressed relatively high levels (0.1–0.2 mg/ml) of the encoded protein in their serum, as assessed by Western blotting and immunoprecipitation.

We should point out that a similar experiment has been performed by Arnold et al. (1988), who generated a transgenic mouse strain that expressed a soluble counterpart of the H-$2K^k$ protein. The analysis of this mouse strain was complicated by several factors: 1) the level of expression of the soluble counterpart of the H-$2K^k$ protein was very low (at the limit of detectability of the assay, about 100 ng/ml); and 2) the transgenic strain was developed in (C57Bl/6 × DBA/2)F_2 embryos, so the analysis of the immunological behavior of the progeny was complicated by multigenic factors.

The results of our studies demonstrate the generation of a mouse strain in a homozygous C57Bl/6 background that expresses high levels in serum

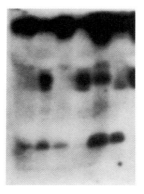

FIGURE 15-5 Immunoprecipitation of the H-2D^d/Q10^b molecule from the serum of DQ2.1, the transgenic founder mouse. Pellets or supernatants from serum precleared with either PBS, 34.2.12 (labelled 34.2.), or 34.5.8 (labelled 34.5.) were analyzed by Western blotting. One microliter of serum was diluted to 10 μl with PBS and added to 10 μl of 34.5.8 (a monoclonal antibody specific for the α2 domain of H-2D^d; Ozato et al. 1982), 34.2.12 (a monoclonal antibody specific for the α3 domain of H-2D^d), or PBS and incubated at 4° C for 18 hours. Twenty-five microliters of protein A-Sepharose in lysis buffer [0.15 M NaCl, 10 mM Tris (pH 7.6), 1.5 mM MgCl$_2$, 0.5% NP40, 1% aprotinin, 0.1% SDS] were added and the mixture incubated at 4° C for 90 min with constant agitation. The mixtures were centrifuged at high speed, the supernatants saved, and the pellets washed three times. The supernatants, pellets, or 1 μl of untreated serum were mixed with 10 μl of nonreducing sample buffer [0.05 M Tris (pH 6.8), 2% SDS, 20% glycerol, 0.3 M iodoacetamide] and subjected to electrophoresis through a 10% SDS-polyacrylamide gel. Subsequent Western blotting and identification of the proteins was carried out as described in the legend to Figure 15-4.

of a soluble counterpart of the H-2D^d molecule. Experiments underway will assess the nature of T-cell function in these animals. Specifically, we are studying the ability of the transgenic animals to mount an anti-H-2D^d allogenic cytotoxic T-lymphocyte response, the ability of these animals to reject or to tolerate skin grafts from donors that express membrane-bound H-2D^d, and their ability to mount H-2D^d-restricted, antigen-specific responses. Furthermore, we expect to evaluate the role of soluble H-2D^d in the generation of class II-restricted, anticlass I responses. In future studies, we hope to compare and evaluate the contribution of the α3 domain of the class I molecule in the generation of T-cell tolerance and education using

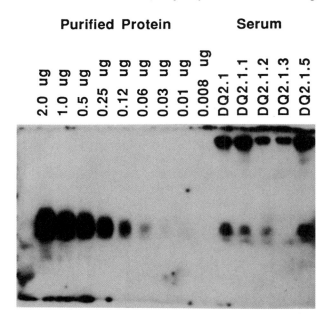

FIGURE 15-6 Quantitative Western blotting analysis of serum from H-2D^d/Q10^b transgenic mice. Methods were as described in the legend to Figure 15-4. From left to right: lanes 1 to 9, twofold dilutions of H-2D^d/Q10^b protein in the indicated amounts; lanes 10 to 14: 1 μl of serum from the founder mouse (DQ2.1) or selected offspring from her first litter (DQ2.1.1, DQ2.1.2, DQ2.1.3, DQ2.1.5).

a newly created H-2D^d chimeric gene that encodes a soluble molecule which consists of $\alpha 1$, $\alpha 2$, and $\alpha 3$ domains of H-2D^d linked only to the carboxy-terminal 27 amino acid residues of Q10^b (S. Kozlowski et al. unpublished observations). A complete functional analysis of the transgenic mouse strain described here, as well as of other related strains currently being produced, should permit a more detailed understanding of the influence of the MHC on the development of T cells in the thymus.

REFERENCES

Arnold, B., Dill, O., Küblbeck, G., Jatsch, L., Simon, M.M., Tucker, J., and Hämmerling, G. J. (1988) *Proc. Nat. Acad. Sci. USA* 85, 2269–2273.

Bjorkman, P. J., Saper, M. A., Samraoui, B., Bennett, W.S., Strominger, J.L., and Wiley, D.C. (1987a) *Nature* 329, 506–512.

Bjorkman, P. J., Saper, M. A., Samraoui, B., Bennett, W.S., Strominger, J.L., and Wiley, D.C. (1987b) *Nature* 329, 512–518.

Evans, G.A., Margulies, D.H., Skykind, B., Seidman, J.G., and Ozato, K. (1982) *Nature* 300, 755–757.

Herrmann, S. H. and Mescher, M. F. (1986) *J. Immunol.* 136, 2816–2825.
Hogan, B., Costantini, F., and Lacy, E. (1986) *Manipulating the Mouse Embryo*, Cold Spring Harbor Laboratory: Cold Spring Harbor.
Kappler, J., Roehm, N., and Marrack, P. (1987) *Cell* 49, 273–280.
Kappler, J., Staerz, U., White, J., and Marrack, P.C. (1988) *Nature* 332, 35–40.
Kisielow, P. Blüthmann, H., Staerz, U.D., Steimetz, M., and von Boehmer, H. (1988) *Nature* 333, 742–746.
Kress, M., Cosman, D., Khoury, G. and Jay, G. (1983) *Cell* 34, 189–196.
Lew, A.M., Maloy, W.L., and Coligan, J.E. (1986) *J. Immunol.* 136, 254–258.
MacDonald, H.R., Schneider, R., Lees, R.K., et al. (1988) *Nature* 332, 40–45.
Maloy, W.L. (1987) *Immunol. Res.* 6, 11–29.
Margulies, D.H., Lopez, R., Boyd, L.F., and McCluskey, J. (1987) *Immunol. Res.* 6, 101–116.
Margulies, D.H. and McCluskey, J. (1985) *Surv. Immunol. Res.* 4, 146–159.
Margulies, D.H., Ramsey, A.L., Boyd, L.F., and McCluskey, J. (1986) *Proc. Nat. Acad. Sci. USA* 83, 5252–5356.
McCluskey, J., Boyd, L.F., Highet, P.F., Inman, J., and Margulies, D.H. (1988) *J. Immunol.* 141, 1451–1455.
McCluskey, J., Germain, R.N., and Margulies, D.H. (1985) *Cell* 40, 247–257.
Ozato, K., Mayer, D., and Sachs, D. (1982) *Transplant.* 39, 113–120.
Schwartz, R.H. (1986) *Adv. Immunol.* 38, 31–201.

PART IV

Use of Transgenics in Medicine

CHAPTER 16

Principles of Gene Transfer and the Treatment of Disease

Evelyn M. Karson

The purpose of this chapter is to present an overview of research into the use of gene transfer for treating human genetic diseases. In contrast to most of the other chapters in this book, this chapter is limited to a discussion of the possibilities for application of gene transfer technology to somatic cells of an organism. Indeed, the difficulties associated with insertional mutagenesis and temporal and tissue specificity are indicators that the technology has not yet reached the point where it is reasonable or ethical to consider correction of disease through the germline, and there remains much that is not understood about the treatment of somatic cells.

The focus of this chapter will primarily be directed at 1) those diseases that are approachable in the near future, 2) the biological criteria for identifying such diseases, 3) the types of cells that might be treatable, 4) experience with animal models, and 5) a novel application of gene transfer technology in cancer research.

There are over 3,000 known genetic diseases. The medical costs as well as the social and emotional costs of genetic disease are enormous. Monogenic diseases account for 10% of all admissions to pediatric hospitals in North America (Hall et al. 1978) and 8.5% of all pediatric deaths (Roberts et al. 1970). They affect at least 1% of all liveborn infants (Denniston 1982) and they cause 7% of stillbirths and neonatal deaths (Mueller et al. 1983).

Those survivors with genetic diseases frequently have significant physical, developmental, or social impairment (Costa et al. 1985). At present, medical intervention provides complete relief in only about 12% of Mendelian single-gene diseases; in nearly half of all cases, attempts at therapy provide no help at all (Hayes et al. 1985). If a safe procedure that could correct genetically defective fetuses, infants, and or children were available, it could have widespread application. Because of the advances in recombinant DNA technology and our understanding of human genetic disease, gene therapy is now becoming feasible (Anderson 1984; Bernstein et al. 1985; Williams et al. 1986; Belmont and Caskey 1986).

16.1 WHICH DISEASES ARE CANDIDATES FOR TREATMENT?

There are several important considerations, both ethical and scientific, in the choice of initial candidates for gene therapy. Because we are introducing a new, unproven technology, the risks are not completely quantifiable. Consequently, only severe disorders, those for which current therapy cannot stop the progression to severe crippling or early death, should be considered for treatment initially. Clearly, the success or failure of the first attempts at any new treatment will influence future use and development of an innovative technology. Scientifically, it is important to begin with the fewest complicating biological variables in order to maximize the likelihood of success in the initial trial. Specific biological characteristics of a genetically treatable disease include the following:

1. The disease should be a recessive disease caused by the absence of a normally functioning copy of a gene. Treatment of autosomal dominant diseases, in which the presence of one abnormal copy of a gene causes disease, would remain unapproachable until methods have been perfected for "turning off" or removing defective genes.
2. The gene involved must be clonable by recombinant DNA techniques.
3. The gene in the normal person must code for an enzyme formed from a single protein that does not need to be located in a highly specialized cellular or subcellular environment to function effectively.
4. The normal function of the cell should not require precise regulation of the amount of the protein produced.
5. It should be theoretically possible to ameliorate or correct the patient's condition by treatment of a target cell that is either directly affected by the disease process or is unaffected but able to produce a new gene product that is accessible to an affected cell. In the latter case, the gene product must be able to function at a distance from the affected tissues (for example, by lowering the concentration of a toxic by-product or by increas-

ing levels of a necessary metabolite), or the target cell should be able to produce and deliver the gene product to affected organs.
6. It should be possible for the target cells to be removed safely from the body for manipulation and then returned.
7. Although not mandatory, it is desirable that the treated cells should have a selective advantage in terms of growth or survival, so that elimination of the patient's remaining untreated cells would not be necessary.

16.1.1 Bone Marrow as a Target Organ

Based on the success of bone marrow transplantation (BMT) in treating a variety of genetic diseases, prospects for the use of bone marrow as a target for gene transfer should be summarized here (see also review by Parkman 1986). Some of the treated cells, the totipotent stem cells, after being removed from the body for processing and then reinfused, may have the capacity for self-renewal throughout the lifetime of the patient. A large number of cells from other organs, many of which cannot propagate and currently cannot be easily transplanted, interact with the many types of blood cells and macrophages that differentiate from the bone marrow cells. Thus, many disorders besides those directly affecting bone marrow-derived cells might be correctable using genetically engineered bone marrow.

For example, the enzymes affected in lysosomal storage diseases have a characteristic that makes them particularly suitable for indirect delivery via gene therapy using BMT; posttranslational modifications add a mannosyl-phosphate carbohydrate side chain that directs the uptake of these enzymes into lysosomes. Liver biopsies obtained from a patient with Hurler syndrome after BMT demonstrated that enzyme from donor Kupffer cells (which are a type of macrophage derived from bone marrow) cleared hepatocytes of mucopolysaccharide (MPS) inclusions. In addition, hepatosplenomegaly and corneal clouding were corrected (Hobbs et al. 1981). Lymphoid cells in vitro have also been shown to transfer lysosomal enzymes directly to fibroblasts (Olsen et al. 1981; Abraham et al. 1985). Brain biopsies from iduronidase-deficient dogs treated by allograft BMT showed clearing of glycosaminoglycans from neurons, glial cells, and perivascular cells (Shull et al. 1987). Reduced levels of cerebrospinal fluid glycosaminoglycans were also noted in the dogs, as well as in pediatric patients with Hurler and Sanfilippo A syndromes after BMT (Whitley et al. 1986a and 1986). Cloning of iduronidase and also iduronate sulfatase (the enzyme that is defective in Hunter syndrome, in which nonneurological symptoms have been treated by BMT) is in progress.

Unless the gene product requires cell-specific, posttranslational modifications and/or direct localization in a specific compartment of the true target cells affected by a particular disease, the transfer of exogenous genes with appropriate regulatory signals might allow bone marrow-derived cells to produce proteins that are not usually expressed. For instance, phenylalanine hydroxylase, the enzyme that fails to function normally in many cases of phenylketonuria, is not ordinarily produced in blood cells. The cloned gene inserted into cultured cells has been shown to produce active protein (Ledley et al. 1986). Because dietary restriction decreases intellectual impairment, it appears that the damage to the central nervous system (CNS) in phenylketonuria is secondary to accumulation of phenylalanine or a toxic by-product rather than to cerebral deficiency of tyrosine. Thus, it is possible that creating a pool of enzymatic activity in the peripheral blood cells might be an effective treatment without more specific targeting of the enzyme.

16.1.2 Other Possible Target Cells

Preliminary experiments in vitro with human cord blood obtained at the time of premature and term deliveries and experiments with fetal sheep and primates (discussed in a later section) suggest that it may be feasible to use circulating hematopoietic progenitor cells as an alternative source of "bone marrow" cells (Karson et al. 1987; Broxmeyer et al. 1989). Recently, infusion of cord blood from an HLA-identical sibling was used in place of BMT to successfully transplant and cure a child with Fanconi's anemia (Gluckman et al. 1989). During the hematopoietic expansion in the fetus, cells migrate from the liver and spleen to the bone marrow, which becomes the major site of hematopoiesis in later life. The treatable cell pool in the fetus can be removed by percutaneous umbilical vessel blood sampling (PUBS) under ultrasound guidance, manipulated in the laboratory, and reinfused by PUBS or ultrasound-guided intraperitoneal injection. This method might be particularly useful in diseases where there is significant deterioration by the time of birth and might possibly permit cells to cross the potential blood-brain barrier. Alternatively, where these factors are less important, it may be advantageous to wait until after birth for treatment. One could aseptically collect fetal cord blood at delivery for treatment and reinfuse the cells by a partial exchange transfusion if necessary. A larger fraction of the blood can be sampled at this time and the risks to the mother and child associated with the actual manipulations are reduced. In either case, repopulation of the hematopoietic system with the "repaired" fetal cells, particularly in cases where these cells have a selective advantage for survival, might be accomplished without the risks associated with radiologic and/or chemotherapeutic ablation, which are usually necessary to achieve engraftment in later postnatal life.

Using gene transfer, other types of tissues may also be exploited as shuttles or generators of the normal gene product in diseased patients. Because of their contiguity with the bloodstream, endothelial cells are a particularly attractive target that would be suitable recipients of genes for enzymes and proteins normally found in the blood, as well as those which can function there as an alternative (such as the examples described above). Endothelial cells could be isolated from the patient, expanded and treated ex vivo, and then returned to the patient either to seed the blood vessel directly or in the form of an artificial graft or shunt. Expression of exogenous genes in cultured rabbit endothelial cells before and after seeding onto synthetic vessel grafts have been demonstrated in vitro (Zwiebel et al. 1989; Zwiebel, personal communication).

Fibroblasts from an individual can also be propagated in vitro and transduced with exogenous genes. Cells can be immobilized on collagen matrices and can then be reimplanted to provide a source of enzyme that either migrates to an affected tissue or can function autonomously. Application of an angiogenesis factor to the matrix can promote vascularization, which will support survival of the implanted cells (Thompson et al. 1988).

16.1.3 The Initial Candidates

In the past, clinical investigators thought that the human genetic diseases that would most likely be the initial candidates for successful treatment by gene therapy would be the hemoglobinopathies, such as beta-thalassemia. Bone marrow is one of the easiest tissues to manipulate outside the body, but hemoglobin is composed of two subunits, which are encoded by genes on two different chromosomes, and hemoglobin itself switches from embryonic to fetal to adult forms during development. Moreover, the levels of each subunit are controlled by complicated regulatory signals.

Currently, it appears that the form of severe combined immunodeficiency disease which results from adenosine deaminase (ADA) deficiency is likely to be the first disease to be treated by gene therapy. Children with ADA deficiency have recurrent infections of the skin, the CNS, and the sinopulmonary and gastrointestinal tracts, as well as a failure to thrive because of chronic diarrhea and malabsorption, and a high incidence of lymphoma. Most individuals who do not have a histocompatible bone marrow donor die in early childhood. Because BMT can be curative and since the other criteria described above are met, it is thought that an "autologous" transplantation of bone marrow cells that have received a copy of the normal ADA gene ex vivo might be the first successful application of gene therapy. In a subsequent section, we will discuss in greater detail the requirements that should be met before a therapeutic protocol for treating patients is actually designed and carried out.

16.1.4 Other Potential Candidates

From a theoretical perspective, what other categories and specific diseases might be candidates for gene therapy in the near future? Table 16-1 lists some of the diseases in which the gene of interest has been cloned and for which there are varying possibilities for successful treatment in the near future. Subset A includes several of the examples given above and other similar diseases for which gene therapy holds much promise. Subset B in-

TABLE 16-1 Prospects for Gene Therapy—Diseases for Which Relevant Genes Have Been Cloned

Likely to Suceed
Adenosine deaminase deficiency
Argininosuccinic aciduria (argininosuccinase)
Citrullenemia (arginosuccinate synthetase)
Gaucher disease type I (glucocerebrosidase)
Phenylketonuria (phenylalanine hydroxylase)
Purine nucleoside phosphorylase deficiency

Possibility of Success
Elliptocytosis 1 (protein 4.1)
Elliptocytosis 2 (spectrin)
Granulocyte actin deficiency

Complex Regulation
Coagulation factors VIII, IX, X, XIIIa
Complement factors C2, C4, C9
Hemoglobinopathies
Hereditary angioneurotic edema (C1 inhibitor)
Thalassemia

cludes examples in which the affected gene codes for a blood cell structural protein. The addition of normal molecules to these cells is likely to ameliorate the abnormal morphology, despite the continued production of abnormal protein (in contrast to the problem of cirrhosis in cases of α1-antitrypsin deficiency, discussed below). In subset C, the problems of complex regulation of the factors involved may complicate the application of gene therapy for some time.

16.1.5 Unlikely Candidates

Regrettably, structural proteins for most tissues, such as muscle, visceral organs, or neurons, probably cannot be supplied via BMT (Table 16–2). The gene for α1-antitrypsin has been cloned and expression of the functional, glycosylated protein in fibroblasts in culture has been achieved (Garver et al. 1987). Only partial correction of the clinical defects may, however, be possible. Theoretically, production of the protein by the alveolar macrophages derived from stem cells (Thomas et al. 1976) might prove to be a useful treatment for the emphysema component of the deficiency disease, but the cirrhosis, which is postulated to result from the deposition of nonsecreted protein, will continue to progress. Indeed, transgenic mice in which the human PiZ mutant isozyme of α1-antitrypsin is expressed show liver histopathology similar to homozygous, affected human patients (DeMayo et al. 1986).

TABLE 16–2 Diseases Unlikely to be Cured by Gene Therapy in the Near Future Because of Targeting Problems

α-1 Antitrypsin deficiency
Carbamyl phosphate synthetase deficiency
Fabry disease (alpha galactosidase)
Fucosidosis (alpha fucosidase)
Gaucher disease types II and III
Hypophosphatasia (alkaline phosphatase)
Metachromatic leukodystrophy, variant (SAP 1)
Lesch-Nyhan disease (HPRT)
Ornithine transcarbamylase deficiency
Propionyl CoA carboxylase deficiency
Sandhoff disease (hexosaminidase A and B)
Tay-Sachs disease (hexosaminidase)

At present it has not definitely been established whether or not bone marrow-mediated gene transfer will be applicable to diseases with significant manifestations in relatively inaccessible body spaces, such as bone matrix or the CNS. The osteopetrosis BMT model suggests that donor osteoclasts are functional for about three to four months after transplantation. Although improved joint mobility was noted in iduronidase-deficient dogs, there were no radiographic changes in children with Hurler syndrome (see above) over the course of approximately one year, and the chondrocytes from one child remained vacuolated 10 months after the transplant. More longitudinal studies are needed to determine whether the deformation of bones and joints and the coarsening of features can be arrested or even reversed. However, no significant data are available regarding possible delivery of correction to osteoblasts or chondrocytes.

CNS manifestations of disease can arise by a variety of mechanisms, only some of which may be correctable with currently anticipated technologies. Damage to the brain and/or neurons may result from physical compression, either intracellularly by direct accumulation within the organelles or cytoplasm of the cells, or extracellularly from adjacent tissues on either side of the blood-brain barrier. Alternatively, the accumulation of toxic metabolites or the absence of a necessary substrate may cause deterioration or developmental failure that may or may not be reversible. CNS problems originating outside the blood-brain barrier are more likely to be reversible with treatment. In addition, it has been hypothesized that microglial cells arise from cells derived from the bone marrow that are capable of crossing the blood-brain barrier (Konigsmark and Sidman 1963; Bartlett 1972). Mice prepared for BMT with lethal total-body irradiation (TBI) have demonstrated the presence of donor-derived microglial cells only after nine months, suggesting a slow turnover of these cells (Ting et al. 1983). It is not clear, however, whether engraftment in the brain can be achieved without using such harsh ablative treatment. As will be discussed below, one reason for attempting to treat the fetus or the newborn is that the blood-brain barrier may be more permeable to treated cells in these younger patients.

Similar considerations apply to other lysosomal storage diseases. Genes for alpha-galactosidase (Fabry disease; Calhoun et al. 1985), alpha-fucosidase (fucosidosis; Fukushima et al. 1985), alpha (Myerowitz and Proia 1984) and beta (O'Down et al. 1985) subunits of β-hexosaminidase (Tay Sachs and Sandhoff diseases), and glucocerebrosidase (Gaucher disease) have been cloned, but most of these diseases are not currently being actively considered for gene therapy in humans because of the questions of access to the CNS. However, after allogenic BMT following total lymphoid irradiation in fucosidase-deficient dogs, 25 to 45% of normal enzyme levels were detectable in the brain and peripheral nervous tissue. In addition, decreased vacuolization was seen in many tissues at postmortem, 10 months after transplantation (Taylor et al. 1986).

16.1.6 Retroviral Vectors As a Gene Delivery System

A retroviral vector is a defective recombinant retrovirus in which all or part of the three structural genes, *gag*, *pol*, and *env*, those needed for viral replication, have been replaced by the gene(s) of interest. This particle retains the ability to infect and integrate into a single target cell and deliver its genetic information, but it has lost the ability to replicate. The advantages and disadvantages of this delivery system are described in other sections of this book and will not be addressed further here. The structure of the Moloney murine leukemia virus and two derivative retroviral vectors (Eglitis et al. 1985; Kantoff et al. 1986), which were used in the experiments described below, are shown in Figure 16-1. Each contains a gene called neo^R, which encodes an enzyme, neomycin phosphotransferase (NPT), that inactivates several antibiotics including one called G418. G418 is toxic to mammalian cells, so that one can select for cells that contain the inserted gene and against cells that do not contain the neo^R gene. The vector N2 contains the neo^R gene controlled by long terminal repeat sequences, whereas the SAX vector, in addition, has the gene for human ADA under the control of the early promotor of the SV40 virus.

In the next section, several experimental protocols using these vectors from the laboratory of W. French Anderson at the National Institutes of Health and collaborators at several other centers will be described. While these animal models are useful for initial testing of many procedures that might be performed in authentic gene therapy for patients in the future, it must be noted that the animals are not enzyme deficient and, thus, are not actually models for any of the genetic diseases that would be treated. In adult animals, endogenous marrow was ablated with radiation to facilitate study of the transplanted, transduced population. It is anticipated that such a procedure will not be necessary prior to treatment of ADA-deficient pa-

16.1 Which Diseases are Candidates for Treatment? 197

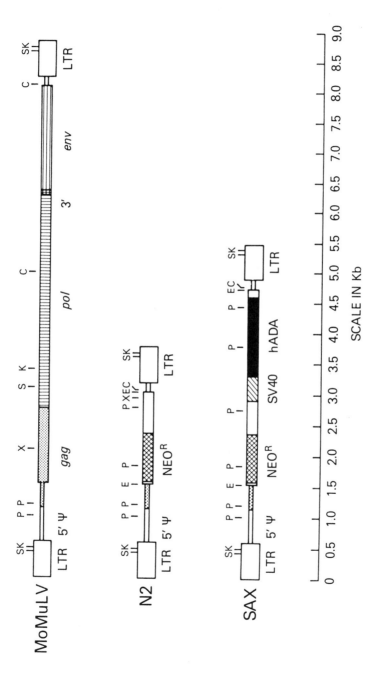

FIGURE 16-1 Structures of a retrovirus and retroviral vectors N2 and SAX. MoMuLV, Moloney murine (mouse) leukemia virus, is the retrovirus from which the two retroviral vectors were constructed. N2 is a shuttle for the *neo*[R] gene and SAX is a shuttle for both *neo*[R] and the human gene for adenosine deaminase (hADA) with the SV40 early promotor.

tients, and perhaps others, in which the treated cells will have a selective advantage.

16.2 BONE MARROW-MEDIATED GENE TRANSFER IN THE MOUSE

The first target cells tested for somatic cell gene transfer in vivo were mouse bone marrow cells. Bernstein and coworkers (Joyner et al. 1983) initially demonstrated retroviral vector-mediated transfer of the neo^R gene into progenitor cells of mouse bone marrow utilizing resistance to G418 in the hematopoietic colony assay, a method for culturing clonal cells in vitro in a semisolid, nutrient medium. Expression of the neo^R gene, which mapped to a common integration site in several blood cell lineages of long-term reconstituted, irradiated mice, was then used to demonstrate the existence of primitive, multipotent, bone marrow cells (Dick et al. 1985; Keller et al. 1985). A number of investigators have now demonstrated retroviral transfer of genes, including human genes for ADA and hypoxanthine phosphoribosyl transferase into mouse hematopoietic lineages in whole animals (Williams et al. 1984 and 1986; Taylor et al. 1986).

16.3 GENE TRANSFER VIA AUTOLOGOUS BONE MARROW TRANSPLANTATION IN THE ADULT PRIMATE

A collaborative, primate bone marrow transplantation program was established at two locations: Memorial Sloan-Kettering Cancer Center, where the primate used was the Cynomolgus macaque, and at the National Institutes of Health, where the primate used was the rhesus monkey (Kohn et al. 1987; Kantoff et al. 1987). The principal vector used in these studies was SAX. Bone marrow cells were aspirated from the animal and transduced either by co-cultivation of the marrow cells with virus-producing cells or by incubation in virus-containing supernatant (Figure 16–2). Prior to peripheral infusion of the treated cells, the monkey received total body irradiation to create "marrow space." After various intervals of time, blood and marrow cells were then sampled and analyzed for vector DNA by Southern blotting and for NPT and human ADA activity. Reconstitution represented the full recovery of all blood lineages. Twelve monkeys were studied (Table 16–3): the best results were with Robert, who became fully reconstituted and demonstrated detectable levels of human ADA activity in his blood cells for several months (Figure 16–3). At the time of maximum expression, the amount of human ADA produced by his cells represented about 0.5% of the endogenous monkey ADA activity. Independent verification that vector DNA was present and expressed in blood was obtained by in situ hybrid-

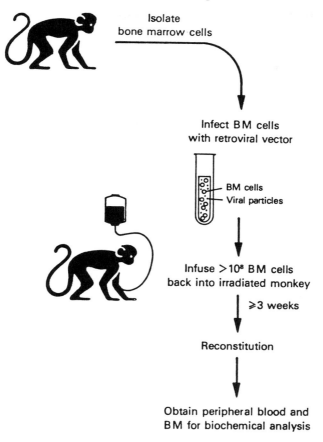

FIGURE 16–2 Protocol for adult primate, bone marrow-mediated, gene transfer.

ization. Approximately one cell in 200 was found to be positive for the neo^R-specific RNA.

We interpreted these data to mean that the level of human ADA activity (and, in addition, of neo^R activity) seen in these monkeys was due to expression of human ADA at levels comparable to endogenous monkey activity, but that only the few cells into which the human gene sequence was successfully transferred were expressing the gene. That is, the overall low level of human ADA activity was due to a low efficiency of transduction of primate hematopoietic cells and was not due to inefficient expression in individual cells. After six months, human ADA activity was not seen, presumably as a result of the loss of the vector-infected population of cells (see the discussion of fluctuating levels of activity in sheep in the next section) or inactivation of the vector genes.

TABLE 16-3 Primate BMT/Gene Transfer: Summary for the h*ADA* Gene

No.	Name[a]	Date	Method[b]	Reconstitution	NPT	ADA	ADA Analysis % A→1	% Endogenous
1	Bill (C)	07/12/85	C	No	Pos	Pos	1	<0.01
2	Mork (R)	09/06/85	C	No	—	—	—	—
3	Mindy (R)	09/06/85	C	No	—	—	—	—
4	Kate (C)	10/29/85	C	No	—	—	—	—
5	Ethel (R)	10/29/85	S	Yes	Neg	Pos	3	<0.01
6	Robert (C)	11/19/85	S	Yes	Pos	Pos	66	0.5
7	Kyle (C)	11/19/85	S	Yes	Pos	Pos	17	0.2
8	Venus (R)	11/20/85	C	No	Pos	Pos	0.5	<0.01
9	George (C)	03/19/86	S	Yes	Neg	Neg	0	0
10	Ken (C)	03/19/86	S	Yes	Neg	Pos	2	<0.01
11	Oppie (C)	08/06/86	S	No	—	—	—	—
12	Barney (C)	08/06/86	S	Yes	Neg	Neg	0	0

[a] C, cynomolgus; R, rhesus monkeys.
[b] Transduction, by either the cocultivation (C) or supernatant (S) method, was performed using the SAX vector. Activity of the *neo*[R] and the human *ADA* genes were measured.

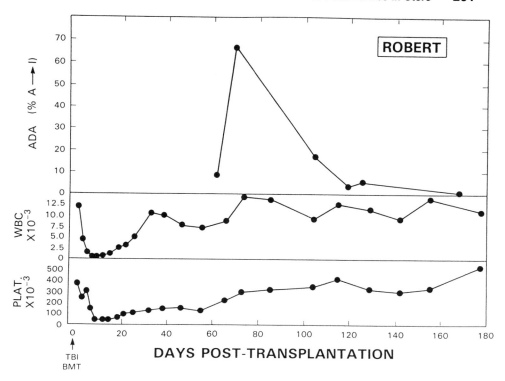

FIGURE 16-3 Posttransplantation course for the monkey Robert, whose bone marrow was transduced with the SAX vector. After six months, human ADA activity was not detectable.

16.4 PROTOCOL FOR GENE TRANSFER INTO FETAL LAMBS IN UTERO

Looking toward the possibility of early treatment of disorders such as Lesch-Nyhan syndrome and Tay Sachs disease and other diseases in which irreversible damage may occur even before birth, a collaborative protocol was established to develop an animal model for gene transfer in utero. Studies were conducted in fetal lambs at 93 to 105 days of gestation (normal term: 145 days) over a two-year period (Kantoff et al. 1989). The overall design of the study was to obtain circulating hematopoietic progenitors from the fetus, insert an exogenous gene(s) via retroviral-mediated gene transfer ex vivo, and reinfuse the treated cells into the donor fetus (autologous transplantation). Direct access to fetal circulation was achieved by hysterotomy for placement of a catheter in the carotid artery of the fetus (Harrison et al. 1980), both for blood withdrawal from the fetus and infusion of maternal blood or transduced fetal hematopoietic cells. The fetuses were then allowed to complete gestation (for 45 to 62 days until birth) and the newborn lambs were examined for evidence of the exogenous gene at various intervals after birth (Figure 16–4). The presence and (or) activity of the exogenous gene in hematopoietic progenitors was assessed by examining (1) an aliquot of fetal cells after gene transfer, and (2) bone marrow and peripheral cells obtained from newborn lambs for resistance to G418 in plasma-clot or methyl cellulose colony assays (Roodman and Zanjani 1979; Ash et al. 1981).

Gene transfer was attempted in 20 animals: 16 received the N2 vector and four received the SAX vector (Table 16–4). As with any surgical procedure involving young fetuses, there was high perioperative mortality. Eight fetuses were lost; 12 fetuses survived the procedure and were born alive. Of the 10 animals that could be analyzed (nine of whom received the N2 vector and one the SAX vector), six were positive for G418-resistant hematopoietic cells and four were negative one week postpartum. Lambs found to be negative for the neo^R gene soon after birth continued to be negative

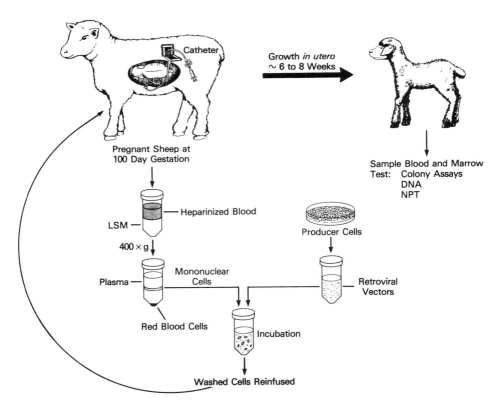

FIGURE 16–4 Schematic diagram of the protocol used for transducing circulating hematopoietic cells from fetal sheep.

TABLE 16–4 Summary of Information About the Fetal Lambs in Which Circulating Hematopoietic Progenitor Cells Were Transduced with Retroviral Vectors

Animal	Vector	Outcome	G418 Resistant Progenitors[a]
2516	SAX	Aborted 1 month later	ND
2517	SAX	Born alive, lost to animal rights' group	ND
2518	N2	Died in utero 3 days later	ND
2519	N2	Born alive 03/20/86: alive and well	Yes
2997	SAX	Aborted	ND
2998	SAX	Born alive, died 1 week later	Yes
2999	N2	Born alive, died during transport	Yes
3000	N2	Born alive, died 1 week later	Yes
3381	N2	Died in utero 4 days later	ND
3382	N2	Born alive, died in transport	No
3383	N2	Mother died 4 days later	ND
3384	N2	Died in utro	ND
3390	N2	Aborted	ND
3391	N2	Born alive, died 2 months later	Yes
3392	N2	Born alive 01/13/87; alive and well	No
3900	N2	Born alive 01/13/87; alive and well	No
4218	N2	Born alive 05/11/87; alive and well	Yes
4261	N	Aborted	ND
4262	N2	Born alive, died 3 days later	ND
4263	N2	Born alive 05/23/87; alive and well	No

[a] ND, cells unavailable or unsuitable for testing because of untimely loss of the animal.

up to several months later. Only two of the lambs that gave G418-resistant colonies at one week after birth remained available for long-term, follow-up studies.

Lambs 2519 and 4218 continued to exhibit G418-resistant colonies throughout the study period. Results of cultures from lamb 2519 are shown in more detail in Figure 16–5. It can be seen that significant fluctuations in the relative incidence of G418-resistant progenitors occurred without a set pattern in this animal; periods of no activity were followed by the reappearance of G418-resistant colonies. A similar picture was found for animal 4218. Re-emergence of hematopoietic cells that express the neo^R gene is consistent with the hypothesis put forth by Mintz et al. (1984) and Lemischka et al. (1986) to explain their data in mice. They postulated that normal hematopoiesis derives from the sequential activation of different clones of stem cells rather than from an averaged contribution of the entire pool of stem cells. Furthermore, they suggested that some stem cells, although replicating to a sufficient extent in vitro to permit proviral integration during the process of transduction by the retroviral vector, can return to a quiescent

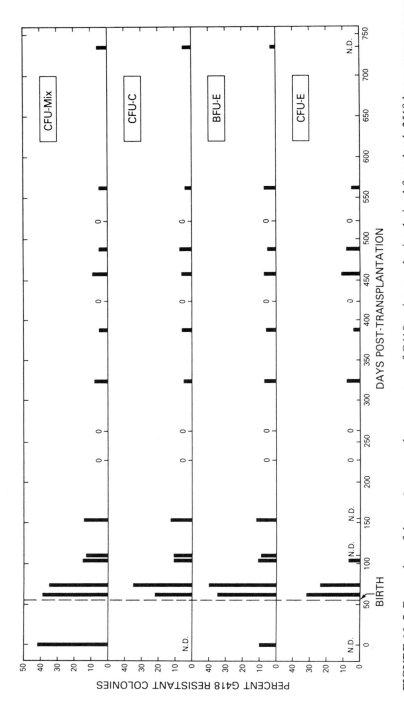

FIGURE 16-5 Expression of the *neo*R gene as the percentage of G418-resistant colonies derived from lamb 2519 bone marrow over time. Each value represents the mean of results from triplicate cultures established with 2 mg/ml G418. The lamb was born 55 days after reinfusion of treated cells. In all cases, values for uninfected, age-matched control lambs were less than 1%.

state upon transplantation back into the animal and, theoretically, they can be reactivated to express their exogenous gene later in the lifetime of the animal. It is also possible to interpret the data in terms of fluctuations in the total population of stem cells (Snodgrass and Keller 1987).

Lamb 2519, the first positive lamb in the series, has been followed continuously for more than two years since the transplant. She was bred to a normal (untreated) ram and was delivered of a healthy lamb at about two years of age (about 26 months after transplantation). Despite the presence of significant numbers of G418-resistant progenitor cells in the mother, who had been treated in utero, no hematopoietic stem cells that expressed the neo^R gene were detected in the second-generation newborn.

The unique feature of this study is the persistence of durably transduced stem cells in vivo (particularly since no cytoablation or positive selection in vivo was used), a result not previously seen in any species except the mouse. The physiology of hematopoietic differentiation in the fetal lamb may help to explain why fetal blood cells are susceptible to gene transfer. During the stage of development at which gene transfer was attempted, two important changes were occurring in the hematopoietic system: extensive growth-associated hematopoietic expansion and a change in the primary site of hematopoiesis, from the liver and spleen to the bone marrow. It is generally believed that the latter process is associated with the seeding of the developing marrow spaces by hematopoietic stem cells delivered via the circulation.

It is therefore possible that a large number of early progenitor cells with a high capacity for self-renewal are in the circulating pool and are thus recoverable by exchange transfusion. These cells (after being transduced and reinfused) would represent a small but significant proportion of the overall pool of stem cells, thereby allowing the detection of progeny of stem cells that carry the vector DNA for many months after transplantation. Furthermore, because of the large numbers of cells delivered as one single bolus, the reinfused cells may have an advantage at "homing" into developing hematopoietic sites in the bone marrow. It is possible, in addition, that the explantation and treatment in vitro of a portion of the total population of hematopoietic stem cells may confer a proliferative or seeding advantage on these cells, creating an overrepresentation in the total marrow. It was previously shown that the transplantation in utero of heterologous hematopoietic cells resulted in the creation of long-term chimera in sheep (Flake et al. 1986) in the absence of cytoablation. Therefore, the reinfused transduced cells may compete favorably for unoccupied spaces in the bone marrow. Nonetheless, we still do not understand why the percentages of G418-resistant colonies on days 62 (one week after birth) and 75 (20 days after birth) were so high.

Successful long-term engraftment of stem cells after gene transfer in vitro has been achieved in the mouse model but not in primate or dog models (Stead et al. 1988) in which cytoablation is used to confer a selective survival and proliferative advantage on infected stem cells. The apparent increased efficiency of gene transfer in this sheep model does not appear to be due primarily to differences between species in susceptibility to viral infection (or in the capacity for expression of vector genes) but rather may be due to an inherent difference between progenitor cells derived from the fetus versus those from the adult. Bone marrow cells of an adult sheep transduced with the N2 virus were transplanted into a heterologous recipient fetal lamb. Hematopoietic chimerism of hemoglobin markers in the newborn lamb was achieved, but no evidence of neo^R gene activity was detected as measured by growth of G418-resistant colonies. Additional studies assessing the efficiency of retroviral vector-mediated transfer of the neo^R gene into adult and fetal hematopoietic sheep cells in vitro, in colony assays suggest that the fetal cells express the exogenous gene as much as twofold more efficiently than adult cells transduced under the same conditions (Ekhterae et al. 1988).

16.5 GENE TRANSFER INTO FETAL HUMAN AND NONHUMAN PRIMATE HEMATOPOIETIC CELLS

Encouraged by the results in fetal sheep, our group initiated several investigations using fetal human and nonhuman primate cells. Because our goal is human gene therapy, we undertook the evaluation in vitro of human hematopoietic cells from residual cord blood from term and premature births. Meanwhile, nonhuman primate models are being developed to test posssible protocols for gene therapy in the future.

16.5.1 Human Cord Blood Cells Are more Receptive to Gene Transfer than Adult Cells

Circulating hematopoietic progenitors in the fetus, which are thought to be a population of stem cells migrating from the liver to the bone marrow (Hassan et al. 1979; Linch et al. 1982; Tchernia et al. 1981; Hann et al. 1983), can be cultured by techniques similar to those used for culture of bone marrow and fetal liver cells (Karson et al. 1987; Broxmeyer et al. 1989). To determine whether these cells might be more readily infected by retroviral vectors and/or better able to express a transferred gene than adult bone marrow cells, nucleated cells from normal adult bone marrow and from residual human fetal cord blood that remained after appropriate clinical specimens had been obtained immediately after delivery were compared in methylcellulose colony assays. In the granulocyte-macrophage (CFU-GM) and burst-forming erythroid (BFU-E) lineages, the fetal progenitor cells appeared to be at least as abundant as progenitors in normal adult marrow when samples were normalized for the number of nucleated cells (Karson et al. 1987). Because the maturation and growth of such fetal cells as well as adult progenitor cells are markedly arrested in culture by G418, the colony assay was also used to demonstrate the transfer and effective expression of the neo^R gene by showing increased survival of the hematopoietic cells transduced with the retroviral vector, N2 (Figure 16-6).

When nucleated, human hematopoietic cells from an adult, a 24-week, and a 29-week fetus were treated in parallel with N2 viral supernatant, a greater proportion of the CFU-GM colonies derived from the fetus were resistant to G418 than were the CFU-GM colonies from the adult (Table 16-5). The trend for fetal-derived BFU-E colonies was similar although the difference between survival of treated and untreated adult human, bone marrow cells was small. These data have been corroborated, however, in other experiments with a much larger number of colonies.

At present, we are unable to determine whether the fetal cells are intrinsically more transducible, perhaps because of the presence of different cell-surface markers, such as amphotropic receptors, or to more rapid cell cycling, or whether transferred foreign genes are either less likely to be inactivated or more likely to be expressed in fetal cells. Although the enzymatic activity of the protein product of the neo^R gene, neomycin phosphotransferase, can be measured directly (Reiss et al. 1984) and the gene fragment can be identified on a Southern blot (Southern 1975), the actual amounts of each which would be present in the number of cells that can be recovered are below the levels detectable by current methods. Studies using the polymerase chain reaction (PCR) method (Saiki et al. 1985) to amplify the neo^R gene sequence for detection are underway.

Regardless of the mechanism responsible for the increased proportion of drug-resistant progenitor cells, our results may have important implications for the potential for human gene therapy initiated during fetal life. Fetal progenitor cells, collected after as little as 19 weeks of gestation, have

TABLE 16-5 Transducibility of Granulocyte-Macrophage Colonies (CFU-GM) Grown from Hematopoietic Progenitors in Fetal Blood or Adult Marrow

Cell Type	mg/ml	No. Surviving/Total		% Surviving		% Difference
		Control	Transduced	Control	Transduced	
Adult	1.2	20/405	23/368	4.94	7.07	2.13
	1.8	0/405	1/184	0.00	1.09	1.09
29-wk fetal	1.2	34/251	60/251	13.55	23.90	10.36
	1.8	2/251	10/251	0.80	3.98	3.19
24-wk fetal	1.2	17/176	76/175	9.66	43.43	33.77
	1.8	2/176	20/175	1.14	11.43	10.29

Mononuclear cells were incubated for 3 hr with retroviral vector or mock cell-free supernatant. The harvested cells were then cultured for 2 wks in semisolid medium that contained nutrients, growth factors, fetal bovine serum, and methyl cellulose with and without G418. Clusters of greater than 10 cells were scored.

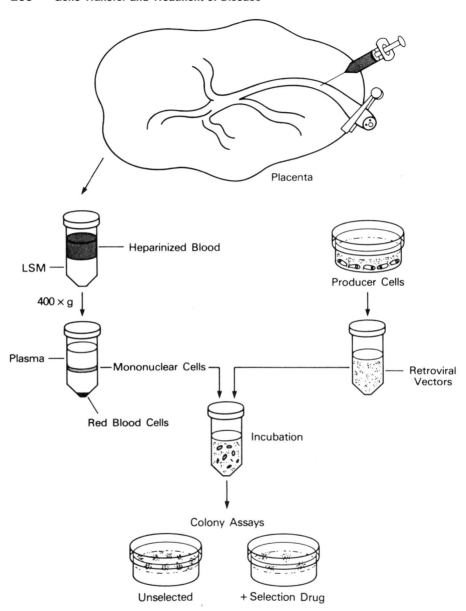

FIGURE 16-6 Procedure for determining the transducibility of hematopoietic progenitor cells from human cord blood in vitro.

been evaluated and are transducible targets for retroviral vector delivery systems; even earlier intervention may also be possible. In cases in which a genetic disease can be diagnosed prenatally, it may be possible to begin therapy early enough to reduce or prevent irreversible damage to some affected organ systems.

16.5.2 Gene Transfer into Primate Cord Blood Cells

Analogous results have been obtained in similar experiments with cells from fetal Cynomolgus and rhesus monkeys in vitro. Three types of experiment are in progress. The transducibility of bone marrow and peripheral hematopoietic progenitors from preterm abortuses, newborns, and adult animals have been compared in colony assays of resistance to G418. As in the case of sheep, survival of up to 25% of colonies is seen in the preterm and newborn specimens, as compared to specimens from adults with which percentages greater than 5% are rarely seen.

In contrast to the protocol for sheep, the protocol for retroviral-mediated gene transfer into autologous cells in utero utilized the less invasive procedure of ultrasound-guided blood sampling and reinfusion. Preliminary studies (Bond et al. 1987; Slotnick et al. 1988) in which heterologous liver hematopoietic cells from a 60-day fetal donor were injected under ultrasound guidance into the peritoneal cavity of a 60-day recipient, demonstrated that chimerism could be established without immunosuppression. Although the original plan was to perform PUBS, the extremely small size of the fetal monkeys necessitated ultrasound-guided cardiac puncture to obtain even a minimal number of cells. Likewise, cells were returned by the intraperitoneal rather than the intravascular route. Small numbers of G418-resistant colonies were isolated from the newborn animals.

Finally, two newborn Cynomolgus monkeys were treated. Because of their small size and the reflex vasoconstriction of their umbilical vessels at birth, even under controlled conditions of delivery by cesarean section, very few cells were obtained. The cells were incubated with growth factors for one to four days prior to exposure to the N2 viral supernatant in an attempt to increase the rate of transduction. Two days after pretreatment with a single dose of busulfan, the autologous cells were reinfused into the femoral vein of each infant. Evaluation of these animals is in progress.

16.6 GENE TRANSFER INTO TUMOR-INFILTRATING LYMPHOCYTES (TIL)

Although the overall efficiency of gene transfer and expression in nonhuman primates has not been high enough to justify the risks of a human trial, an opportunity to utilize the system of retroviral-mediated gene transfer into human cells for a somewhat different purpose has arisen. A stable intra-

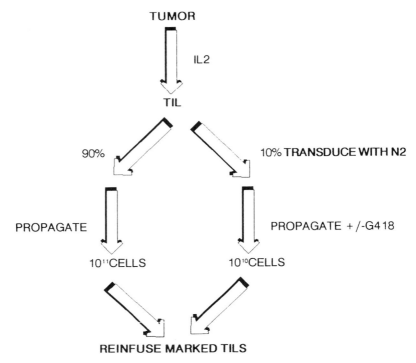

FIGURE 16-7 Schematic diagram of the procedure to be followed with TIL, with N2 as a marker for the infused cells.

cellular marker was needed to trace tumor-infiltrating lymphocytes (TIL) to learn why only some terminal cancer patients are helped by this new experimental therapy. TIL, one of the body's natural defenses against cancer, are lymphocytes that recognize tumor antigens and can be isolated from a patient's tumor after surgery. The therapy involves growing a large number of these cells ex vivo and reinfusing them to attack remaining tumor cells. For successful application of the retroviral technology to insert a particular gene for use as a marker, demonstration of the sustained expression of the gene is unnecessary. Also, because the cells are grown extensively in culture after transduction and can be evaluated for possible recombinant replication-competent virus, the potential hazards are less imposing. The protocol, which involves transducing a portion of the TIL population with the vector N2 (Figure 16-7), was approved in 1989, and analysis of the findings in the first five patients is in progress at the time of this writing.

REFERENCES

Abraham, D., Muir, H., Olsen, I., and Winchester, B. (1985) *Biochem. Biophys. Res. Commun.* 129, 417–425.

Anderson, W.F. (1984) *Science* 226, 401–409.

References

Ash, R.C., Detrick, R.A., and Zanjani, E.D. (1981) *Blood* 58, 309.
Bartlett, P.F. (1972) *Proc. Natl. Acad. Sci. USA* 79, 2722–2725.
Belmont, J.W., and Caskey, C.T. (1986) *Ann. Clin. Res.* 18, 322–326.
Bernstein, A., Berger, S., Huszar, D., and Dick J. (1985) in *Genetic Engineering: Principles and Methods* (Setlow, J., and Hollaender, A., eds.), Vol. 7, p. 235, Plenum, New York.
Bond, S.J., Harrison, M.R., Crombleholme, T.M., et al. (1987) *Blood* 70, 290a.
Broxmeyer, H.E., Douglas, G.W., Hangoc, G., et al. (1989) *Proc. Natl. Acad. Sci. USA* 86:3828–32.
Calhoun, D.H., Bishop, D.F., Bernstein, H.S. et al. (1985) *Proc. Natl. Acad. Sci. USA* 82, 7364–7368.
Costa, T., Scriver, C.R., and Childs, B. (1985) *Am. J. Med. Genet.* 21, 231–242.
DeMayo, J.L., Sifers, R.N., Carlson, J.A., et al. (1986) *Am. J. Human Genet.* 39, supplement, A195.
Denniston, D. (1982) *Ann. Rev. Genet.* 16, 219–255.
Dick, J.E., Magli, M.C., Huszar, D., Phillips, R.A., and Bernstein, A. (1985) *Cell* 42, 71–79.
Eglitis, M.A., Kantoff, P., Gilboa, E., and Anderson, W.F. (1985) *Science* 230, 1395–1398.
Ekhterae, D., Crumbleholme, T., Karson, E., et al. (1988) *Blood* 72, 386a.
Flake, A.W., Harrison, M.R. Adzick, N.C., and Zanjani, E.D. (1986) *Science* 233, 776–778.
Fukushima, Hl., de Wet, J.R., and O'Brien, J.S. (1985) *Proc. Natl. Acad. Sci. USA*, 82, 1262–1265.
Garver, R.I., Chytil, A., Karlsson, S., et al. (1987) *Proc. Natl. Acad. Sci. USA* 84, 1050–1054.
Gluckman, E., Broxmeyer, H.A., Auerbach, A.D., et al. (1989) *N. Engl. J. Med.* 321:1174–1178.
Hall, J.B., Powers, E.K., McIlvaire, R.T., and Ean, V.H. (1978) *Am. J. Med. Genet.* 1, 417–436.
Hann, I.M., Bodger, M.P., and Hoffbrand, A.V. (1983) *Blood* 62, 118–123.
Harrison, M.R., Jester, J.A., and Ross, N.A. (1980) *Surgery* 88, 174.
Hassan, M.W., Lutton, J.D., Levere, R.D., Rieder, R.F., and Cederqvist, L.L. (1979) *J. Haematol.* 41, 477–484.
Hayes, A., Costa, T., Scriver, C.R., and Childs, B. (1985) *Am. J. Med. Genet.* 21, 243–255.
Hobbs, J.R., Hugh-Jones, K., Barrett, A.J., et al. (1981) *Lancet* II, 709–712.
Joyner, A., Keller, G., Phillips, R.A., and Bernstein, A. (1983) *Nature* 305, 556–558.
Kantoff, P.W., Flake, A.W., Eglitis, M.A., et al. (1989) *Blood* 73, 1066–1073.
Kantoff, P.W., Gillio, A. McLachlin, J., et al. (1987) *J. Exp. Med.* 166, 219–234.
Kantoff, P.W., Kohn, D.B., Mitsuya, H., et al. (1986) *Proc. Natl. Acad. Sci. USA* 83, 6563–6567.
Karson, E.M., Eglitis, M.A., Kantoff, P., and Anderson, W.F. (1987) Retroviral mediated gene transfer into fetal hematopoietic cells, presented at *Am. Coll. Obstet. Gynecol.* 37th Annual Meeting, Las Vegas.
Keller, G., Paige, C., Gilboa, E., and Wagner, E.F. (1985) *Nature* 318, 149–154.
Kohn, D.B., Kantoff, P.W., Eglitis, M.A., et al. (1987) *Blood Cells* 13, 285–298.
Konigsmark, B.W., and Sidman, R.L. (1963) *J. Neuropath. Exp. Neurol.* 22, 643–648.
Ledley, F.D., Grenett, H.E., McGinnis-Shelmutt, M., and Woo, S.L.C. (1986) *Proc. Natl. Acad. Sci. USA* 83, 409–413.

Lemischka, I.R., Raulet, D.H., and Mulligan, R.C. (1986) *Cell* 45, 917–927.
Linch, D.C., Knott, L.J, Rodeck, C.H., and Huehns, E.R. (1982) *Blood* 59, 976–979.
Mintz, B., Anthony, D., and Litwin, E.D. (1984) *Proc. Natl. Acad. Sci. USA* 81, 7835–7839.
Mueller, R.F., Sybert, V.P., Johnson, J., Brown, Z.A., and Chen, W.J. (1983) *New Eng. J. Med.* 309, 586–590.
Myerowitz, R., and Proia, R.L. (1984) *Proc. Natl. Acad. Sci. USA* 81, 5394–5398.
O'Down, B.F., Quan, F., Willard, H.F., et al. (1985) *Proc. Natl. Acad. Sci. USA* 82, 1184–1188.
Olsen, I., Dean, M.F., Harris, G., and Muir, H. (1981) *Nature* 291, 244–247.
Parkman, R. (1986) *Science* 232, 1373–1378.
Reiss, B., Sprengel, R., Will, H., and Schaller, H. (1984) *Gene* 30, 211–218.
Roberts, D.F., Chavez, J., and Court, S.D.M. (1970) *Arch. Dis. Childhood* 45, 33–38.
Roodman, G.D., and Zanjani, E.D. (1979) *J. Lab. Clin. Med.* 94, 699.
Saiki, R.K., Scharf, S., Faloona, F., et al. (1985) *Science* 230, 1350–1354.
Shull, R.M., Hastings, N.E., Selcer, R.R., et al. (1987) *J. Clin. Invest.* 79, 435–43.
Slotnick, R.N., Crombleholme, T., Anderson, J.Z., et al. (1988) *Am. J. Human Genet.* 43, A133.
Snodgrass, R., and Keller, G. (1987) *EMBO J.* 6, 3955.
Southern, E.M. (1975) *J. Mol. Biol.* 98, 503–517.
Stead, B., Kwod, W.W., Storb, R., and Miller, D.A. (1988) *Blood* 71, 742.
Taylor, R.M., Farrow, B.R.H., Stewart, G.J., and Healy, P.J. (1986) *Lancet* II, 772–774.
Tchernia, G., Mielot, F., Coulombel, and L. Mohandas, N. (1981) *J. Lab. Clin. Med.* 97, 322–331.
Thomas, E.D., Ramburg, R.E., Sale, G.E., Sparks, R.S., and Golde, D.W. (1976) *Science* 192, 1016–1018.
Thompson, J.A., Anderson, K.D., DiPietro, J.A., et al. (1988) *Science* 241, 1349–1352.
Ting, J.P-Y., Nixon, D.F., Weiner, L.P., and Frelinger, J.A. (1983) *Immunogenetics* 17, 295–301.
Whitley, C.B., Belani, K.G., Ramsay, N.K.C., Kersey, J.H., and Krivit, W. (1986a) *Am. J. Human Genet.* 39, supplement, A87.
Whitley, C.B., Ramsay, N.K., Kersey, J.H., and Krivit, W. (1986b) *Birth Defects: Original Article Series*, Vol. 22, Number 1, 7–24.
Williams, D.A., Lemischka, I.R., Nathan, D.G., and Mulligan, R.C. (1984) *Nature* 310, 476–480.
Williams, D.A., Orkin, S.H., and Mulligan, R.C. (1986) *Proc. Natl. Acad. Sci. USA* 83, 2566–2570.
Zwiebel, J.A., Freeman, S.F., Kantoff, P.W., et al. (1989) *Science* 243, 220–222.

CHAPTER 17

Transgenic Mice Carrying HIV Proviral DNA

David S. Pezen
John M. Leonard
Jan W. Abramczuk
Malcolm A. Martin

The biology and molecular structure of the human immunodeficiency virus (HIV), the etiological agent in the acquired immunodeficiency syndrome (AIDS), has been determined predominantly through the use of in vitro techniques. However, the study of viral replication in vitro may not fully reveal the complex interactions between the virus and its host in the natural disease state. The development of an animal model for the HIV life cycle may bring us closer to understanding these interactions. Unfortunately, at this time no good animal model exists for this disease and, even though chimpanzees are susceptible to infection by HIV (Alter et al. 1984; Frances et al. 1984; Nara et al. 1987), the disease develops only in humans.

Obviously, an animal model simulating all phases of HIV infection would provide the most information about the biology of HIV in vivo. Such a model might be generated by the production of transgenic mice capable of expressing the human CD4 protein. The CD4 protein is the principle determinant of susceptibility to infection by HIV (Dalgleish et al. 1984; Klatzmann et al. 1984; McDougal et al. 1986; Maddon et al. 1986) and is expressed at the cell surface of human T-helper cells and macrophages. Even though previous work has shown that no intrinsic barrier exists in mouse cells for the assembly of infectious virions after transfection of an infectious molecular clone of HIV (Adachi et al. 1986), several lines of mouse cells that synthesize human CD4 have been shown to be refractory to infection by the virus (Maddon et al. 1986). Viral particles bind to mouse cells expressing CD4 but fail to enter them. For this reason, transgenic mice expressing human CD4 will probably not be a useful model for HIV disease.

In contrast, transgenic animals containing HIV proviral DNA have the potential to provide a model of those steps in viral infection that occur subsequent to the integration of the HIV provirus into the DNA of the host: the synthesis of viral RNA and proteins and their subsequent assembly into virions in vivo. We took this latter approach and have succeeded in generating several transgenic lines that yield HIV virions. In the case of one strain, a spontaneous and fatal disease occurs that mimics several features of human AIDS.

One-cell embryos from 8- to 12-week old FVB/N mice were microinjected with 75 to 150 copies of HIV proviral DNA derived from the infectious molecular plasmids pNL4-3 (Adachi et al. 1986) or pNL432 (Strebel et al. 1987). pNL4-3 was digested with both *Sma*I and *Nru*I to release a 12.4-kb segment that contained the 5' cellular flanking region (1.2 kb), the HIV provirus (9.7 kb), and the 3' flanking cellular region (1.5 kb). pNL432 was linearized with *Aat*II, which cleaves once within pUC 18 and generates a 15.1-kb fragment that contains the sequences listed above plus plasmid sequences (2.7 kb). After microinjection, the embryos were inserted into the oviducts of surrogate FVB/N female mice. These mice were then housed in a stainless-steel glovebox system within a BL4 facility. All transgenic animals were maintained in the glovebox system for the duration of the experiment. All containment practices and procedures followed in the BL4 facility were reviewed and approved by the Biosafety Committee of the National Institutes of Health.

Sixty-four of the microinjected ova were successfully carried to term. Thirteen of the 64 animals contained 0.4 to more than 64 copies of HIV proviral DNA per haploid genome, as determined by slot-blot analysis. Southern blot analysis indicated that 12 of the 13 founder animals carried full-length HIV proviruses arranged as head-to-tail oligomers. Seven of these 12 founder (F_0) animals transmitted the transgene to their progeny. During a 10-month observation period, all F_0 animals appeared healthy by the criteria of normal fertility, growth, development, and body weight. In addition, no HIV was ever recovered from the blood of any F_0 animal after cocultivation with phytohemagglutinin-stimulated, human peripheral blood lymphocytes (PBLs) during a 17-day incubation period. Only one of the seven founder animals, F_0 number 13, synthesized serum antibodies against HIV as determined by enzyme-linked-immunosorbent assay (ELISA). Subsequent immunoblotting demonstrated reactivity with gp120 envelope and p64 reverse transcriptase.

One of the seven founder animals, animal number 13, produced progeny that developed a characteristic and fatal syndrome in 45% of the pups upon mating with a nontransgenic FVB/N male. This founder animal unexplainably was and still remains free of disease. The progeny, which contain two copies of proviral DNA at one integration site, were readily distinguishable from their unaffected littermates by their small size (50–80% of normal) and the presence of a skin disease characterized by dry, scaling lesions, which affected the tail, paws, and ears. Over the next one to two weeks of life, all of the affected animals nursed normally and displayed no evidence of diarrhea, respiratory distress, gross neurological dysfunction, or other readily observable clinical abnormalities. Shortly before death, however, the affected animals became sluggish and developed ruffled fur. All affected animals either died or were killed by day 25 after birth because of their moribund state. Attempts to breed these F_1 animals have not been possible because of death before sexual maturity.

Of the 41 F_1 animals born to founder 13 and examined by slot-blot analysis, 23 were found *not* to carry the HIV transgene. All 23 animals appeared healthy with none of the abnormalities seen in their transgenic littermates. Of the remaining 18 F_1 animals, 12 contained proviral DNA and exhibited growth retardation, skin changes, or died spontaneously. DNA was unavailable from the remaining six animals because of their unexpected death and the subsequent autolysis of tissues. Five of these deaths occurred in the first F_1 litter, alerting us to the existence of a potentially interesting phenotype. The sixth death, which occurred in the third F_1 litter, which also exhibited the characteristic dermatologic abnormalities. Since no animal that has died and whose DNA has been analyzed failed to contain proviral DNA, we have assumed that these six animals also contained HIV proviral DNA (Table 17–1).

TABLE 17–1 Mortality/Morbidity of HIV Transgenic Mice

Characteristic	Strain Number			
	13	16	62	38, 39, 42, 64
Number of litters	4	3	3	14
Total live births	41	35	32	96
Moribund or spontaneous deaths	18/41 (44%)	7/35 (20%)	4/32 (13%)	1/96 (1%)
DNA isolated	12/18	4/7	2/4	1/1
Positive for HIV proviral DNA	12/12 (100%)	4/4 (100%)	1/2 (50%)	0/1 (0%)

Necropsies were performed on eight affected strain 13 F_1 mice and six nontransgenic, aged-matched littermates that served as controls. In every case, the transgenic animal exhibited runting and presented with a severely roughened, scaling, thickened tail, particularly on the ventral surface (Figure 17–1). The four paws, the inner surface of the ears, and the tips of the nose were similarly affected. The diameter of the tail was twice that of controls, particularly in the proximal, ventral portion. Similarly, the paws were enlarged to about twice their width and ears were twice their normal thickness. Skin with hair appeared normal except at the base of the tail. At autopsy, enlarged (up to 3 × 4 mm) axillary, cervical, inguinal, mesenteric, and thoracic lymph nodes were present. No lymphadenopathy was identified in any of six unaffected littermates. The spleens of the affected animals were two to three times larger than normal, as determined by cell count. In addition, thymuses were smaller in severely affected mice than in normal mice. Cutaneous masses in the submandibular area, resembling trichofolliculomas, were noted in three of the affected animals.

Histologic examination of the skin demonstrated that the grossly apparent lesions of the tail, paws, and ears of strain 13 F_1 mice were characterized by a striking epidermal hyperplasia (Figure 17–2A to 2C). Dyskaryotic cells and mitotic figures were numerous throughout the epidermis,

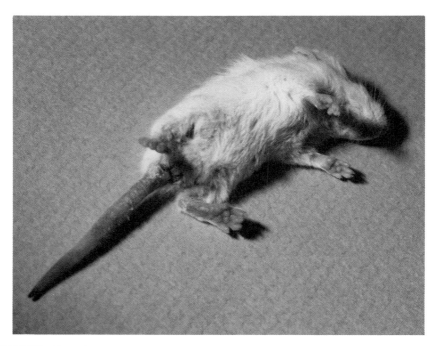

FIGURE 17–1 Twenty-day-old transgenic mouse. Note thickened, scaly tail, ears, and paws.

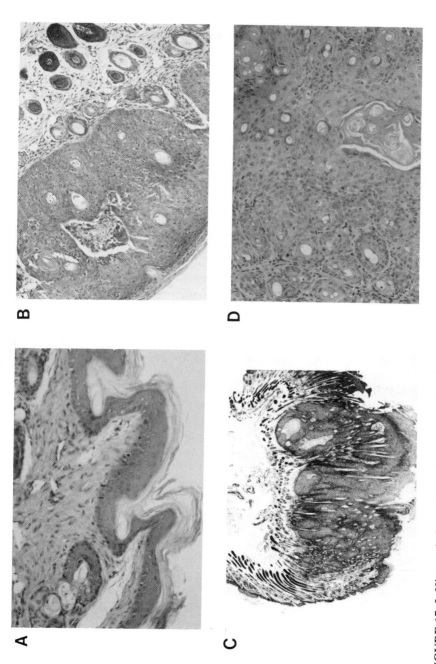

FIGURE 17-2 Histopathology of skin from transgenic mice. (A) Tail skin from transgenic mouse with marked epidermal hyperplasia, hyperkeratosis and parakeratosis. (B) Tail skin from control mouse with normal epidermis. (C) Skin from the base of the tail of an affected animal. Note the transition from normal thickness of the epidermis on the right to markedly hyperproliferative epidermis in the center. (D) A nodular lesion removed from the submandibular region that consists of proliferating keratinocytes and hairshafts.

particularly in the paw, and were associated with marked acanthosis and parakeratosis. The granular layer of the epidermis was absent in the severely affected ventral surface of the tail skin. Examination of the masses recovered from the submandibular area in three of the mice demonstrated nodular proliferation of mature keratinocytes interspersed with hair shafts as well as associated hyperkeratosis and parakeratosis as shown in Figure 17-2D.

In the lungs of the affected animals, the most characteristic and consistent pulmonary lesions were perivascular cuffs around small arterioles throughout the parenchyma, unassociated with any other pathologic changes (Figure 17-3A). The cuffs were composed primarily of tightly packed lymphoid cells with varying numbers of polymorphonuclear cells and eosinophils (Figure 17-3B). The lymphoid cells were mainly small lymphocytes with scattered lymphoblasts; mitotic figures were rare. Animals killed at a later stage of development (up to 24 days) displayed a progression of the pulmonary abnormalities with accumulation of lymphocytes, polymorphonuclear leukocytes, and macrophages within the perivascular interstitium. In no case was there evidence of a bronchopneumonia suggestive of the presence of a bacterial or viral pathogen. Specific staining for fungi, acid-fast bacilli, and pneumocystis was negative.

The lymph nodes of the affected animals were found to be grossly enlarged upon histologic examination, which revealed only reactive changes and no well-formed germinal centers (Figure 17-3C). The histological features were indistinguishable from those seen in the very small lymph nodes present in control animals. In the spleen, the borders of the periarteriolar lymphoid sheaths were obscured by a large number of lymphocytes which filled the red pulp. Neutrophilic and lymphocytic infiltrates were found in the liver, primarily in the portal triads and occasionally in the sinusoids (Figure 17-3D). Hyperplasia of Kupffer cells was also present in the livers of all affected mice. In the thymus, severe cortical atrophy was observed only in the most severely affected animals. No pathology was observed in the brain, small intestine, eye, kidney, or bone marrow.

In situ hybridization was carried out on several specimens of tissue from the affected mice using sense and antisense probes. Sections from affected tail and ear skin and a few scattered cells in the liver and the gastrointestinal tract reacted specifically with antisense HIV probes and failed to hybridize with sense probes. Reactivity in the skin sections was localized to the outer two-thirds of the epidermis and was associated both with hair follicles and with cells interspersed throughout the stratum spinosum (Figure 17-4). No hybridization was detected in sections taken from the brain, lung or kidney. High background in sections from lymph nodes and spleens made it impossible to interpret the results of hybridization with samples of these tissues.

To determine whether infectious HIV particles were produced in the tissues of affected mice, sections (2-5 mm^3) of brain, lung, kidney, skin, liver, spleen, and lymph node were aseptically removed from affected and

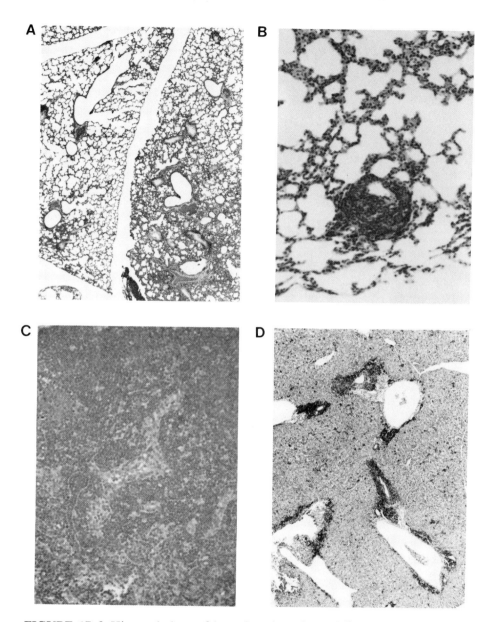

FIGURE 17-3 Histopathology of lung, lymph node, and liver. (A) Lung from a transgenic mouse that shows small, perivascular lymphoid aggregates. (B) Higher magnification of one such aggregate composed mainly of lymphocytes. (C) An enlarged lymph node from an affected animal lacking germinal centers. (D) Inflammatory cell infiltrates present in the portal triads and to a lesser extent in the sinusoids of the lobules of the liver of a transgenic animal.

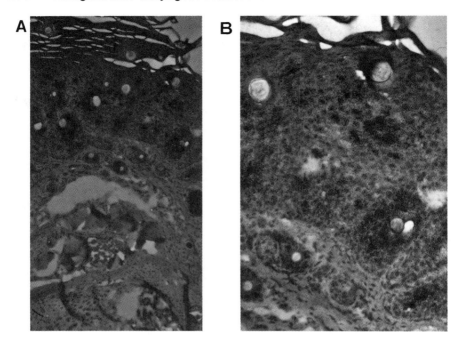

FIGURE 17-4 (A) In situ hybridization of transgenic mouse tail with HIV-specific riboprobes. Note the reactivity associated with both hair follicles and individual cells throughout the stratum spinosum. (B) High-power view.

normal animals at autopsy and cocultured with PBLs or A3.01 cells (Folks et al. 1985), a CD4+ continuous human T lymphocyte line that is susceptible to infection by HIV (Figure 17-5A). Virus was recovered from skin, lymph node, or spleen specimens from five of five affected mice but not from tissues of nontransgenic littermates. Lysates were subsequently prepared from A3.01 cells infected with isolates of spleen, lymph node, or skin and analyzed by immunoblotting with pooled serum from HIV-infected individuals (Figure 17-5B). The reactive bands generated by the mouse-derived viral isolates comigrated with authentic HIV proteins.

The status of the immune system of the affected strain 13 F_1 animals was also examined. The most consistent immunologic finding was splenomegaly (spleens two to three times larger than those of unaffected littermates) and lymphadenopathy, both of which were observed in all affected animals. Analysis of lymph node cells from diseased animals by a fluorescence-activated cell sorter showed an increase in both the number and proportion of Lyt-2+ cells, as well as a reduction in the percentage but *not* in the number of the L3T4+ subset of T cells (Figure 17-6A and 6B). B cells from the affected animals were increased in both number and proportion.

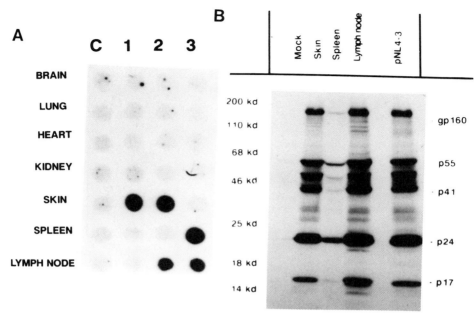

FIGURE 17-5 Recovery of HIV from transgenic mouse tissues. (A) Tissues from transgenic mice (lanes 1, 2, and 3) and from a nontransgenic littermate (lane C) were collected under sterile conditions at necropsy and cultured with phytohemagglutinin-stimulated human peripheral blood lymphocytes (2×10^6 cells) maintained in RPMI media plus 10% fetal calf serum. Supernatants were collected and assayed for reverse transcriptase (RT) activity. The RT activity of 10-μl aliquots of culture medium at three weeks is shown in the autoradiogram. (B) Immunoblot analysis of virus recovered from transgenic mice. Note that reactive bands from lysates of A3.01 cells infected with equivalent amounts of virus (as determined by RT activity) from tail, skin spleen, and lymph nodes of affected transgenic mice comigrate with those of the parental virus (pNL4-3).

In the spleen, the relative proportions of B and T cells were comparable to those in normal mice. However, due to the splenomegaly present in affected mice, the total number of T and B cells was two to three times greater in the affected mice. The proliferative response of splenic T cells to concanavalin A was moderately but consistently reduced in affected animals, with the greatest disparity found in severely affected mice more than 20 days old (Figure 17-7).

As has been mentioned previously, only one of the seven founder animals (13) gave rise to the phenotype described. However, two of the other six strains that transmitted full-length copies of HIV proviral DNA to their progeny had very high mortality rates. As seen in Table 17-1, only one premature death in 96 F_1 offspring of four founder animals was recorded.

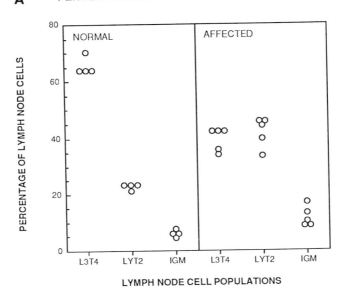

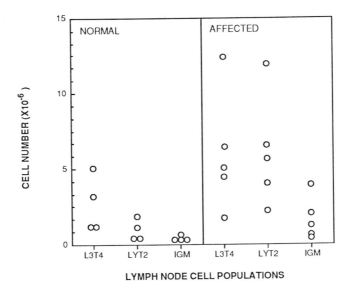

FIGURE 17-6 Characterization of lymphocytes from lymph nodes of transgenic mice. Inguinal, axillary, and cervical lymph nodes of individual transgenic and control mice were pooled and analyzed for lymphocyte populations using a fluorescence-activated cell sorter. Data are expressed as (A) the relative proportions of L3T4, Lyt-2, and IgM bearing cells, and as (B) the total numbers of these populations in the combined lymph node population. Each data point represents the result from an individual mouse.

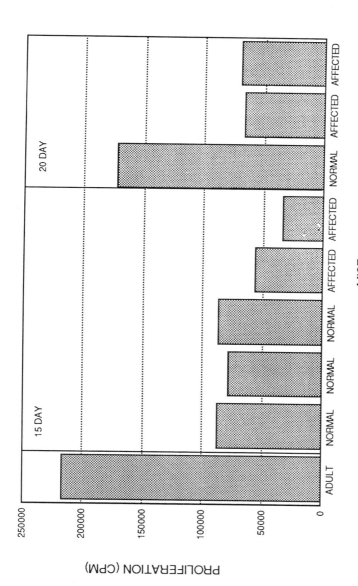

FIGURE 17-7 Proliferative response of splenic lymphocytes to concanavalin A. Cultured splenocytes from 15- and 20-day-old affected mice and their nontransgenic littermates were incubated with concanavalin A. Proliferation was determined by measurement of incorporation of ^{3}H-thymidine. Note that the disparity in the proliferative response between affected and normal animals increased by day 20, primarily as a result of the increased response of the older normal animal, which is approaching adulthood.

In contrast, seven of 35 F_1 mice from founder 16 and four of 32 F_1 mice from founder 62 either died unexpectedly or became moribund and were sacrificed. Chi-square analysis showed that the increased mortality rates of F_1 progeny from founders 16 and 62 were statistically significant at $P \leq 0.0002$ and 0.004, respectively, as compared to the rates for the 96 progeny of founders 38, 39, 42, and 64. DNA prepared from four of the seven strain 16 F_1 animals contained HIV proviral DNA, but because of autolysis due to spontaneous deaths no DNA was obtained from the three remaining mice.

The expression of intact HIV provirus in transgenic F_1 progeny of strain 13 results in a consistent and characteristic set of histopathologic abnormalities: hyperproliferative skin, perivascular pulmonary lymphoid infiltrates, splenomegaly, and lymphadenopathy, as well as growth retardation and premature death, all of which occur in the absence of Leu3 or T4 cell depletion. The major histologic abnormalities in the affected animals were found in the skin and lungs. While each of the skin lesions is characterized by epidermal hyperplasia, it is difficult to equate them to analogous lesions found in humans. The pulmonary interstitial lymphoid infiltrates observed in the mice are similar to the lesions present in cases of nonspecific interstitial pneumonitis in adult patients with AIDS. (Suffredini et al. 1987). Infiltrates in both the transgenic mice and AIDS patients occur in the absence of identifiable pulmonary pathogens. Thus, these lesions may represent a direct response to the presence of HIV gene products.

The reduced size of the thymus in all affected animals and the complete thymic involution in two severely affected animals are reminiscent of the precocious thymic involution observed in AIDS patients (Grody et al. 1985; Joshi et al. 1986). However, thymic involution is also seen as a nonspecific response to stress in mice. The observed splenomegaly and lymphadenopathy could very well be the result of a chronic immune response to the endogenous production of viral proteins. A similar immunopathological condition has been described in mice chronically infected with lymphocytic choriomeningitis virus (Doherty and Zinkernagel 1974). Lymphadenopathy, hyperplasia of subsets of B and T cells, and decreased proliferative responses of splenic T cells have also been reported in virus-infected neonatal mice (Buchmeier et al. 1980). Thus, the immunologic alterations present in the strain 13 mice are similar to those seen in mice challenged with various viral or allogenic stimuli.

It should be noted that the introduction of the HIV provirus into the germline of mice bypasses the early events in the life cycle of the virus, including the binding of viral particles to its receptor, the human CD4 molecule. Thus, a spreading infection in these newborn mice may not account for the observed abnormalities because of the absence of human CD4. The disease syndrome that we observed probably results from the intracellular production of HIV proteins and/or particles. A different mechanism, which is potentially responsible for the disease phenotype, is the interruption of an essential mouse gene as a result of the integration of proviral DNA.

However, insertional mutations in transgenic animals tend to be recessive traits, which are expressed in offspring homozygous for the defective gene (Jaenisch 1988); affected strain 13 progeny are in a uniformly heterozygous state.

Although the pathogenic mechanism(s) responsible for the syndrome observed in strain 13 is not clear, it should be noted that the disease was associated with only low levels of HIV gene activity. Virus was recovered from the diseased tissue only after two to three weeks of cocultivation with the CD4-positive A3.01 cell line. A characteristic of many lentivirus infections, including infection by HIV, is the disparity between profound clinicopathological findings and low levels of viral expression in affected tissues. One explanation for the dissociation of HIV expression and severe clinical disease in both transgenic mice and humans is that the synthesis of very small amounts of viral proteins or progeny particles may be sufficient to elicit the production of cellular proteins that may cause deleterious effects. For instance, a low level of HIV expression in macrophages might upset the regulation of the synthesis or secretion of factors that mediate chemotaxis, inflammation, or cell growth, resulting in the fatal disease observed. The epidermal proliferation observed in the affected mice could also represent the indirect effects of HIV expression in cells of the monocyte, macrophage lineage (Morhenn 1988).

It may be possible to use the mice that carry the HIV provirus to determine cellular and viral factors involved in the regulation of its transcription in vivo. The transcriptional regulation of genes involves the interaction of *trans*-acting factors with specific *cis*-acting DNA sequences (Cedar 1988). The *tat* gene product of HIV, for example, is a *trans*-acting factor that enhances viral gene expression through interaction with the *trans*-activating responsive region located in the long terminal repeat (LTR). Cellular regulatory proteins affecting other elements in the viral LTR could be involved in reactivating viral expression (Jones et al. 1986; Nabel and Baltimore 1987). Still unknown are the factors that repress the expression of quiescent, integrated copies of the proviral DNA present in our six other transgenic strains. The absence of detectable HIV expression in these animals could be due to the configuration of the cellular DNA that surrounds the provirus and/or the need for cellular factors to be present for the initial expression of the provirus to occur. Methylation of DNA, for instance, has been shown to affect gene structure and to limit access of specific sequences to *trans*-acting factors necessary for expression (Cedar 1988). 5-azacytidine (5-aza C), a potent demethylating agent, can activate such methylated genes. In fact, injection of 5-aza C into mice that contain highly methylated, nonexpressing Moloney murine leukemia proviral DNAs induced expression of viral RNA in the thymus, spleen, and liver (Jaenisch et al. 1985). Treatment with 5-aza C also restored expression of a nonexpressing HIV LTR-CAT construction, which had been stably integrated into Vero or murine cells (Bednarik et al. 1987). Activation of HIV has also been achieved by treating cells with mitomycin C, a cross-linker of DNA and ultraviolet light (Valerie et al. 1988). We are presently attempting to induce viral expression in the progeny of the six nonexpressing strains of transgenic HIV mice using 5-aza C and mitomycin C. Preliminary results indicate that virus can be recovered from a number of these animals after treatment with either drug.

Transgenic mice may serve as valuable tools in elucidating the pathogenic mechanisms that underlie HIV-related diseases in humans. Furthermore, the administration of agents prior to the onset of symptoms in the strain 13 animals could result in the identification and development of novel compounds that abrogate the onset or progression of the disease. Additional transgenic animals containing subgenomic and mutagenized HIV proviruses are being constructed in an attempt to elicit disease without production of infectious virus. Such mice may prove invaluable in demonstrating the relationship between specific HIV protein(s) and disease.

REFERENCES

Adachi, A., Gendelman, H.E., Koenig, S., et al. (1986). *J. Virol.* 59, 284–291.
Alter, H.J., Eichberg, J.W., Masur, H., et al. (1984). *Science* 226, 549–552.
Bednarik, D.P., Mosca, J.D., and Raj, N.B. (1987). *J. Virol.* 61, 1253–1257.
Buchmeier, M.J., Welsh, R.M., Dutko, F.J., and Oldstone, M.B. (1980). *Adv. Immunol.* 30, 275–331.
Cedar, H. (1988) *Cell* 53, 3–4.
Dalgleish, A.G., Beverly, P.C., Clapham, P.R., et al. (1984) *Nature* 312, 763–767.
Doherty, P.C., and Zinkernagel, R.M. (1974) *Transpl. Rev.* 19, 89–120.
Folks, T., Benn, S., Rabson, A., et al. (1985) *Proc. Natl. Acad. Sci. USA* 82, 4539–2543.
Francis, D.P., Feorini, P.M., Broderson, J.R., et al. (1984) *Lancet 2*, 1276–1277.
Grody, W.W., Fligiel, S., and Naeim, F. (1985) *Am. J. Clin. Pathol.* 84, 85–95.
Jaenisch, R. (1988) *Science* 240, 1468–1474.
Jaenisch, R., Schnieke, A., and Harbers, K. (1985). *Proc. Natl. Acad. Sci. USA* 82, 1451–1455.
Jones, K.A., Kadonaga, J.T., Luciw, P.A., and Tjian, R. (1986). *Science* 232, 755–759.
Joshi, V.V., Oleske, J.M., Saad, S., et al. (1986) *Arch. Pathol. Lab. Med.* 110, 837–842.
Klatzmann, D., Barrie-Sinoussi, F., Nugeyre, M.T., et al. (1984) *Science* 225, 59–63.
Maddon, P.J., Dalgleish, A.G., McDougal, J.S., et al. (1986) *Cell* 47, 333–348.
McDougal, J.S., Kennedy, M.S., Sligh, J.M., et al. (1986) *Science* 231, 382–385.
Morhenn, V.B. (1988) *Immunol. Today* 9, 104–107.
Nabel, G., and Baltimore D. (1987) *Nature* 326, 711–713.
Nara, P.L., Robey, W.G., Arthur, L.O., et al. (1987) *J. Virol.* 61, 3173–3180.
Strebel, K., Daugherty, D., Clouse, K., et al. (1987) *Nature* 328, 728–730.
Suffredini, A.F., Ognibene, F.P., Lack, E.E., et al. (1987) *Ann. Int. Med.* 107, 7–13.
Valerie, K., Delers, A., Bruck, C. et al. (1988) *Nature* 333, 78–81.

CHAPTER 18

Apolipoprotein A-I Gene Expression in Transgenic Mice

Annemarie Walsh
Yasushi Ito
Jan L. Breslow

The synthesis of apolipoprotein A-I (apo A-I), the major structural protein of high-density lipoprotein (HDL), should be an important determinant of HDL-cholesterol (HDL-C) levels. To study the effect of increased synthesis of apo A-I on HDL levels, we introduced the human gene for apo A-I and various lengths of the 5' flanking sequence into transgenic mice. In each of the five transgenic lines that were established, the human gene for apo A-I was expressed in liver but not in the small intestine, the two normal sites of synthesis. Animals from one transgenic line that had sufficient human apo A-I in their plasma to increase the size of the pool of apo A-I were

---

We wish to thank Dr. Elizabeth Lacy for instructing us in the transgenic mouse technique, Dr. Eliot Brinton for valuable advice, Ms. Sylvia White and Ms. Margaret Timmons for their technical assistance, and Ms. Lorraine Duda for her expertise in preparing the manuscript. This research was supported in part by grants from the National Institutes of Health, which included a Clinical Nutrition Research Unit grant and a General Clinical Research Center grant; by an American Heart Association-New York City Affiliate grant; as well as by a general grant from the Pew Trusts. Dr. Yasushi Ito is a fellow supported by the H.O. West Foundation, Parenteral Drug Association, Inc., and Suntory Fund for Biomedical Research. Dr. Jan L. Breslow is an Established Investigator of the American Heart Association.

studied in detail to document the effects of the oversynthesis of apo A-I on HDL-C levels.

18.1 MATERIALS AND METHODS

18.1.1 Production of Transgenic Mice
Fertilized eggs from the oviducts of superovulated F_1 females (C57BL/6J × CBA/J) were microinjected with an aliquot of 1–3 μg/ml of one of three DNA fragments derived from a human apo A-I genomic clone that contained 5.5 kb, 4.5 kb, or 256 bp of 5' flanking sequence. After injection, one- or two-cell embryos were surgically transferred to the oviducts of surrogate females who carried them to term. Founder animals were bred to (C57BL/6J × CBA/J) F_1 animals and transgenic lines were established.

18.1.2 Analysis of DNA
A small piece of tail, 1–2 cm in length, was excised from 3- to 4-week-old mice and DNA was extracted with an Applied Biosystems 340A Nucleic Acid Extractor. Southern blot analysis was performed using a nick-translated genomic DNA probe for human apo A-I (specific activity $2-5 \times 10^8$ cpm/μg).

Total RNA was isolated by the guanidine thiocyanate method (Chirgwin et al. 1979). Slot-blot analysis was performed using a nick-translated 0.7-kb *Sac*I-*Pst*I fragment from a genomic clone for human apo A-I for hybridization with human apo A-I mRNA or a 345-bp fragment of 3' cDNA from mouse to probe for mouse apo A-I mRNA.

18.1.3 Quantification of Levels of Apo A-I by ELISA and SDS-PAGE
Plasma was prepared from blood collected during excision of the tail tip or from the retroorbital plexus. Quantitation of levels of human apo A-I by sandwich enzyme-linked immunosorbent assay (ELISA; Denke and Breslow 1988) was performed using an antibody with less than 0.01% cross-reactivity with mouse apo A-I. Total levels of plasma apo A-I (human plus mouse) were quantitated by densitometric scanning of gels after nonreducing SDS-PAGE of plasma proteins.

18.1.4 Analyses of Plasma Lipid and Lipoprotein
Mice at approximately three months of age were fasted overnight and 600 μl of blood was obtained by retroorbital bleeding. After separation of plasma by centrifugation, lipoprotein fractions were prepared by sequential ultracentrifugation in an Airfuge (Beckman). Cholesterol and triglycerides were

measured enzymatically (with reagents from Boehringer Mannheim) in the total plasma and in the lipoprotein subfractions. Electrophoresis of lipoproteins was performed by applying 2 µl of whole plasma to a 0.6% agarose gel (100 × 120 mm² gel support film, Biorad). After electrophoresis, the gel was fixed and then stained with 0.23% Sudan Black in 60% ethanol.

The distribution by size of HDL particles was assessed by electrophoresis on native polyacrylamide gradient gels (GGE) of the total lipoprotein fraction with density less than 1.21 g/ml. Western blotting of HDL separated by GGE was used to determine the distribution of human apo A-I in HDL particles of different sizes.

18.1.5 Turnover of Apo-1 In Vivo

Human and mouse plasma were obtained from a single, healthy donor and from pooled sera, respectively. HDL, with a density of 1.063–1.21 g/ml, was prepared by sequential ultracentrifugation; it was delipidated; and apo A-I was isolated by ion-exchange chromatography (on a Mono Q column, FPLC system, Pharmacia, NJ), which was followed by gel-filtration chromatography (on Superose 12, Pharmacia). Iodination of apo A-I was carried out by Bilheimer's modified version (Bilheimer et al. 1972) of the McFarlane method. An aliquot of 8.5 µg of human apo A-I and 17 µg of mouse apo A-I, corresponding, respectively, to approximately 0.15 and 0.3% of the total pool of plasma apo A-I in the mouse, was injected into each of four transgenic mice via the external jugular vein. Ten microliters of whole blood was obtained serially by retroorbital bleeding under methoxyfluorane anesthesia for measurements of radioactivity.

18.2 RESULTS

Three fragments of DNA containing the entire human gene for apo A-I but with differing amounts of 5' and 3' flanking sequence, were microinjected into fertilized eggs from (C57BL/6J × CBA/J) F_1 mice (Figure 18-1). Of the 61 animals born, nine had integrated the human gene for apo A-I and seven of these expressed the human protein in their plasma. Five transgenic lines were established with mean plasma levels of human apo A-I that varied from 0.07–339 mg/dl (Table 18-1).

An examination of the tissue specificity of expression of the gene for apo A-I in each of the five transgenic lines produced unexpected results. Most mammalian apo A-I is synthesized principally in liver and intestine with minor rates of synthesis in other tissue. Slot-blot analysis (Figure 18-2), utilizing a mouse 3' cDNA probe with little cross-reactivity to human apo A-I mRNA, showed that approximately equal levels of mouse apo A-I were present in the liver and intestines of control and transgenic mice. However, using a genomic probe for human 3' apo A-I with little cross-

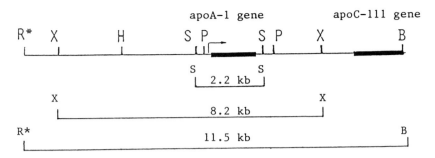

FIGURE 18-1 Restriction map of constructions of the human gene for apolipoprotein used for microinjection. A restriction map is shown of a cloned portion of the human genome that contains the 1863-bp gene for apo A-I (thick bar). The direction of transcription is shown by the arrow. This region also contains a portion of the apo CIII gene (thick bar) 2.5 kb downstream from the apo A-I gene. The apo CIII gene is transcribed from the opposite DNA strand. The three pieces of DNA from which transgenic lines were produced are indicated. The 2.2-kb fragment extends between the *Sma*I site 256 bp 5′ to the gene and a site 80 bp 3′ to the gene. The 8.2-kb fragment extends between the *Xmn*I site 4.5 kb 5′ to the gene and the *Xmn*I site 1.5 kb 3′ to the gene. The 11.5-kb fragment extends between a false *Eco*RI site that is 5.5 kb 5′ to the gene and a *Bam*HI site that is 3.8 kb 3′ to the gene. A summary of the results of microinjection is shown below the restriction map.

reactivity to mouse apo A-I mRNA, we saw expected levels of human apo A-I mRNA in the liver in each of the five lines that carried the transgene. However, none of the lines showed intestinal expression even though expression in inappropriate tissue was observed in the testes, ovaries, heart, and spleen of individual animals.

Mice from the Tg(OHSA-AI)179 line with a mean level of human apo A-I of 245 mg/dl were used to study the effects of oversynthesis of apo A-I on plasma lipoprotein levels (Figure 18-3). In transgenic animals, total levels of apo A-I (mouse plus human) were 381 ± 43 mg/dl significantly

TABLE 18-1 Summary of Transgenic Lines Generated with the Human Gene for APO A-I

Fragment Size (kb)	Transgenic Lines Generated[a]	Concentration of Human Apo A-I in Transgenic Line (Mean ± SD mg/dl)	
2.2	Tg(OHSA-AI)427	339 ± 69	n = 10[b]
8.2	Tg(YHSA-AI)139	2.5 ± 2.1	n = 5
	Tg(OHSA-AI)145	0.07 ± 0.03	n = 5
	Tg(OHSA-AI)149	94 ± 27	n = 7
11.5	Tg(OHSA-AI)179	245 ± 86	n = 31

[a]The recommended nomenclature for transgenic mice, as outlined by Peters (1985), is used.
[b]n represents the number of transgenic animals used to obtain the mean value of the concentration from ELISA data.

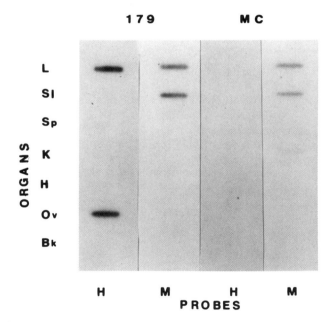

FIGURE 18-2 Tissue-specific expression of the human apo A-I transgene. Total RNA was isolated from liver (L), small intestine (SI), kidney (K), spleen (Sp), heart (H), and ovary (Ov) of an F_1 progeny of transgenic animal 179. Ten micrograms of RNA was loaded in each slot. After electrophoresis, the gel was blotted onto nitrocellulose and allowed to hybridize with either a nick-translated genomic probe of human apo A-I (H) or a mouse apo A-I cDNA probe (M). RNA from organs of a control mouse (MC) and a blank control (BK) were included for purposes of comparison. In the transgenic 179 line, the human probe detected mRNA for human apo A-I in the liver but not in the intestines, whereas the mouse probe detected mRNA for mouse apo A-I in both tissues. Line 179 also shows evidence of expression of human apo A-I in the ovary.

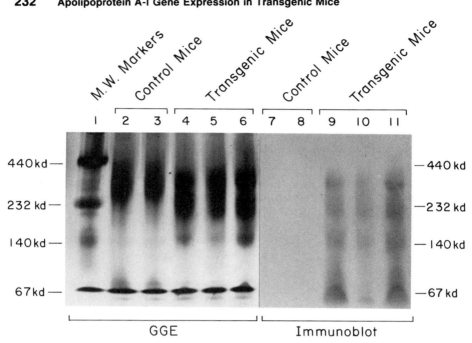

FIGURE 18-3 The spectrum of sizes of HDL particles and the distribution of human apo A-I in control and transgenic mouse lipoproteins. Electrophoresis on nondenaturing gradient gels and immunoblotting of the lipoprotein fraction (d <1.21 g/ml) from control and transgenic mice are shown. Samples were subjected to electrophoresis on an 8.25% native polyacrylamide gradient gel. Silver staining of the gel is shown in lanes 2 to 6 and an immunoblot made from the same gel with an antibody specific for human apo A-I is shown in lanes 7 to 11. Lane 1 contains molecular weight markers as indicated. Note that there are smaller HDL particles in samples from transgenic mice and that the antibody specific for human apo A-I reacts with all subfractions of HDL. The antibody did not react at all with lipoproteins from the control mouse.

different from the levels of 153 ± 17 mg/dl in control mice. The levels of total cholesterol and HDL-C were significantly higher in transgenic animals and, although the increment was small, the transgenic animals had significantly higher levels of VLDL- and LDL-cholesterol. Total triglyceride levels or those in the lipoprotein subfraction were not significantly different between the groups of animals. As shown in Table 18-2, there was a highly significant difference in the molar ratio of HDL-C to apo A-I between the transgenic and the control mice. Since this ratio reflects the size of the HDL particle, it appears that the increase in levels of HDL in the transgenic mice occurs in the smaller HDL particles. To confirm this observation, the distribution of HDL particles by size was examined by electrophoresis on native polyacrylamide GGE. In Figure 18-3, lanes 2 to 6, the HDL in control

TABLE 18-2 Total Levels of APO A-I and Lipoprotein Profiles in Control and Transgenic Mice

	Total A-I mg/dl	Cholesterol (mg/dl)				Triglycerides (mg/dl)				HDL-C/A-I (MR)[a]
		Total	VLDL	LDL	HDL	Total	VLDL	LDL	HDL	
Control (mean ± SD)	153±17	78±13	4±1	20±4	55±11	44± 7	12±5	6± 4	27±3	25±2
Transgenic (mean ± SD)	381±43	125±12	8±3	27±5	90± 7	54±19	16±7	13±10	28±5	17±1
P value	0.0001	0.0001	0.01	0.03	0.0001	NS	NS	NS	NS	0.0001

Differences between control and transgenic mice were tested by Student's t-test.
[a]MR indicates the molar ratio.
[b]NS, not significant.

mice is seen to migrate more slowly than the 232-kd molecular weight marker, whereas a substantial fraction of the HDL in the transgenic animals migrates more rapidly. Since this difference in size could be the result of differential metabolism of human apo A-I in the mouse, an immunoblot was prepared. As seen in lanes 9 to 11 of Figure 18-3, the human apo A-I is spread over the entire range of sizes of HDL particles and is not just in the smaller-sized species present in the transgenic animals. This result suggests that human apo A-I is metabolized in the same way as mouse apo A-I. Furthermore, studies of the turnover of radioiodinated mouse and human apo A-I in vivo showed them to have an identical rate of catabolism.

18.3 DISCUSSION

The introduction of the human gene for apo A-I into mice resulted in five transgenic lines of mice that contained the human protein in their plasma. The amount of human protein differed greatly between the various lines. This variation was not a function of the size of the 5' flanking region of the apo A-I gene in the particular fragment injected, but was probably due to factors specific to the site of integration of the DNA. In two of the lines, the amount of human protein produced only minimally increased the size of the total apo A-I pool. However, in three of the lines, the human protein was present at approximately 60 to 200% of normal levels of apo A-I in the mouse and added significantly to the size of the total apo A-I pool. These mice made it possible for us to study the effect of excess plasma apo A-I, presumably due to oversynthesis, on levels of lipoprotein. The major change was an increase in HDL-C levels. HDL particles occur over a wide range of densities and sizes and, from results of compositional analysis and gradient gel electrophoresis, it appeared that the increase in levels of HDL-C in the transgenic animals was in small HDL particles, similar to the dense human HDL_3 particles.

The other major finding reported here is that in vivo only 256 bp of the 5' flanking sequence are required for liver expression of the gene for apo A-I while not even 5.5 kb allowed expression in the intestines. This result is compatible with results of studies of transient expression in the human hepatoma cell line HepG2, in which it was shown that liver-specific, *cis*-acting, transcriptionally active regions are located within the 256 bp 5' to the apo A-I gene. However, similar studies in the intestinal cell line CaCO2 identified a transcriptionally active region required for intestinal expression, which was within 2 kb of the origin of the 5' flanking sequence (Sastry et al. 1988). Our results in vivo suggest that the region important for intestinal expression may be elsewhere.

REFERENCES

Bilheimer, D.W., Eisenberg, S., and Levy, R.I. (1972) *Biochim. Biophys. Acta* 260, 212–221.
Chirgwin, J., Przybyla A., MacDonald, R., and Rutter, W.J. (1979) *Biochemistry* 18, 5294–5299.
Denke, M.A., and Breslow, J.L. (1988) *J. Lipid Res.* 29, 963–969.
Peters, J. (1985) in *Mouse News Letter*, 72, p. 27, Oxford University Press, Oxford, England.
Sastry, K.N., Seedorf, U., and Karathanasis, S.K. (1988) *Mol. Cell. Biol.* 8, 605–614.

CHAPTER 19

Effects of Human Growth Hormone on Reproductive and Neuroendocrine Functions in Transgenic Mice

A. Bartke, J.G.M. Shire, V. Chandrashekar
R.W. Steger, A. Mayerhofer, A.G. Amador
P. Bain, K. Tang, J.S. Yun, T.E. Wagner

Reproductive competence of transgenic animals is an obvious concern in any attempt to utilize transgenic technology for improvement of productivity of swine, sheep, or other domestic animals. In an attempt to uncover mechanisms that may account for reproductive abnormalities in animals that express various growth-promoting genes (see Chapters 20 and 21), we are studying reproductive and neuroendocrine functions in mice that express genes for human growth hormone (hGH), bovine GH (bGH), or a variant of hGH (hGH-B). In this chapter, we describe results obtained in mice that express and transmit a mouse metallothionein-I (mMT)/hGH gene construct, and some comparisons are made with results obtained with other lines of transgenic mice.

These studies were supported by the National Institutes of Health through grants HD20001 (A.B.) and HD09042 (T.E.W.) and by the Wellcome Research Travel Grant (J.G.M.S.). We thank S. Hodges and T. Began for their assistance and M. Hilt for typing the manuscript.

19.1 PRODUCTION OF TRANSGENIC MICE AND THEIR CHARACTERISTICS

Fertilized mouse eggs for microinjection were recovered in cumulus from the oviducts of C57BL/6J × C3H/J (B6C3) females that had mated with males several hours earlier. Approximately 1,000 copies of the blunt-ended mMT/hGH fusion gene were microinjected into the male pronucleus of each fertilized egg. Microinjected eggs were implanted into the oviducts of one-day pseudopregnant foster mothers of the ICR strain and were carried to term (Wagner et al. 1981). Five transgenic animals produced in this fashion included one fertile male that was positive for expression of hGH, as assessed by DNA hybridization analysis. This mouse had one fusion gene incorporated into his genome and he transmitted mMT/hGH gene constructs to approximately half his progeny (Selden et al. 1989). Transgenic mice used in the present study were produced by mating transgenic male descendants of this particular animal with B6C3 F_1-hybrid females purchased from the Jackson Laboratory (Bar Harbor, ME).

Although transmission of the transgene in the resulting line generally followed the expected pattern of a dominant autosomal trait, the ratio of transgenic to phenotypically normal progeny was consistently different from 1:1 with a slight shortfall of transgenic males, a greater shortfall of transgenic females, and a significant sex difference in the ratio of transgenic to normal animals ($P < 0.01$). Similar departures from the expected transmission ratio were observed in other lines of transgenic mice in breeding colonies at both Edison Center and Southern Illinois University and were particularly striking in the case of mMT/hGH-B animals (Yun, unpublished observations; Shire, unpublished observations). Analysis of DNA extracted from tail tissue of phenotypically normal females by DNA hybridization indicates that only phenotypically transgenic animals are "DNA-positive." This result strongly suggests that departures from the expected phenotypic ratios in the progeny are due to a shortfall in the number of animals that inherit the mMT/hGH transgene rather than to a failure of phenotypic expression of the gene construct in some of the animals. We have no explanation for these findings but, from the analysis of sizes of litters of females mated to normal and transgenic males, we suspect that increased mortality of transgenic embryos or fetuses is not involved.

Mice that express the mMT/hGH transgene (hereafter referred to as "transgenic mice") are of normal size and appearance at birth and during early postnatal growth. However, soon after weaning (at the age of approximately four weeks), they can be distinguished from their normal littermates by the appearance of their fur. They show a modification of the normal agouti phenotype, which resembles a mild version of that found in Tabby (Ta) homozygotes and hemizygotes and in crinkled and downless homozygotes. They have reduced amounts of hair behind the ears and on the abdomen, although their fur does contain some zig-zag hairs. Quantitative measurements on the skin of adult transgenics show a reduction in the density of hair follicles. There are no differences in the number of secondary vibrissae. There appear to be minor differences from the tail rings found in normal mice, but the tails are not hairless as they are in Ta mice (but not Ta^J).

Accelerated growth usually starts during the fourth week of postnatal life, presumably corresponding to the stage of development at which hepatic production of insulin-like growth factor-1 (IGF-1 = somatomedin C) begins to be responsive to the stimulatory effects of GH (Matthews et al. 1988). The increased growth rate is maintained and adult transgenic mice weigh at least 50% more than their normal siblings. Thus, the effects of the mMT/hGH transgene on somatic growth in this line of mice closely resemble those described in another line that expresses the same gene construct (Palmiter et al. 1983) and those in animals that express genes for rGH, bGH, hGH-B, or hGRF (Matthews et al. 1988; Selden et al. 1988). Presumably, the effects of both ectopic secretion of heterologous GH and overproduction of mGH by the pituitary in situ on somatic growth are mediated predominantly, if not exclusively, by increased production of IGF-1, and the observed 50–100% increase in body weight of transgenic mice represents the maximal stimulation of growth by this mechanism. However, a recent histopathological study of mice that expressed genes for bGH, hGRF, or hIGF-1 suggests that only some of the effects of an excess of GH in transgenic mice can be ascribed to elevated levels of IGF-1 (Quaife et al. 1989).

Analysis of DNA and transcripts of the gene for hGH provides evidence that a single copy of the mMT/hGH gene is stably incorporated into the chromosomal complement of these animals and is expressed primarily in the liver, kidneys, and intestines (Selden et al. 1989). This pattern of expression of a transgene that contains a mMT promoter is consistent with observations in other lines of transgenic mice (Palmiter et al. 1983; Shea et al. 1987).

Serum levels of hGH in adult transgenic mice used in our study were somewhat higher in males than in females: 8.34 ± 1.62 versus 4.49 ± 0.66 ng/ml; $P < 0.05$. Use of the same commercial RIA kit for measurement of hGH (Hybritech Inc., San Diego, CA) in samples of plasma, rather than serum, resulted in detection of somewhat lower levels of hGH, 4.8 ± 0.4 ng/ml in males and 3.8 ± 0.3 ng/ml in females (Bartke et al. 1988; Bartke, unpublished observations).

The average lifespan of transgenic mice is reduced, with transgenic mice appearing to age faster than their littermates. In one-year-old mice, the incidence of overt mammary cancer in untreated transgenic females was 6/18 while it was 0/13 in normal littermates ($P < 0.05$). Mammary cancers were also found in two males that carried the transgene. Externally palpable tumors in the axillary, thigh, and neck regions were common in both sexes.

19.2 REPRODUCTIVE AND NEUROENDOCRINE FUNCTIONS IN TRANSGENIC FEMALES

Transgenic (mMT/hGH) females are sterile. Female sterility was also noted in another, independently derived line of mMT/hGH transgenic mice (Hammer et al. 1985; Hammer, personal communication), as well as in transgenic mice that expressed the rat gene for GH (Hammer et al. 1984). In our line of transgenic animals, young adult females (2–3 months old) have ovaries that appear normal in terms of gross histological structure and numbers of pre-antral and antral follicles (Mayerhofer et al. 1988). At this age, transgenic females exhibit cyclic variations in ovarian function and appear to ovulate normally, as indicated by changes in the cellular composition of vaginal smears and the presence of copulatory plugs when they are housed with males. However, copulation is not followed by pseudopregnancy or pregnancy. Instead, the animals usually mate again five to seven days later. The length of the estrous cycle is significantly increased (6.23 ± 0.39 versus 4.37 ± 0.20 days; $P < 0.01$) in agreement with an average interval of 6.5 days between consecutive vaginal plugs (Bartke et al. 1988).

In middle-aged, seven-month-old, transgenic females, the percentage of pre-antral follicles is reduced, the percentage of antral follicles is increased, and conspicuous clusters of lipid-filled interstitial cells resemble changes seen in ovaries of very old animals (Mayerhofer et al. 1988; E. Anderson, personal communication). At this stage, most transgenic females are anovulatory, but they can be induced to ovulate and mate by injections of pregnant mare's serum gonadotropin (PMSG) and human chorionic gonadotropin (hCG). Positive proof of the ability of the ovaries of transgenic females to produce normal ova was obtained by transplantation of their ovaries into normal, ovariectomized hosts and successful pregnancy in these animals (Yun and Wagner 1987).

From the observations summarized above, we suspected that sterility of transgenic females might be due to luteal failure. In the normal mouse, copulation leads to activation of the corpora lutea followed by increases in peripheral levels of progesterone, which serve to prepare the uterus for implantation and to prevent ovulation of a new crop of ova several days later. In transgenic mice, this process appeared to have failed since after mating the animals did not become pregnant and continued to cycle. In support of this interpretation, we found that replacement of luteal function by treatment with 1 mg progesterone per day starting on the day of mating resulted in maintenance of pregnancy in five of seven treated mice (Bartke et al. 1988 and 1989).

In the mouse, functional activation of corpora lutea and maintenance of luteal function depend on mating-induced stimulation of the release of endogenous prolactin (PRL). Therefore, we suspected that luteal failure in transgenic females reflects reduced secretion of PRL. This possibility was substantiated by the demonstration that release of PRL in vivo in ovariectomized transgenic females is severely attenuated. Furthermore, treatment with PRL-secreting ectopic pituitary transplants is very effective in maintaining pregnancy in these animals, and the activity of dopaminergic neurons, which inhibit the release of PRL, is significantly increased in adult transgenic females (Bartke et al. 1988). Evidence for altered dopaminergic control of release of PRL in transgenic mice was obtained by measuring the turnover of dopamine (DA) in the median eminence, a region that contains terminal fields of the tuberoinfundibular dopaminergic (TIDA) system and where DA is released into the blood supply of the anterior pituitary. Presumably, hGH activates this neuronal system via its inherent lactogenic activity in the mouse (Forsyth et al. 1965; Hartree et al. 1965) and this activation inhibits the synthesis and release of PRL and leads to luteal failure. Circulating levels of hGH are presumably inadequate to substitute for the mating-induced stimulation of the release of PRL. However, we cannot discount the possibility that infertility of transgenic females is due to altered ratios of gonadotropins to PRL or of LH to follicle-stimulating hormone (FSH) (Yoshinaga and Fujino 1979) and that treatment with PRL-secreting pituitary transplants corrects this abnormality.

The importance of the lactogenic activity of hGH in suppressing the fertility of transgenic mice is underscored by the observation that most mice that express the mMT/bGH gene are fertile (Yun and Wagner, unpublished observations). bGH is somatotropic but not lactogenic in the mouse.

Lactational performance of transgenic mice rendered fertile by ectopic pituitary transplants is extremely poor. Only three of nine pituitary-grafted females that delivered live pups were able to raise them to weaning. Maternal behavior seemed normal in all females but in most litters pups appeared to have no milk in their stomachs and died within 48 hours after delivery (Bartke et al. 1988 and 1989; Bartke, unpublished observations). We have no explanation for this unexpected finding.

19.3 REPRODUCTIVE AND NEUROENDOCRINE FUNCTIONS IN TRANSGENIC MALES

Almost all mMT/hGH transgenic males from the line used for these studies are fertile. Daily production of sperm, determined from the content of late spermatids in the testes and the epididymal reserves of sperm, are increased while production of sperm per gram of testicular tissue is not affected (Amador et al. 1988). We have examined the weights of several endocrine and target organs in male mice transgenic for one of four different GH gene constructs. All four kinds of transgenic males were heavier ($P < 0.0001$) than normal adults of the same stock. The mean weights of the organs shown in Figure 19–1 were all significantly larger ($P < 0.001$) for the transgenic

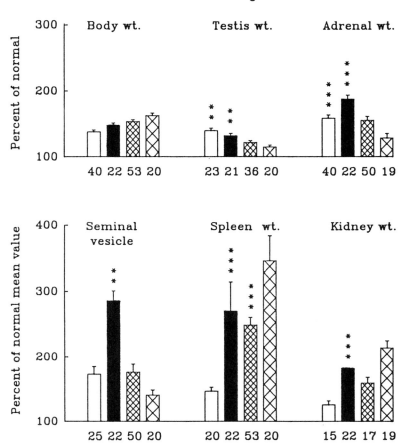

FIGURE 19-1 The mean (± SE) body weight and organ weights of transgenic male mice aged between 12 and 36 weeks, expressed as a percentage of the mean values for normal mice of the same stock. The four gene constructs were: mMT/hGH-A, open box; mMT/hGH-B, filled box; mMT/bGH, close cross-hatching; PEPCK/bGH, open cross-hatching. The numbers below each column give the numbers of mice used for measurements. The asterisks show significant differences in the mean positions of the allometric relationships: ** $P < 0.01$; *** $P < 0.001$.

mice than for their normal adult littermates. The weights of these organs were not correlated with age, which ranged from 12 to 36 weeks.

We carried out analyses of covariance on logarithmically transformed weights, as did Shea et al. (1987) in their important study of mice transgenic for rat GH. Such an allometric analysis enables an increase in the weight of an organ that simply reflects the general growth pattern of that organ to be distinguished from a disproportionate enlargement of that particular organ in response to specific effects of excess hormone. Organs that showed

19.3 Reproductive and Neuroendocrine Functions in Transgenic Males

such disproportionate increases in transgenic mice are indicated in Figure 19-1. The testis and the adrenal glands, both steroidogenic tissues, showed similar patterns of disproportionate enlargement in the mice with GH transgenes but only proportionate growth in the transgenics with bGH gene constructs. The seminal vesicles of mice that carried the hGH-B gene were disproportionately heavy but those of mice with the hGH gene were not, at least in mice less than nine months of age. However, greatly enlarged seminal vesicles, weighing up to 3.1 g, were found in older mice with the hGH gene, while seminal vesicles weighing up to 2.9 g were found in hGH-B mice aged 12 months. The coagulating glands showed an even greater age-related increase in weight in hGH males. In such mice, aged between 9 and 13 months, the combined weight of the coagulating glands and the seminal vesicles showed a highly significant ($P < 0.001$) and disproportionate increase in weight, reaching almost 7.5 g in one transgenic mouse, which weighed 68 g. The spleen showed proportionate growth in the mice transgenic for the hGH gene. In contrast, in both hGH-B and mMT/bGH transgenics, the spleen showed very marked enlargement, with significant differences in both the slope and the position of the allometric growth lines, as did the rGH transgenics studied by Shea et al. (1987). The weight of the thymus, despite significant negative regression with age, as well as with body weight, showed a similar pattern to that of the spleen. The thymus was significantly larger ($P < 0.0001$), in absolute terms, in the three kinds of transgenics with enlarged spleens and did not differ in size between hGH transgenic mice and their normal littermates. Kidney growth was, in general, proportionate as in the case of mice transgenic for rGH except in the case of the mice with the hGH-B transgene.

The specific effects on organ growth described here were not due simply to increased levels of particular classes of GH molecules but also reflect induced alterations in other components of the endocrine system that regulate the development and growth of these organs. While circulating levels of GH were raised in transgenic mice of all four stocks, levels of endogenous PRL were depressed in mice with the hGH transgene and levels of LH were raised. The hGH protein has some PRL-like actions in mice whereas the hGH-B protein is not known to have such effects; the bovine molecule is considered to be free of such effects. PRL-like effects may well prove to be involved in the disproportionate growth of both the testes and the adrenal glands since both are responsive to PRL (Colby 1979; Bartke 1980). PRL also acts on the growth and function of male accessory reproductive glands (Bartke 1980; Prins 1987) and induces significant enlargement of the seminal vesicles in male mice (Nonomura et al. 1985; Bartke et al. 1987). Therefore, stimulation of the growth of the seminal vesicles in transgenic mMT/hGH males is most likely due to the lactogenic activity of hGH in this species (Forsyth et al. 1965; Hartree et al. 1965). Extreme hypertrophy of the seminal vesicles in aging transgenic males may be related to a similar tendency in inbred C57BL mice (Nonomura et al. 1985), that is, in one of the parental strains that contribute to the genetic background of these animals.

The observed differences between the stocks that express the two GH genes in the pattern of onset of excessive stimulation of the seminal vesicles may reflect differences in the spectrum of stimulatory activities of homologous and heterologous molecules with PRL-like activities. Similar questions are raised by the pattern of responses found in the immune system; both GH and PRL are considered to have trophic effects (Berczi 1988), yet no disproportionate effects were seen in mice that expressed the hGH gene.

A complicating factor in this study, and in many other studies of transgenics, is that the genetic background on which the transgenes were segregating was not itself uniform. The background approximated that of an F_2 cross between the two strains used to produce the F_1 eggs in which the gene constructs were originally incorporated. Such variation in the genetic background of both transgenic and normal mice increases the "noise" against which any effects of a transgene have to be measured. Variation present in the background of the transgenes described here includes that affecting the altered sensitivity of C57BL mice to androgens (Bartke 1974; Shire 1981), the increased sensitivity of their seminal vesicles to PRL (Nonomura et al. 1985), adrenal weight (Shire 1981), and immunological characteristics (Crichton and Shire 1982).

Plasma levels of PRL are significantly reduced ($P < 0.001$) in transgenic males (Chandrashekar et al. 1988). This reduction is almost certainly due to the lactogenic activity of hGH present in the circulation of these animals and, presumably, it is mediated via activation of TIDA neurons. We were unable to obtain direct support for this possibility from measurements of the rate of turnover of DA in the median eminence of these animals (R.W. Steger and A. Bartke, unpublished observations) and, therefore, we suspect that stimulation of this neuronal system during prepubertal or pubertal development of transgenic males leads to compensatory adjustments of the sensitivity of the TIDA neurons to hGH and of the lactotrophs to DA. Results obtained in chronically hyperprolactinemic rats and in seasonally hypoprolactinemic hamsters provide precedents for such a possibility (Steger et al. 1983; Morgan et al. 1985). Suppression of the release of PRL in mice that express GH genes might also be mediated by an increase in the secretion of somatostatin (Patel and Srikant 1986).

Plasma levels of LH are significantly increased in our transgenics ($P < 0.001$; Chandrashekar et al. 1988). It is noteworthy that stimulation of the release of LH after acute administration of LHRH is attenuated and the postcastration increase in plasma levels of LH is severely reduced or absent (Chandrashekar et al. 1988). Moreover, the inhibitory effect of exogenous testosterone on plasma levels of LH in castrated males is much less pronounced in transgenic animals than in their normal siblings (Chandrashekar et al. 1988). Clearly, the hypothalamic mechanisms that control the secretion of LH are profoundly affected. Measurements of the turnover of norepi-

19.3 Reproductive and Neuroendocrine Functions in Transgenic Males

nephrine in the median eminence of these animals suggest that both the increases in basal plasma levels of LH and the suppression of the increase in levels of LH in response to castration are due to alterations in the noradrenergic activity of the hypothalamic neurons that are responsible for the control of the release of LHRH (R.W. Steger and A. Bartke, unpublished observations). Increases in hypothalamic noradrenergic activity and in plasma levels of LH were previously described in DBA/2 male mice with experimentally induced chronic hyperprolactinemia (Bartke et al. 1987).

Plasma levels of FSH are normal (Chandrashekar et al. 1988) and the responses of these levels to castration and exogenous testosterone are only slightly reduced. The divergent effects of the expression of a transgene on levels of LH and FSH are consistent with the existence of different mechanisms for the control of the synthesis and release of these gonadotropins (McCann 1980).

In spite of a significant increase in plasma levels of LH, the concentration of testosterone in the plasma of transgenic mice is not altered (Chandrashekar et al. 1988). Moreover, testicular levels of testosterone, release of testosterone by testicular fragments incubated in Krebs-Ringer bicarbonate buffer, and stimulation of the production of testosterone by hCG in vivo and in vitro appear to be entirely normal (Amador et al. 1988). The concentration of hCG (LH) receptors in the testis is also not altered (Amador et al. 1988). This finding is in sharp contrast to the situation in hyperprolactinemic DBA/2 mice, in which numbers of testicular receptors for LH are significantly depleted (Klemcke and Bartke 1981). Since plasma levels of LH are significantly elevated in both transgenic and hyperprolactinemic mice (Klemcke and Bartke 1981; Bartke et al. 1987; Chandrashekar et al. 1988), these results serve to emphasize major differences between the endocrine functions in these animals. Clearly, the effects of the transgene cannot be explained solely in terms of lactogenic activity of hGH in the mouse, and endocrine findings in transgenic males cannot be interpreted as yet another case of hyperprolactinemia. For example, maintenance of normal levels of testicular receptors for LH in transgenic mice but not in hyperprolactinemic mice (Klemcke and Bartke 1981; Amador et al. 1988) may be due to a direct stimulatory effect of IGF-1 on the binding of LH by testicular Leydig cells (Lin et al. 1986). Thus, stimulation of production of IGF-1 due to somatotropic activity of hGH may prevent at least one of the consequences of its lactogenic action.

In the course of routine maintenance of our breeding colony, we have noticed that most transgenic males cease to sire litters when they are approximately one year old, that is, at an age when most normal males are still fully competent to reproduce. Autopsy of sterile transgenic males revealed no gross abnormalities of the reproductive system and an abundance of motile spermatozoa in the vas deferens. When spermatozoa removed from the vasa deferentia and caudae epididymidis of four sterile transgenic males were used for surgical insemination of normal pseudo-pregnant fe-

TABLE 19-1 Mount Latency in Normal and Transgenic Males in the Presence of Ovariectomized Females Injected with Estradiol Plus Progesterone, Determined by Direct Observation (means ± SE)

Type of Male (number)	Mount Latency (minutes)[a]
Normal Males	
Young, naive (8)	28.1 ± 9.0
Transgenic Males	
Young, naive (10)	23.3 ± 8.2
Middle-aged, fertile (6)	29.2 ± 10.6
Middle-aged, sterile (4)	16.0 ± 14.5

[a]Animals that did not mount were assigned a value of 60 minutes (duration of test).

males (two females per male), seven of eight females became pregnant. This evidence for continued presence of functionally normal spermatozoa in males that were no longer able to impregnate females suggested the possibility of abnormalities in copulatory behavior. When old transgenic males were paired for one hour with ovariectomized females brought into behavioral estrus by injections of estradiol and progesterone, both fertile and sterile males mounted vigorously with latencies that suggested no abnormalities in sexual motivation ("libido") that could be related to age or fertility (Table 19-1). However, none of the sterile transgenic males ejaculated during the test or when left overnight with sexually receptive test females. When these males were housed with normal intact females for two weeks and the females were checked daily for the presence of vaginal plugs, none were found. From these findings, we conclude that age-related loss of fertility in transgenic males is due to their inability to ejaculate.

19.4 CONCLUSIONS

Expression of the hGH gene in transgenic mice leads to sterility of females and premature reproductive aging in males. Both sexes exhibit multiple alterations in neuroendocrine functions. Studies of domestic species reviewed elsewhere in this volume indicate that abnormalities of reproductive function may be common in transgenic animals that express various growth-promoting genes. It is unclear which of the findings in mMT/hGH mice may be applicable to other GH genes or to other species because hGH is lactogenic in the mouse and, therefore, capable of suppressing the endogenous release of PRL, and PRL is absolutely required for luteal function in this species (Bartke 1973). To answer this question and to elucidate further details of the mechanism of action of GH on reproductive and neuroendocrine functions, we are studying mice that express bGH or hGH-B ("variant") genes. Results obtained to date demonstrate the existence of reproductive and endocrine abnormalities in these animals even though neither bGH nor hGH-B appear to be lactogenic in the mouse.

REFERENCES

Amador, A.G., Johnson, L., Mayerhofer, A., et al. (1988) *Program of the 10th Annual. Testis Workshop*, Abstract. 70, p. 90.
Bartke, A. (1973) *Biol. Reprod.* 9, 379–383.
Bartke, A. (1974) *J. Endocrinol.* 60, 145–148.
Bartke, A. (1980) *Fed. Proc.* 39, 2577–2581.
Bartke, A., Morgan, W.W., Clayton, R.N., et al. (1987) *J. Endocrinol.* 112, 215–220.
Bartke, A., Steger, R.W., Hodges, S.L., et al. (1988) *J. Exp. Zool.* 248, 121–124.
Bartke, A., Tang, K., Bain, P., Yun, J.S., and Wagner, T. (1989) *J. Cell. Biochem.*, Suppl. 13B, p. 168, Abstract F100.
Berczi, I. (1988) *J. Immunol. Immunopharm.* 8, 186–194.
Chandrashekar, V., Bartke, A., and Wagner T.E. (1988) *Endocrinology.* 123, 2717–2722.
Colby, H.D. (1979) *Endocrinology.* 104, 1299–1303.
Crichton, D.N., and Shire, J.G.M. (1982) *Genet. Res.* 39, 275–285.
Forsyth, I.A., Folley, S.J., and Chadwick, A. (1965) *J. Endocrinol.* 31, 115–126.
Hammer, R.E., Brinster, R.L., Rosenfeld, M.G., Evans, R.M., and Mayo, K.E. (1985) *Nature* 315, 413–416.
Hammer, R.E., Palmiter, R.D., and Brinster, R.L. (1984) *Nature* 311, 65–67.
Hartree, A.S., Kovacic, N., and Thomas, M. (1965) *J. Endocrinol.* 33, 249–258.
Klemcke, H.G., and Bartke, A. (1981) *Endocrinology* 108, 1763–1768.
Lin, T., Haskell, J., Vinson, N., and Terracio, L. (1986) *Endocrinology* 119, 1641–1647.
Matthews, L.S., Hammer, R.E., Brinster, R.L., and Palmiter, R.D. (1988) *Endocrinology* 123, 433–437.
Mayerhofer, A., Bartke, A., and Wagner, T.E. (1988) *Anat. Rec.* 222, 62A–63A.
McCann, S.M. (1980) *Neuroendocrinology* 31, 355–363.
Morgan, W.W., Bartke, A., and Herbert, D.C. (1985) *Brain Res.* 335, 330–333.
Nonomura, M., Hoshino, K., Harigaya, T., Hashimoto, H., and Yoshida, O. (1985) *J. Endocrinol.* 107, 71–76.
Palmiter, R.D., Norstedt, G., Gelinas, R.E., Hammer, R.E., and Brinster, R.L. (1983) *Science* 222, 809–814.
Patel, Y.C., and Srikant, C.B. (1986) *Ann. Rev. Physiol.* 48, 551.
Prins, G.S. (1987) *Endocrinology* 120, 1457–1464.
Quaife, C.J., Mathews, L.S., Pinkert, C.A., et al. (1989) *Endocrinology* 124, 40–48.
Selden, R.F., Wagner, T.E., Blethen, S., et al. (1988) *Proc. Natl. Acad. Sci. USA* 85, 8241–8245.
Selden, R.F., Yun, J.S., Moore, D.D., et al. (1989) *J. Endocrinol.* 122, 49–60.
Shea, B.T., Hammer, R.E., and Brinster, R.L. (1987) *Endocrinology* 121, 1924–1930.
Shire, J.G.M. (1981) *Symposia Zoo. Soc. London* 47, 547–574.
Steger, R.W., Bartke, A., Goldman, B.D., Sores, M.J., and Talamantes, F. (1983) *Biol. Reprod.* 29, 872–878.

Wagner, T.E., Hoppe, P.C., Jollick, J.D., et al. (1981) *Proc. Natl. Acad. Sci. USA* 78, 6376–6380.
Yoshinaga, K., and Fujino, M. (1979) *Maternal Recognition of Pregnancy; CIBA Foundation Series*, 64, 85–110.
Yun, J.S., and Wagner, T.E. (1987) *The 10th Korea Symposium on Science and Technology* 3-1, 279–282.

PART V

Use of Transgenics in Animal Agriculture and Other Animals Used As Research Models

CHAPTER 20

Enhanced Growth Performance in Transgenic Swine

Carl A. Pinkert
David L. Kooyman
Timothy J. Dyer

In 1980 and 1981, six laboratories reported the successful transfer of foreign DNA into fertilized mouse ova (Brinster and Palmiter 1986). The potential of these studies was dramatically illustrated by the transfer of genes that caused accelerated growth in mice (Palmiter et al. 1982 and 1983). In general, the mice that integrated and expressed a growth hormone (GH) gene construct, which was composed of either the gene for rat growth hormone (rGH) or human growth hormone (hGH) regulated by a metallothionein (MT) promoter sequence, grew to almost twice the size, by 7 to 12 weeks of age, of the sex-matched littermate controls.

The success of the experiments with mice led to similar research aimed at enhancing growth performance in species of domestic animals raised for food (Table 20–1). Using gene constructs that caused accelerated growth in mice, these investigators were unable to detect enhanced growth of a similar

---

The authors gratefully acknowledge T.E. Wagner, R.L. Brinster, R.D. Palmiter, V.G. Pursel, S.H. Holtzman, M. Wieghart, J.J. Kopchick, R.W. Hanson, and M.M. McGrane for useful discussions and assistance. Supported in part by USDA 87-CRSR-2-3223 (C.A.P.) and funds from DNX Corporation.

TABLE 20-1 Gene Transfer Experiments to Improve Growth Performance in Swine

Gene	Ova Injected no.	Offspring no.	Offspring %	Integration no.	Integration %	Transgene Expression no.	Transgene Expression %	Germ Line Transmission	Reference
MT-hGH	268	15	6	1	0.4	—	—	—	Brem et al. 1985
MT-hGH	2035	192	9	20	1.0	11/18	61	5/6	Hammer et al. 1985
MT-pGH	423	17	4	6	1.4	1/6	17	2/2[a]	Vize et al. 1988
MT-bGH	2198	149	7	11	0.5	8/11	73	2/3	Pursel et al. 1987
PRL-bGH	289	23	8	4	1.3	2/4	50	4/4	Polge et al. 1989
PEPCK-bGH	1057	124	12	7	0.6	5/7	71	1/3	Wieghart et al. 1988, 1990
MLV-rGH	170	15	9	1	0.6	1/1	100	—	Ebert et al. 1988
MT-hGRF	2627	234	9	8	0.3	2/8	25	1/1	Pinkert et al. 1987b; Pursel et al. 1989b
Alb-hGRF	968	132	14	5	0.5	3/3	100	—	Pursel et al. 1989a

Gene constructions consisted of the regulatory sequences of metallothionein (MT), prolactin (PRL), Moloney murine leukemia virus (MLV), albumin (alb) or phosphoenolpyruvate carboxykinase (PEPCK) fused to structural genes for growth hormone (GH) or growth hormone releasing factor (GRF) of rat (r), human (h), bovine (b) or porcine (p) origin. Transgenic efficiency is based on the number of transgenic pigs produced and the percentage of transgenic pigs per injected ovum.

[a]Only pigs that did not express transgene mRNA exhibited germline transmission.

magnitude in the transgenic domestic animals that were produced. Efforts continue toward development of a greater understanding of the cascade of events involved in animal growth and the effective regulation of animal development through genetic engineering.

20.1 DISCUSSION

The potential of transgenic livestock with enhanced growth performance was suggested by pioneering experiments in 1982. Working with GH genes, Palmiter and coworkers (1982 and 1983) published their initial results on the significantly enhanced growth of transgenic mice. The first construct that they described included an rGH gene fused to a mouse MT-1 regulatory sequence (enhancer/promoter). Later, similar results were obtained with the MT regulatory sequence fused to a human or a bovine GH structural gene (Palmiter et al. 1983; Hammer et al. 1985). It is noteworthy that the genomic rGH and hGH genes, under the control of their own (endogenous) regulatory sequences, did not enhance growth of mice in early experiments (Wagner et al. 1983; Hammer et al. 1984). This negative result suggested that foreign genes for GH with endogenous promoters were subject to the normal regulatory mechanisms that influence a normal pattern of growth. The MT regulatory sequence was necessary to drive accelerated transcription of the gene, which resulted in elevated levels of GH peptide in the blood and enhanced growth in the mice. In addition, other regulatory sequences, namely, those associated with genes for albumin (Pinkert et al. 1987a), murine leukemia virus (Ebert et al. 1988), and phosphoenolpyruvate carboxykinase (PEPCK; McGrane et al. 1988), were successfully used to drive expression of a GH gene with resultant enhancement of the growth of transgenic animals. Recently, Behringer et al. (1988) succeeded in using a 310-bp 5' flanking sequence from rGH fused to a structural gene for hGH, to target expression of hGH to the pituitary.

While the increased body size of mice transgenic for GH genes is important, perhaps of greatest interest to livestock producers are the various morphological characteristics associated with the MT-GH transgenic mice. These characteristics include increased feed efficiency and growth rate, increased lean-muscle mass, and reduced fat. It has been estimated that a modest 10% increase in feed efficiency and growth rate could equate to an annual saving of one billion dollars to swine producers in the United States (Pinkert 1987).

In addition to the work with transgenic mice, studies of exogenous administration of GH to livestock further demonstrate the potential impact of GH-transgenic livestock on the agriculture industry. The exogenous administration of GH to pigs can increase deposition of protein and muscle mass while simultaneously causing a decrease in fat deposition both in the muscle and subcutaneous regions (Turman and Andrews 1955; Chung et al. 1985). Exogenous administration of GH to dairy cattle has also been shown to increase milk production dramatically (Peel et al. 1981 and 1983). GH is secreted from acidophiles in the anterior pituitary gland in response to the hypothalamic hormone; GH releasing factor (GRF). Exogenous administration of GRF to pigs (Kraft et al. 1985; Etherton and Walton 1986), cattle (Moseley et al. 1984), sheep (Hart et al. 1985), and rats (Spiess et al. 1983) stimulated release of GH from the pituitary with all the subsequent growth-performance characteristics discussed above.

In contrast to the favorable results obtained with GH-transgenic mice and exogenous administration of GH or GRF to large animals, results from studies of GH-transgenic pigs are less than optimal (Pursel et al. 1989b; Pinkert et al. 1990). There are a number of reasons why microinjection of GH genes into pig eggs has been disappointing. One problem faced is the relatively opaque cytoplasm in zygotes (Wagner et al. 1984). This problem was overcome to a limited degree by high-speed centrifugation to stratify cytoplasmic lipids and allow visualization of the pronuclei for microinjection (Wall et al. 1985). In addition, the relative timing of ovulation in superovulated commercial pigs is not precise and the conditions for culture in vitro of one-cell pig eggs are not well defined. Therefore, reasonable rates of conception with microinjected eggs necessitate the use of both one- and two-cell eggs with rapid transfer of microinjected ova to recipient females (Rexroad and Pursel 1988). Additionally, because it is difficult to regulate the levels of transgene-encoded protein in vivo, the potential phenotypic side-effects of pharmacological levels of GH in transgenic pigs were observed. The side effects reported have included altered endocrine profiles and metabolism, insufficient thermoregulatory capacity, joint pathology (lameness and arthritis), low libido and fertility, as well as an increased susceptibility to pneumonia (Pursel et al. 1989a and b; Rexroad and Pursel 1988). Similar pathology was also observed in pigs subjected to chronic administration of exogenous porcine GH (pGH) (Machlin 1972). It has been suggested that the growth performance and deleterious effects observed in transgenic pigs resulted from inadequate dietary intake and altered metabolic requirements of the transgenic pigs (Pursel et al. 1989a and b). The production of an MT-pGH transgenic pig with a higher feed efficiency and growth rate than those of control littermates with no phenotypic abnormalities has been reported (Vize et al. 1988). Unfortunately, this pig died of pneumonia at 18 weeks of age.

As in the case of exogenous administration of GH, it has been established that GH-transgenic pigs do exhibit significantly greater feed efficiency and less back fat (Pursel et al. 1989a and b; Vize et al. 1988; Wieghart et al. 1988 and 1990). The process of growth and development is a well-orchestrated and intricate cascade of events. Therefore, one cannot become discouraged by the problems associated with the present state of transgenic pig production. The great potential for benefit to both the consumer and producer of pork in the United States warrants continued research into the development of a suitable GH-transgenic pig.

20.2 PROSPECTS

In the future, three benefits from production of transgenic pigs are envisioned: (1) enhancement of production characteristics; (2) molecular farming; and (3) use of such pigs as models of human disease. Enhancement of production characteristics will necessitate tighter control of gene regulation in the production of transgenic swine. More precise control of transgenes, by a regulator such as that of PEPCK, may allow regulation in vivo of structural genes that has been unavailable to date (McGrane et al. 1988).

Beyond the area of enhanced growth performance, the possibility of decreasing an animal's susceptibility to disease offers another potential mechanism for improving production efficiency (Storb et al. 1986; Pinkert et al. 1989). In addition to enhancement of production characteristics and molecular farming (Van Brunt 1988), further projects with transgenic pigs may include production of models for human diseases. The pig has been used extensively as an alternative to human subjects because of the many physiological similarities between pigs and humans (Pond and Houpt 1978). Models for cardiac and metabolic disorders or cancer may be best addressed in a transgenic pig.

In order to enhance growth performance in swine using the transgenic model, a number of approaches may be utilized in the future. These approaches may involve new genes of choice as well as improved methodology; for example, the targeting of genes with greater specificity and improvement of techniques for gene transfer hold great promise (Pinkert et al. 1990).

A gene with greater specificity that may prove useful is that for insulin-like growth factor-1 (IGF-1). IGF-1 is a GH-dependent peptide that has potent mitogenic effects on a variety of cell types (Zapf and Froesch 1986; Claustres et al. 1987). Although its role in vivo is not fully defined, IGF-1 is produced in certain tissues in response to GH and is thought to be a mediator of the actions of GH. The induction of production of IGF-1 is most pronounced in liver tissue but induction also occurs in extrahepatic tissues, such as heart, thymus, diaphragm, and skeletal muscle (Roberts et al. 1986 and 1987; Hynes et al. 1987; Murphy et al. 1987). The largest relative induction (before versus after stimulation by GH) occurs in skeletal muscle. Such induction may indication potential for the use of IGF-1 in transgenic modeling to enhance muscle growth, decrease fat content in muscle, or affect muscle development in some other manner, in particular if IGF-1 can be expressed in a tissue-specific and developmentally appropriate manner. To facilitate such work, the genes for human, rat, and porcine IGF-1, as well as the gene for the human IGF-1 receptor have recently been cloned (see

Tavakkol et al. 1988). The use of IGF-1 in transgenic studies may further refine the developmental regulation of growth.

With regard to phenotypic alteration of mammals using gene transfer technology, researchers have been able to circumvent specific genetic defects in mice, for example, deficiencies in the immune response (LeMeur et al. 1985; Pinkert et al. 1985; Yamamura et al. 1985), a demyelinating phenotype (Readhead et al. 1987), dwarfism (Hammer et al. 1984), reproductive dysfunction (Mason et al. 1986), and β-thalassemia (Costantini et al. 1986). In these examples, the endogenous genes were still present. The transgenes were not targeted to replace the endogenous defective genes. Therefore, the transgenes and endogenous "defective" genes are able to segregate in future generations and the genetic defects are maintained in the gene pool.

Since the first stem-cell-derived transgenic animal was produced by Robertson et al. (1986), this field of research has seen rapid advance. Perhaps the most exciting work to date has involved site-directed mutagenesis via homologous recombination (Mansour et al. 1988), which has potentiated realistic approaches to gene therapy. Such techniques may allow researchers to replace endogenous genes for GH with altered genes that express peptides with higher affinity for GH receptors but remain under endogenous control.

20.3 CONCLUSION

It is currently unacceptable to allow transgenic food animals to enter the food chain. All animals that are the product of gene transfer experiments (whether or not they are transgenic) must be destroyed and are further banned from entering any food chain. Thus, independent of patent legislation, the net cost of such research is prohibitive. In the future, researchers will produce transgenic food animals that are significantly more efficient and of greater benefit to consumers. Perhaps transgenic food animals will then gain public acceptability. For the present, however, we continue our attempts at making existing techniques and future discoveries in this area more efficient, economically feasible, and acceptable to society.

REFERENCES

Behringer, R.R., Mathews, L.S., Palmiter, R.D., and Brinster, R.L. (1988) *Gene Develop.* 2, 453–461.

Brem G., Brenig, B., Goodman, H.M., et al. (1985) *Zuchthygiene* 20, 251–252.

Brinster, R.L., and Palmiter, R.D. (1986) *Harvey Lect.* 80, 1–38.

Chung, C.S., Etherton, T.D., and Wiggins, J.P. (1985) *J. Anim. Sci.* 60, 118–130.

Claustres, M., Chatelain, P., and Sultan, C. (1987) *J. Clin. Endocrinol. Metab.* 65, 78–82.

Costantini, F., Chada, K., and Magram, J. (1986) *Science* 233, 1192–1194.

Ebert, K.M., Low, M.J., Overstrom, E.W., et al. (1988) *Mol. Endocrinol.* 2, 277–283.
Etherton, T.D., and Walton, P.E. (1986) *J. Anim. Sci.* 63 (Suppl. 2), 76–88.
Hammer, R.E., Brinster, R.L., and Palmiter, R.D. (1985) *Cold Spring Harbor Symp. Quant. Biol.* 50, 379–387.
Hammer, R.E., Palmiter, R.D., and Brinster, R.L. (1984) *Nature* 311, 65–67.
Hammer, R.E., Pursel, V.G., Rexroad, C.E. Jr., et al. (1985) *Nature* 315, 680–683.
Hart, I.C., Chadwick, P.M.E., James, S., and Simmonds, A.D. (1985) *J. Endocrinol.* 105, 189–192.
Hynes, M.A., Van Wyk, J.J., Brooks, P.J., et al. (1987) *Mol. Endocrinol.* 1:233–242.
Kraft, L.A., Baker, P.K., Ricks, C.A., et al. (1985) *Dom. Anim. Endocrinol.* 2, 133–137.
LeMeur, M., Gerlinger, P., Benoist, C., and Mathis, D. (1985) *Nature* 316, 38–42.
Machlin, L. J. (1972) *J. Anim. Sci.* 35, 794–797.
Mansour, S.L., Thomas, K.R., and Capecchi, M.R. (1988) *Nature* 336. 348–352.
Mason, A.J., Pitts, S.L., Nikolics, K., et al. (1986) *Science* 234, 1372–1378.
McGrane, M.M., deVente, J., Yun, J., et al. (1988) *J. Biol. Chem.* 263, 11443–11451.
Moseley, W.M., Krabill, L.F., Friedman, A.R., and Olson, R.F. (1984) *J. Anim. Sci.* 58, 430–435.
Murphy, L.J., Bell, G.I., Duckworth, M.L., and Friesen, H.G. (1987) *Endocrinology* 121, 684–691.
Palmiter, R.D., Brinster, R.L., Hammer, R.E., et al. (1982) *Nature* 300, 611–615.
Palmiter, R.D., Norstedt, G., Gelinas, R.E., Hammer, R.E., and Brinster, R.L. (1983) *Science* 222, 809–814.
Peel, C.J., Bauman, D.E., Gorewit, R.C., and Snifen, C.J. (1981) *J. Nutr.* 111, 1662–1671.
Peel, C.J., Fronk, T.J., Bauman, D.E., and Gorewit, R.C. (1983) *J. Dairy Sci.* 66, 776–786.
Pinkert, C.A. (1987) *Proc. U.S. Anim. Health Assn.* 91, 129–141.
Pinkert, C.A., Dyer, T.J., Kooyman, D.L., and Kiehm, D.J. (1990) *Dom. Anim. Endocrinol.* 7, 1–18.
Pinkert, C.A., Manz, J., Linton, P., Klinman, N.R., and Storb, U. (1989) *Vet. Immunol. Immunopathol.* 23, 321–332.
Pinkert, C.A., Ornitz, D.O., Brinster, R.L., and Palmiter, R.D. (1987a) *Genes Develop.* 1, 268–276.
Pinkert, C.A., Pursel, V.G., Miller, K.F., Palmiter, R.D., and Brinster, R.L. (1987b) *J. Anim. Sci.* 65 (Suppl. 1), 260 (Abstract).
Pinkert, C.A., Widera, G., Cowing, C., et al. (1985) *EMBO J.* 4, 2225–2230.
Polge, E.J.C., Barton, S.C., Surani, M.A.H., et al. (1989) in *Biotechnology in Growth Regulation* (Heap, R.B., Prosser, C.G., and Lamming, G.E., eds.), pp. 189–199, Butterworths, London.
Pond, W.G., and Houpt, K. A. (1978) *The Biology of the Pig*, Cornell University Press, London.
Pursel, V.G., Miller, K.F., Bolt, D.J., et al. (1989a) in *Biotechnology of Growth Regulation*, (Heap, R.B., Prosser, C.G., and Lamming, G.E., eds.), pp. 181–188, Butterworths, London.
Pursel, B.G., Pinkert, C.A., Miller, K.F., et al. (1989b) *Science* 244, 1281–1288.
Pursel, V.G., Rexroad, C.E., Bolt, D.J., et al. (1987) *Vet. Immunol. Immunopathol.* 17, 303–312.

Readhead, C., Popko, B., Takahashi, N., et al. (1987) *Cell* 48, 703–712.
Rexroad, C.E. Jr., and Pursel, V.G. (1988) *11th Int. Congr. Anim. Reprod. and A.I.* Dublin, 5, 29–35.
Roberts, C.T., Brown, A.L., Graham, D.E., et al. (1986) *J. Biol. Chem.* 261, 10025–10028.
Roberts, C.T., Laskey, S.R., Lowe, W.L., Seaman, W.T., and LeRoith, D. (1987) *Mol. Endocrinol.* 1, 243–248.
Robertson, E., Bradley, A., Kuehn, M., and Evans, M. (1986) *Nature* 323, 445–448.
Spiess, J., Rivier, J., and Vale, W. (1983) *Nature* 303, 532–533.
Storb, U., Pinkert, C., Arp, B., et al. (1986) *J. Exp. Med.* 164, 627–641.
Tavakkol, A., Simmen, F.A., and Simmen, R.C.M. (1988) *Mol. Endocrinol.* 2, 674–681.
Turman, E.J., and Andrews, F.N. (1955) *J. Anim. Sci.* 14, 7–18.
Van Brunt, J. (1988) *Bio/Technology* 6, 1149–1154.
Vize, P.D., Michalska, A.E., Ashman, R., et al. (1988) *J. Cell Sci.* 90, 295–300.
Wagner, E.F., Covarrubias, L., Stewart, T.A., and Mintz, B. (1983) *Cell* 35, 647–655.
Wagner, T.E., Murray, F.A., Minhas, B., and Kraemer, D.C. (1984) *Theriogenology* 21, 29–45.
Wall, R.J., Pursel, V.G., Hammer, R.E., and Brinster, R.L. (1985) *Biol. Reprod.* 32, 645–651.
Ward, K.A., Franklin, I.R., Murray, J.D., et al. (1986) *Proc. World Cong. on Genetics Applied to Livestock Prod.*, 3, XII 6–21.
Wieghart, M., Hoover, J.L., Choe, S.H., et al. (1988) *J. Anim. Sci.* 66 (Suppl. 1), 266 (Abstract).
Wieghart, M., Hoover, J.L., McGrane, M.M., et al. (1990) *J. Reprod. Fertil. Suppl.*, 41, in press.
Yamamura, K., Kikutani, H., Folsom, V., et al. (1985) *Nature* 316, 67–69.
Zapf, J., and Froesch, E.R. (1986) *Hormone Res.* 24, 121–130.

CHAPTER 21

Production of Sheep Transgenic for Growth Hormone Genes

Caird E. Rexroad, Jr.

Injection of lambs with growth hormone (GH) increases feed efficiency and decreases the fat content of the carcass (Wagner and Veenhuizen 1978; Muir et al. 1983; Johnson et al. 1985; Pell et al. 1987). Introduction of transgenes to increase plasma levels of GH holds promise as a method for improving growth characteristics of sheep. Hammer et al. (1985) reported the introduction of a human GH fusion gene into the genome of a ewe lamb by microinjection of DNA into the pronuclei of fertilized ova. Subsequently, several investigators have reported the introduction of genes that increase circulating levels of GH in sheep (Pursel et al. 1987; Rexroad et al. 1988a, 1988b, and 1989; Ward et al. 1986). In this chapter, a discussion is presented of the integration, expression, and transmission of growth-related transgenes in sheep and an attempt is made to outline the possibilities and limitations of the insertion of genes for improved growth, as defined by these experiments.

21.1 INSERTION OF GENES

21.1.1 Microinjection of Genes

One limitation to the transfer of genetic material by pronuclear microinjection in farm animals is that yolk obstructs the view of pronuclei. In spite of the yolk, Hammer et al. (1985, 1986), using differential interference-contrast microscopy, reported that pronuclei were visible in about 80% of fertilized, one-cell, sheep eggs. Even so, pronuclei were not as clearly visible as those in mouse ova.

Poor visibility of sheep pronuclei may account in part for the low rate of production of transgenic lambs, particularly when this rate is compared to the rate of production of transgenic mice. The first attempt to produce transgenic sheep resulted in only 0.1% of the microinjected eggs yielding transgenic lambs (Hammer et al. 1985). More recently, 4.45% of eggs injected by the same group of researchers yielded transgenic lambs (Rexroad et al. 1988b; Table 21-1). The efficiency of production of transgenic sheep may improve with increased experience at microinjection. Other factors may have contributed to this improved efficiency, including the nature of the DNA that was microinjected.

21.1.2 Survival of Microinjected Embryos

Microinjection reduced the ability of fertilized one- and two-cell ova to develop to blastocysts in both pigs and sheep (Hammer et al. 1986). The reduction was greater in sheep than in pigs. Rexroad and Wall (1987) showed that each step of the microinjection process reduced the ability of one-cell sheep ova to develop to blastocysts. Immediate transfer resulted in the development of 86% of these ova into blastocysts. Injection of embryos with buffer resulted in development of 42 and 28% of ova into blastocysts in two studies, whereas injection with DNA resulted in only 18% of ova becoming blastocysts. In most studies of the production of transgenic sheep, the percentage of microinjected embryos that gave rise to offspring was below 20% (see Table 21-1). Three or four injected embryos have been transferred to

TABLE 21-1 Transfer of Genes into Sheep by Microinjection

Fusion Gene	Injected Embryos Transferred No.	Offspring		Transgenic Offspring		Transgenic Offspring Expressing	
		No.	%[a]	No.	%	No.	%
MThGH[b]	1032	73	7.1	1	0.10	—	—
MTbGH[c]	842	47	5.6	2	0.24	2	100
MTOgh[d]	436	27	6.2	1	0.23	0	—
MTHgrf[e]	435	63	14.5	9	2.07	1/7	14
TFbGH[f]	247	42	17.1	11	4.45	3/11	27

[a]Percent of embryos transferred.
[b]Hammer et al. 1985.
[c]Hammer et al. 1986; Pursel et al. 1987.
[d]Ward et al. 1986.
[e]Rexroad et al. 1988a and 1989.
[f]Rexroad et al. 1988b.

each recipient to ensure that a high percentage of recipients becomes pregnant (Hammer et al. 1985 and 1986; Pursel et al. 1987; Rexroad et al. 1989).

21.2 GROWTH-RELATED GENES FOR INSERTION INTO OVA

The first transgenic sheep contained a fusion gene that consisted of the mouse metallothionein-1 (mMT) promoter ligated to the coding sequence for human GH (Hammer et al. 1985). In several studies with sheep, the mMT promoter was joined to coding sequences for a GH gene or a gene for human GH releasing factor (hGRF) (see Table 21-1). Use of the MT promoter in a fusion gene permitted the secretion of GH without the normal feedback inhibition of the secretion of GH on the hypothalamo-pituitary axis. Furthermore, evidence suggested that intact genes for GH might not be functional. In mice made transgenic with an intact gene for GH from rat, no rat-specific GH was detectable in blood and growth was unaltered (Hammer et al. 1984).

mMT was ligated to coding sequences for hGH, bGH (bovine growth hormone), and hGRF to increase secretion of growth hormone in sheep (Hammer et al. 1985; Pursel et al. 1987; Rexroad et al. 1989; see Table 21-1). An Australian group successfully used a sheep metallothionein (MT) promoter ligated to the coding sequence for sheep GH to produce transgenic sheep (Ward et al. 1986; see Table 21-1). Another construct used to cause overproduction of GH in sheep involved the mouse transferrin promoter (Rexroad et al. 1988b), which caused expression of bGH in sheep. Tissue specificity and induction of the transgenes have not been well characterized.

21.3 EXPRESSION OF TRANSGENES

21.3.1 Proportion of Expressing Transgenics

Expression of transgenes in domestic animals has been reported in only a few cases (see Table 21-1). Expression of transgenes varied from about 50% for transgenic sheep that expressed MThGH or MTbGH to about 14% for transgenic sheep that expressed MThGRF (Hammer et al. 1985; Rexroad et al. 1988a and 1989). Poor expression of the MThGRF transgene may be related to its high content of cDNA or perhaps to low activity of the mMT promoter in species other than the mouse. The MThGRF gene was a minigene consisting of genomic DNA with one intron and cDNA, while the MThGH and MTbGH genes contained genomic DNA.

21.3.2 Levels of Expression of Transgenes

Expression of MTbGH in lambs resulted in one lamb having up to 718 ng/ml of GH in its blood, while expression of TFbGH resulted in concentrations of up to 206 and 289 ng/ml of GH, respectively, in the blood of two transgenic lambs (Rexroad et al. 1988b, 1989). The use of a sheep MT promoter ligated to the structural gene for sheep GH resulted in very high concentrations of GH (900–30,000 ng/ml) in the blood of four transgenic sheep (Ward et al. 1989). These observations suggest that species homology of the promoter may be important for expression of transgenes.

21.4 PHYSIOLOGY OF INTEGRATED GENES

In a MThGRF transgenic lamb with high plasma concentrations of hGRF, plasma concentrations of GH were not consistently elevated. Instead, the concentrations cycled through periods of very high and near-normal levels (Rexroad et al. 1989). Varying levels of ovine GH (oGH) may have resulted from the lamb becoming refractory to stimulation by hGRF of the production of GH. When the expressing MThGRF transgenic lamb and nonexpressing MThGRF transgenic lambs were challenged with injections of hGRF, the expressing lamb failed to respond to exogenous hGRF whereas the nonexpressing lambs responded. These observations agree with the finding of Della-Ferra et al. (1986), who observed that lambs became refractory to hGRF after multiple injections.

Introduction of genes for bGH into two sheep resulted in increased plasma concentrations of GH and insulin-like growth factor-1 (IGF-1) but failed to alter growth or feed efficiency (feed/gain; Rexroad et al. 1988b). In addition, glucose in the plasma of the lambs was elevated, suggesting a diabetogenic effect of the elevated levels of GH. Both lambs died before they reached 13 months of age and had high urinary glucose at death. Similarly, introduction of oGH genes into sheep with an MT promoter resulted in increased production of heat by the lambs, abnormalities in joints, and death prior to one year of age in three lambs (Ward et al. 1989). These observations indicate that overproduction of GH in sheep can be very detrimental to their health and that attempts to alter circulating levels of GH should be aimed at causing only small changes in production of GH or perhaps increases in levels that are periodic rather than chronic.

21.5 TRANSMISSION OF TRANSGENES

Transgenic sheep have transmitted the transgenes to their offspring. One ram transmitted a nonexpressing MThGRF gene to four of five fetuses; the copy number in the fetuses was the same as the copy number in the ram (Rexroad et al. 1989), suggesting that the gene had integrated into a single chromosome. Clark et al. (1987) reported the transmission of non-growth-related transgenes to offspring for two generations in sheep.

21.6 CONCLUSION

Growth-promoting genes can be inserted into domestic animals by microinjection into pronuclei. The transgenes are stably integrated and can be transmitted to progeny, usually on a single chromosome. Transgenes can be expressed, but the percentage of animals that expresses the transgene may depend on the gene construct used. Nonexpressed genes have little if any effect on physiology, but expressed genes for GH and GRF have profound effects, which are generally detrimental. Attempts to alter growth characteristics of sheep through production of transgenics is limited by our ability to regulate expression of transgenes. Considerable effort needs to be expended to determine those regulatory sequences that might aid in the control of gene expression and to characterize in greater detail the physiological effects of the GH molecule.

REFERENCES

Clark, A.J., Simons, P., Wilmut, I., and Lathe, R. (1987) *Trends Biotechnol.* 5, 20–24.
Della-Fera, M.A, Buonomo, F.C., and Baile, C.A. (1986) *Domes. Anim. Endocrinol.* 3, 153–164.
Hammer, R.E., Palmiter, R.D., and Brinster, R.L. (1984) *Nature* 311, 65–67.
Hammer, R.E., Pursel, V.G., Rexroad, C.E., Jr., et al. (1986) *J. Anim. Sci.* 63, 269–278.
Hammer, R.E., Pursel, V.G., Rexroad, C.E. Jr., et al. (1985) *Nature* 315, 680–683.
Johnson, I.D., Hart, I.C., and Butler-Hogg, B.W. (1985) *Anim. Prod. Sci.* 41, 207.
Muir, L.A., Wien, S., Duquette, P.F., Rickes, E.L., and Cordes, E.H. (1983) *J. Anim. Sci.* 56, 1315.
Pell, J.M., Blake, L.A., Elcock, C., et al. (1987) *J. Endocrinol.* 112, No. 63 (Abstract).
Pursel, V.G., Rexroad, C.E. Jr., Bolt, D.J., et al. (1987) *Vet. Immunol. Immunopathol.* 17, 302–312.
Rexroad, C.E. Jr., Behringer, R.R., Bolt, D.J., et al. (1988b) *J. Anim. Sci.* 66 (Suppl. 1), 267 (Abstract).
Rexroad, C.E. Jr., Nammer, R.E., Bolt, D.J., et al. (1989) *Mol. Reprod. Develop.* 1, 164–169.
Rexroad, C.E. Jr., Pursel, V.G., Hammer R.E., et al. (1988a) in *Beltsville Symposia in Agricultural Research [12], Biomechanisms Regulating Growth and Development* (Steffens, G.L., and Rumsey T.S., eds.), p. 97, Kluwer Academic Publishers, Boston.
Rexroad, C.E., Jr., and Wall, R.J. (1987) *Theriogenology* 27, 611–619.
Wagner, J.F., and Veenhuizen, E.L. (1978) *J. Anim. Sci.* 47–397 (Abstract).
Ward, K.A., Franklin, I.R., Murray, J.D., et al. (1986) *Proc. 3rd World Congr. Genet. Appl. Livestock Prod.* 12, 6–21.
Ward, K.A., Nancarrow, C.D., Murray, J.D., et al. (1989) *J. Cell. Biochem.* 13B, 164 (Abstract).

CHAPTER 22

Production of Transgenic Cattle by Pronuclear Injection

Kenneth R. Bondioli
Karen A. Biery
Keith G. Hill
Karen B. Jones
Franko J. De Mayo

Pronuclear injection of one-cell embryos has been used extensively to produce transgenic mice and, to a lesser extent, rabbits, pigs, and sheep. One-cell bovine embryos were injected with a pRSVCAT construct (see below) and fetal tissues were collected for analysis of integration and expression. Integration was found to have occurred in seven fetal samples and expression of the transgene was observed in placental tissue of one sample.

Germline genetic transformation of laboratory mice was first reported by Gordon et al. (1980). Reports from several other laboratories followed quickly, establishing the fact that cloned, foreign DNA would be stably integrated into the genome and that Mendelian germline transmission occurred after microinjection into the pronuclei of fertilized mouse embryos

---

The authors wish to acknowledge the efforts of the staff of the commercial embryo transfer unit of Granada Genetics, Inc., in the collection and transfer of embryos used in these studies, and they thank Deborah Sheehan for her help in the preparation of this manuscript.

(Brinster et al. 1981; Costantini and Lacy 1981; Gordon and Ruddle 1981; Wagner et al. 1981).

The method of choice for the transfer of cloned DNA into mammalian embryos is pronuclear injection. The technique for pronuclear injection of mouse embryos has been described by several authors (Gordon et al. 1980; Hogan et al. 1986; Brinster et al. 1985; Palmiter and Brinster 1986; Hammer et al. 1986; Murray et al. 1988). The production of transgenic rabbits, pigs, sheep, and cattle by pronuclear injection has also been reported (Hammer et al. 1985 and 1986; Brem et al. 1986; Church et al. 1986; Ward et al. 1986; Nancarrow et al. 1987; Simons et al. 1988; Biery et al. 1988).

A number of variables that affect the success rate of pronuclear injection have been studied by Brinster et al. (1985). Walton et al. (1987) have studied specific variables associated with pronuclear injection that affect the lysis of mouse and sheep embryos. The production of transgenic cattle by pronuclear injection has been reported by two laboratories (Church et al. 1986; Biery et al. 1988). In this chapter we describe our experiences with the production of transgenic cattle by pronuclear injection. Some of these data have been previously reported in abstract form (Biery et al. 1988).

22.1 MATERIALS AND METHODS

22.1.1 Collection of Embryos

One-cell bovine embryos were collected from excised oviducts of Holstein and cross-bred, beef-type donors after induction of superovulation by injections of follicle-stimulating hormone (FSH). The superovulation regimen consisted of eight injections of FSH (FSH-P, Schering) at 12-hour intervals over a five-day period (treatment with FSH was initiated on the evening of day 1). A 25-mg dose of prostaglandin F2f (Lutalyse, Upjohn) was given with the fifth and sixth injections of FSH and 4,000 IU of human chorionic gonadotropin (HCG, Butler) was administered at noon or during the evening of day 5 (44 or 48 hours after the first injection of prostaglandin). The cows were artificially inseminated on the morning and evening of day 6, and ova were collected during the morning of day 7 (36 to 40 hours after administration of HCG). Ova were recovered after surgical removal of the ovaries and oviducts by flushing the excised oviducts with Dulbecco's modified phosphate-buffered saline supplemented with 0.4% bovine serum albumin, 100 units/ml penicillin and 100 μg/ml streptomycin. Ova remained in this medium throughout the injection procedure.

22.1.2 Pronuclear Injection

One-cell embryos were centrifuged at 15,000 $\times$ g to aid in the visualization of pronuclei (Wall et al. 1985). Injections were performed under an inverted microscope (Zeiss or Nikon) equipped with Nomarski optics at 400$\times$ magnification. Embryos were placed in a drop of 3–5 μl of medium on a glass depression slide. A 2-μl drop of diluted DNA was placed next to this drop and both drops were covered with paraffin oil. Manipulations were carried out with the aid of Leitz or Narishige micromanipulators. Embryos were held by a fine, polished, holding pipet (outer diameter approximately 150 μm) and the plasma and nuclear membranes were penetrated with an injection pipet (tip diameter approximately 1–2 μm). DNA was injected with the aid of an oil-filled microsyringe (Stolting) until expansion of the pronucleus was visible (25% expansion or less). In the majority of embryos, the most visible pronucleus was injected. If both pronuclei were equally visible, the larger pronucleus was injected.

22.1.3 Preparation of DNA

The plasmid pRSVCAT, containing the bacterial gene for chloramphenicol acetyltransferase (CAT) under the control of the Rous sarcoma virus long terminal repeat (Gorman et al. 1982a), was linearized by digestion with restriction endonuclease *Nde*I (New England Bio Labs). The DNA fragment was then purified by extraction with phenol:chloroform:isoamyl alcohol (25:24:1) and precipitated with ethanol. The purified fragment was resuspended in 10 mM Tris (pH 7.5), 0.25 mM EDTA. The concentration of the DNA was then determined spectrophotometrically, and the appropriate concentrations of DNA were made. The DNA was diluted for injection in 10 mM Tris (pH 7.5), 0.25 mM EDTA. The majority of the embryos were injected with DNA at a concentration of 2 ng/μl. In three separate experiments the concentration of injected DNA was varied to determine the effect of concentration of DNA on the viability of embryos and the rates of incorporation.

In the first experiment, concentrations of DNA of 1, 2, and 4 ng/μl were injected. In the second experiment, 2, 25, and 50 ng/μl were injected, and in the third experiment, 2, 50, and 100 ng/μl of DNA were injected. Within each experiment different embryo donors and different injection technicians were equally represented in all treatments. The three experiments were conducted at different times, and, therefore, embryo donors and injection technicians varied between the experiments.

22.1.4 Culture and Transfer of Embryos

After pronuclear injection all embryos were cultured to the morula or blastocyst stage. Culture from the one-cell to the morula or blastocyst stage was conducted either in the ligated sheep oviduct or in vitro. Those embryos cultured in the sheep oviduct were transferred (either with or without embedding in double-layer cylinders of agar) into ewes whose oviducts had been ligated at the uterotubal junction according to procedures described by Willadsen (1979). Embryos remained in the sheep oviduct for six days, at which time they were flushed from the oviducts and evaluated for development to the morula or blastocyst stage. Those embryos cultured in vitro were placed in 20-μl drops that consisted of a 1:1 mixture of Brinster's Mouse Ovum Culture Medium (BMOC) III medium (Brinster 1972) and bovine oviductal fluid. Two to three microliters of bovine oviductal epithelial cells were added to each drop and drops were covered with silicon oil (Dow Corning).

Embryos were cultured at 38°C in an atmosphere of 5% CO_2 and 95% air for seven days and evaluated for development to the morula or blastocyst stage. In experiments where the effect of various concentrations of DNA was investigated, the method of culture was the same for all treatments.

All embryos that developed to the morula or blastocyst stage were transferred by a nonsurgical technique to synchronous recipients (the age of embryos was assumed to be seven days in all cases). In most cases only one embryo was transferred to each recipient but in a few cases two or three embryos were transferred to a single recipient in the same uterine horn.

22.1.5 Collection and Analysis of Tissue

Established pregnancies were carried to 60 to 65 days of gestation, at which time fetuses were surgically removed with a sample of placental tissue.

Fetal and placental tissue were separated and both were snap-frozen in liquid nitrogen. Tissue remained frozen until analysis for incorporation and expression of the transgene.

Fetal and placental tissue was homogenized in STE [0.15 M NaCl, 20 mM Tris (pH 7.8), 10 mM EDTA, that contained 1% sodium dodecylsulfate and 0.2 mg/ml proteinase K (Boehringer Mannheim Biochemicals)]. The homogenate was incubated at 37°C overnight. DNA was then purified by extraction of the homogenate with an equal volume of phenol:chloroform:isoamyl alcohol (25:24:1), followed by precipitation with ethanol. The DNA was resuspended in STE and subjected to a second precipitation. The DNA was finally resuspended in 10 mM Tris (pH 7.8), 1 mM EDTA.

In order to check for the presence of pRSVCAT in the bovine DNA, the isolated DNA was subjected to Southern blot analysis (Southern, 1975). Ten micrograms of bovine DNA was digested with restriction endonuclease *Eco*RI (Boehringer Mannheim Biochemicals) and separated by agarose gel electrophoresis. The DNA was then transferred to nitrocellulose paper and probed with ^{32}P-labeled pRSVCAT. The pRSVCAT was labeled using a nick-translation kit (Bethesda Research Laboratories). Bovine DNA isolated from uterine tissue served as the negative control. pRSVCAT was added to bovine DNA at one copy and 10 copies per haploid genome to serve as positive controls.

Digestion of pRSVCAT with *Eco*RI should yield fragments of 2.5, 2.1, and 0.34 kb in size. If the *Nde*I digest of pRSVCAT DNA was integrated as a head-to-tail concatamer, this pattern should be observed. If the trans-

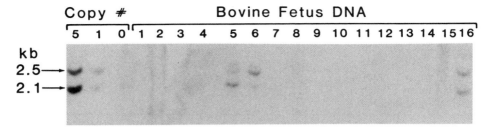

FIGURE 22-1 Southern blot of bovine fetal DNA. Southern analysis was performed as described in the text. Three lanes on the left were loaded with control bovine DNA with pRSVCAT added at the copy number indicated. Fetal samples in lanes 5, 6, and 16 demonstrate integration of the pRSVCAT construct.

gene was integrated as a single copy, only the 2.1- and 0.34-kb bands would be expected. All integrations of bovine DNA observed were of multiple copies. The 0.34-kb fragment was too small to be detected by our blotting procedure (Figure 22-1).

In order to examine expression of the transgene, fetal limb and placental tissues from four positive and two negative samples were assayed for the presence of CAT activity. The tissues were homogenized in 0.25 M Tris (pH 7.5) at 4°C in a Brinkman Polytron. The homogenates were denatured at 65°C for 5 minutes and centrifuged for 10 minutes in an Eppendorf microfuge. The concentrations of protein in the supernatant were determined with a Biorad Protein Assay kit by the Bradford-Lowry procedure. CAT activity was determined in aliquots of supernatant that contained 160 μg of total protein (Gorman et al. 1982b). The assay was incubated at 37°C overnight with 0.5 μCi dichloroacetyl 1-1-2-^{14}C-chloramphenicol, (50 mCi/mmol) (New England Nuclear) and 1.3 mM acetylcoenzyme A. Samples were placed on a thin-layer chromatography (TLC) plate and chromatographed in a 19:1 mixture of chloroform and methanol. The TLC plate was then dried and exposed to X-ray film.

22.2 RESULTS AND DISCUSSION

The procedures for superovulation and collection described above yield approximately 10 ova per collection, of which approximately 70% are fertilized (data not shown). Approximately 90% of the ova collected were one-cell embryos with the remaining 10% being two- to four-cell embryos. Results for embryos injected with the pRSVCAT construct are presented in Table 22-1.

The difference between the proportion of embryos that were fertilized and the proportion of embryos injected (70% fertilized versus 55% injected) can be attributed primarily to the collection of very late pronuclear embryos

TABLE 22-1 Pronuclear Injection of Bovine Embryos with the RSVCAT Construct

No. of ova collected	6207
No. of ova injected	3398 (55%)
No. of ova developing	561 (17%)
No. of fetuses collected	206 (37%)
No. of transgenic fetuses	7 (3.4%)

in which nuclear membranes have started to break down prior to pronuclear fusion.

Microinjection of bovine embryos greatly reduces their viability. Nonmanipulated one-cell embryos cultured in either the sheep oviduct or the culture system in vitro, as described above, develop to the morula or blastocyst stage at a rate of 50-60% (data not shown). This reduction in survival of embryos is similar to that observed for microinjected embryos of pig and sheep (Hammer et al. 1986; Rexroad and Wall 1987; Rexroad and Pursel 1988).

The overall efficiency of integration (seven transgenics from 3398 injected embryos equals 0.2%) is considerably lower than that typically obtained with mouse embryos (Brinster et al. 1981; Brinster et al. 1985) and is somewhat lower than that observed with sheep (Hammer et al. 1985; Nancarrow et al. 1987; Murray et al. 1988; Rexroad and Wall 1987) and pigs (Brem et al. 1986; Hammer et al. 1985; Rexroad and Pursell 1988). The lower overall efficiency of gene incorporation was due both to the low proportion of injected embryos that produced fetuses (6%) and to the low proportion of collected fetuses in which the transgene was incorporated. The results from the DNA concentration experiments are presented in Table 22-2.

Injection of embryos with concentrations of DNA greater than 2 ng/μl did not increase the rates of integration of the DNA. These results are similar to those obtained by Brinster et al. (1985) with mouse embryos. They reported an increase in integration frequency when concentrations of DNA were increased from 0.01 ng/μl to 1 ng/μl, but rates of integration did not increase with concentrations of DNA of above 1 ng/μl. The results in Table 22-2 show that we obtained a total of four transgenic fetuses. The remaining three transgenic fetuses shown in Table 22-1 were from embryos injected with a concentration of DNA of 2 ng/μl.

Viability of embryos varied considerably between the three experiments, a result that is not surprising since embryo donors and injection technicians differed between the experiments. Within each experiment, however, viability was consistently reduced when concentrations of DNA greater than 4 ng/μl were injected. This observation is also in agreement with those of Brinster et al. (1985).

TABLE 22-2 Effect of the Concentrations of Injected DNA on Survival of Embryos and Frequency of Integration

DNA Concentration	No. Injected	No. Developing	(%)	No. Fetuses	% of Transferred	% of Injected	No. Transgenic
Experiment I							
1 ng/μl	120	27	(23)	12	44	10.0	1
2 ng/μl	120	23	(19)	14	61	11.6	2
4 ng/μl	120	21	(18)	11	52	9.2	0
Experiment II							
2 ng/μl	124	21	(17)	7	33	5.6	0
25 ng/μl	118	14	(12)	4	29	3.4	1
50 ng/μl	100	6	(6)	3	50	3.0	0
Experiment III							
2 ng/μl	55	16	(29)	6	38	10.9	0
50 ng/μl	60	11	(18)	3	27	5.0	0
100 ng/μl	52	8	(15)	3	38	5.7	0

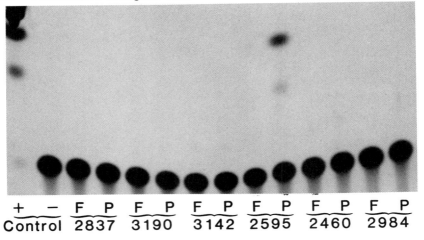

FIGURE 22-2 Autoradiograph of TLC plate from assay of chloramphenicol acetyltransferase (CAT) activity. CAT activity was assayed in fetal and placental tissue as described in the text. Sample number 2595 showed evidence of CAT activity in the placental tissue.

When four of the samples that were positive for integration were assayed for expression of the transgene, one sample (sample no. 2595) showed evidence of expression in the placental tissue (Figure 22-2). No expression was observed in fetal limb tissue from this same sample. The reason for detectable expression in the placental tissue but not in the fetal tissue is unknown.

These results demonstrate that injection of linearized DNA constructs into the pronuclei of bovine embryos can result in integration of the DNA in fetal tissue. This integration occurred without rearrangement of the gene construct and the transgene was capable of being expressed in the placental tissue of one conceptus.

REFERENCES

Biery, K., Bondioli, K., and DeMayo, F. (1988) *Theriogenology* 29, 224 (Abstract).
Brem, G., Breneg, B., Goodman, H., et al. (1986) *Proc. 3rd World Congr. Genet. Appl. Livestock Prod.*, pp. 45–60, Lincoln, NE.
Brinster, R. (1972) in *Growth, Nutrition and Metabolism of Cells in Culture, Vol. II* (Rothblat, G. and Cristofalo, V., eds.), pp. 251–286, Academic Press Inc., New York.
Brinster, R., Chen, H., Trumbauer, M., et al. (1981) *Cell* 27, 223–231.
Brinster, R., Chen, H., Trumbauer, M., Yagle, K., and Palmiter, R. (1985) *Proc. Natl. Acad. Sci. USA* 82, 4438–4442.
Church, R., McRae, A., and McWhir, J. (1986) *Proc. 3rd World Congr. Genet. Appl. Livestock Prod.*, pp. 133–138, Lincoln, NE.

Costantini, F., and Lacy, E. (1981) *Nature* 294, 92–94.
Gordon, J., and Ruddle, F. (1981) *Science* 214, 1244–1246.
Gordon, J., Scangos, G., Plotkin, D., Barbosa, J., and Ruddle, F. (1980) *Proc. Natl. Acad. Sci. USA* 77, 7380–7384.
Gorman, C., Merling, G., Willingham, M., Paston, I., and Howard, B. (1982a) *Proc. Natl. Acad. Sci. USA* 79, 6777–6781.
Gorman, C., Moffat L., and Howard, B. (1982b) *Mol. Cell. Biol.* 2, 1044–1051.
Hammer, R., Pursel, V., Rexroad, C., Jr., et al. (1985) *Nature* 315, 680–683.
Hammer, R., Pursel, V., Rexroad, C., Jr., et al. (1986) *J. Anim. Sci.* 63, 269–278.
Hogan, B., Costantini, F., and Lacy, E. (1986) *Manipulating the Mouse Embryo: A Laboratory Manual*, Cold Spring Harbor Laboratory, Cold Spring Harbor, NY.
Murray, J., Nancarrow, L., and Ward, K. (1988) in *Proc. 11th Inter. Congr. Anim. Reprod., A.I.*, Dublin, Ireland, 5, 20–27.
Nancarrow, C., Marshall, J., Murray, J., Hayelton, I., and Wood, K. (1987) *Theriogenology* 27, 263 (Abstract).
Palmiter, R., and Brinster, R. (1986) *Annu. Rev. Genet.* 20, 465–499.
Rexroad, C., Jr., and Pursel, V. (1988) *Proc. 11th Inter. Congr. Anim. Reprod.*, A.I., Dublin, Ireland, 5, 29–35.
Rexroad, C., Jr., and Wall, R. (1987) *Theriogenology* 27, 611–619.
Simons, J., Wilmut, I., Clark, A., et al. (1988) *Biotechnology* 6, 179–183.
Southern, E. (1975) *J. Mol. Biol.* 98, 508–518.
Wagner, T., Hoppe, P., Jollick, J., et al. (1981) *Proc. Natl. Acad. Sci. USA* 78, 6376–6380.
Wall, R., Pursel, V., Hammer, R., and Brinster, R. (1985) *Biol. Reprod.* 32, 645–651.
Walton, J., Murray, J., Marshall, J., and Nancarrow, C. (1987) *Biol. Reprod.* 37, 957–967.
Ward, K., Franklin, I., Murray, J., et al. (1986) *Proc. 3rd World. Congr. Genet. Appl. Livestock Prod.*, pp. 6–21, Lincoln, NE.
Willadsen, S. (1979) *Nature* 277, 298–300.

CHAPTER 23

Methods for the Introduction of Recombinant DNA into Chicken Embryos

John J. Kopchick, Ed Mills, Charles Rosenblum,
Joyce Taylor, F. Macken, F. Leung,
Jim Smith, Howard Chen

Introduction of recombinant DNA into the pronucleus of fertilized mouse eggs by microinjection or into early mouse embryos by use of murine retroviral vectors is a routine procedure in many laboratories (Palmiter and Brinster 1986). Also, direct microinjection of recombinant DNA into the fertilized eggs of rabbits, sheep, and pigs has resulted in the production of germline transgenic animals (Hammer et al. 1985). This technology, employing recombinant DNA, may be of significant value in agricultural animal husbandry.

We have been interested in generating transgenic poultry, in particular chickens, that can transmit recombinant DNA to their offspring. However, introduction of recombinant DNA into the germline of chickens involves a technology different from that used for insertion of genes into mammalian

---

We would like to thank all individuals at Merck, Sharpe and Dohme Research Laboratories, Rahway, NJ, and at Hubbard Farms, Walpole, NH, who participated in these studies. Also, thanks are due to Diana Guthrie for her help in preparing the manuscript.

embryos. Avian ova are telolecithal and are normally fertilized approximately 30 minutes after ovulation. Cell division occurs in the oviduct for approximately 20 hours before oviposition.

At this time, the embryo is comprised of approximately 60,000 pluripotent cells (Spratt and Haas 1961), which are collectively called the blastoderm (Kochav et al. 1980; Eyal-Giladi and Kochav 1976; Mindur et al. 1985). The presence of a large yolk and multiple pronuclei makes direct microinjection of DNA impractical.

The data presented below show our attempts to produce transgenic chickens using the following strategies: introduction of recombinant DNA into unfertilized chicken ova; transfection of recombinant DNA into day 1 blastoderms; and the use of a replication competent, transformation defective avian retrovirus in attempts to infect chicken precursor germ cells. The latter technique was successful in the production of a commercial strain of germline transgenic chickens.

23.1 MATERIALS AND METHODS

23.1.1 Surgical Procedure for Introduction of Recombinant DNA into Unovulated Follicles

Adult Hubbard Leghorn females were anesthetized with Vetalar (Parke-Davis) (1 ml/100 g body weight). The feathers were removed from the left, dorsolateral thoracic region ventrally from the backbone to the tip of the ribs. The area was disinfected with chlorhexidine and an incision made with a no. 22 surgical blade. The intercostal muscles were separated from the sixth and seventh vertebral ribs. A Weitlaner retractor (4 1/2 inches, 2 × 3 prongs) was used to open the body cavity. A probe was used to puncture the air sac and expose the avian follicles. Individual, maturing, ovarian follicles were manipulated and injected with 1 to 25 µg pBGH-4 DNA (Figure 23-1) as near to the oocyte as possible through a 30-gauge, 1-inch needle.

The musculature of the body cavity was closed with eight to 10 sutures. Zinc bacitracin was applied and the skin was closed with 16 to 20 sutures. After suturing, zinc bacitracin was applied to the entire area.

After surgery, Combiotic (Pfizer, 2 ml per bird) was injected for three days postsurgery to prevent infection and to promote the healing process.

23.1.2 Transfection of Day-1 Blastoderms

Eggs (day-1 blastoderms) were collected immediately after laying and stored at 15°C. A portion of the shell (approximately 3 × 3 mm) was removed without disturbing the membrane. After localization of the blastoderm, 1 to 100 µg of pBGH-4 DNA was injected in the immediate area of the blastoderm through glass capillary pipettes. The opening in the shell was then

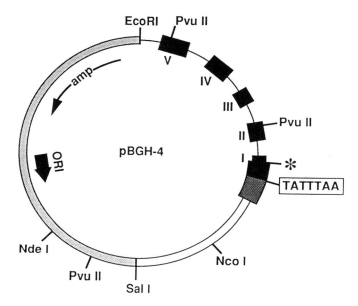

FIGURE 23-1 pBGH-4 DNA. The *bGH* gene, containing five exons (black boxes) and four intervening sequences, was ligated to the long terminal repeat (LTR) of Rous sarcoma virus (RSV). The location of the RSV "TATA box" is shown. The two DNA molecules were joined at the *Bst*EII site of RSV and the *Bam*HI site of *bGH*, which is indicated by an asterisk. The DNA was inserted between the *Sal*I and *Eco*RI sites of pBR322 (dots).

covered with cellophane tape and sealed with paraffin wax. Eggs were incubated at 38°C until hatching. This protocol did not effect hatchability.

23.1.3 Cell Culture and Propagation of Virus

Primary chicken embryo fibroblasts (CEF) were derived from Spafas C/E day-10 embryos. CEF were dispersed by trypsinization and then plated in Ham's F10-medium 199 (1:1) that contained 5% Tryptose phosphate broth and 4% newborn calf serum.

Avian leukosis virus was propagated after transfection of CEF with cloned viral DNA that encoded a transformation-defective Schmidt-Ruppin A strain of Rous sarcoma virus (RSV), obtained from Steve Hughes, National Cancer Institute NCI (Figure 23-2). The *src* gene in this virus was deleted and replaced by a unique site of cleavage for the restriction enzyme *Cla*I. Virus was propagated for 7 to 14 days. Medium was removed from the cultures of CEF and cellular debris removed by centrifugation (10,000 × g, 10 min.). Viral titers were determined by end-point dilution assay and aliquots of the medium were frozen at −70°C (Lannett and Schmidt 1980).

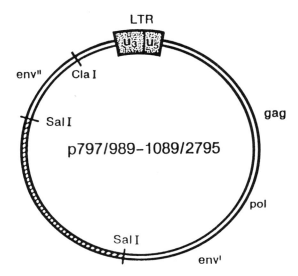

FIGURE 23-2 pSRA DNA. Plasmid p797/989-1089/2795 was obtained from S. Hughes (National Cancer Institute, Frederick, MD) and has been renamed pSRA. The entire RSV genome has been inserted at the *Sal*I site in pBR322 (hatching). A unique *Cla*I site is found following the *env* gene. The RSV LTR and viral *gag, pol*, and *env* genes are indicated. The *env* gene has been separated by the pBR322 insert (hatching).

23.1.4 Recombinant DNA

DNA used for transfection of day-1 blastoderms and unfertilized ova contained the bovine gene for growth hormone (bGH) ligated to the long terminal repeat (LTR) of RSV and was termed pBGH-4 (see Figure 23-1). This plasmid DNA directs expression of bGH in cultured mammalian and avian cells (Kopchick et al. 1985). The antibody used to detect bGH in culture fluids or in serum does not react with endogenous chicken growth hormone. pBGH-4 DNA was introduced into day-1 chicken blastoderms as either covalently closed, circular supercoils, or as linear molecules (after cleavage with *Nde*I) in the presence or absence of diethylaminoethyl (DEAE)-Dextran (200 μg/ml) or as calcium phosphate precipitates (Kopchick et al. 1985). The amount of DNA used was 5, 50, or 100 μg. The volume of the injected material ranged from 10 to 50 μl in a TE (10 mM Tris, 1 mM ethylenediaminetetra-acetic acid (EDTA), pH 7.5) buffer. The DNA was placed near or under the blastoderm by injection through a glass capillary pipette.

23.1.5 Preparation of Chicken Genomic DNA

Chicken genomic DNA, derived from various tissues, was prepared by homogenization of 50 mg of tissue or 50 μl of packed red blood cells in a 3 ml solution of proteinase K (0.5 mg/ml), 1% sodium dodecyl sulphate (SDS),

5 mM EDTA, and 10 mM Tris, pH 7.5. The solution was incubated overnight at 42°C. After two extractions with equal volumes of phenol and chloroform, DNA was precipitated with ethanol at −20°C. Precipitated DNA was collected by centrifugation, washed with 70% ethanol, and suspended in 2.0 ml of sterile H_2O.

23.1.6 Digestion with Restriction Endonucleases and Hybridization of DNA

Genomic DNA (10 µg) was treated with *Hind*III or *Pvu*II (50 units) overnight at 37°C and resolved by electrophoresis on a 1% agarose gel. After electroblotting to Gene Screen Plus membranes (Dupont/NEN) at 42 volts in 0.33 × tris, acetate, EDTA (TAE) (0.04 M Tris-acetate, 0.002 M EDTA, pH 7.5) membranes were exposed to a ^{32}P-labeled DNA hybridization probe (6×10^5) cpm/ml of hybridization solution, according to the suggestion of the manufacturer.

23.1.8 Hybridization Probes

For detection of the bGH gene in chicken genomic DNA, a *Pvu*II DNA fragment from pbGH-4 was used (see Figure 23–1). For detection of exogenous or endogenous proviral DNA sequences in chicken genomic DNA, a 250-bp fragment derived from the U_3 region of an exogenous viral clone, SR-RSV-A, or a 146-bp fragment derived from the U_3 region of an endogenous viral clone, RAV-O was used. The exogenous and endogenous cloned viral DNA was kindly provided by L. Crittendon (USDA, East Lansing, MI). Probes with specific activities of at least 1×10^9 cpm/µg were used in the analyses.

23.1.8 Titering of Viruses

Suspensions of viruses were rapidly thawed at room temperature and 10-fold dilutions were prepared in cold cell-culture medium. One milliter of each dilution was inoculated into each of six primary cultures of CEF (in 60-mm dishes). Cultures were incubated at 37°C in an atmosphere of 5% CO_2 and 95% air for five days, at which time cells were transferred to fresh medium and grown for an additional five days. Then the supernate from lysed cells was assayed for p27 antigen by enzyme-linked immunosorbent assay (ELISA). Antibodies raised in rabbits against p27 and the same antibodies conjugated to horseradish peroxidase, which were used in the assay, were obtained from Spafas (Storrs, CT). Viral titers were calculated as described elsewhere (Lannett and Schmidt 1980).

23.1.8.1 Assays of Virus in Blood, Meconium, and Cloacal Swabs
Isolation of virus from 1-day-old chicks was achieved by expressing meconium

from the chick into a tube that contained 1.5 ml of cell culture medium supplemented with five times the normal levels of antibiotics. Samples were frozen at $-70°C$. Samples were thawed, centrifuged ($10,000 \times g$), and 0.4 ml of supernatant was inoculated into primary cultures of CEF derived from Spafas C/E embryos. Cultures were incubated at 37°C in an atmosphere of 5% CO_2 and 95% air for five days, transferred to fresh medium, and grown for an additional five days. Supernatants from lysed cells were assayed for p27 antigen by ELISA.

Older birds were screened for virus by analysis of either blood samples or cloacal swabs. Cloacal swabs were placed in 1.5 ml of cold cell-culture medium that contained five times the normal levels of antibiotics and frozen at $-70°C$. Both blood and cloacal-swab samples were inoculated into primary cultures of CEF and cultures were treated as described above.

23.1.8.2 Injection of the Blastoderm with Retrovirus Eggs to be injected were collected and stored at 15°C (large end up) for seven to eight days to allow the air cell to form and the blastoderm to rotate to the large end of the eggs.

The air cell was outlined in pencil and a small section (approximately 1 cm²) of shell was removed. Eggs were examined, and all eggs in which the blastoderm could be seen through the membrane were saved for injection.

Approximately 10^5-10^6 viral particles were injected adjacent to and slightly below the blastoderm. Volumes of 10–50 µl were injected using glass pipettes or repeating micropipettes with a 27-gauge needle. All injection procedures were carried out in biological safety cabinets.

After injection, eggs were sealed with cellophane tape and the large ends dipped in melted paraffin wax. Eggs were set in a commercial incubator and, at hatching, chicks were screened for virus in samples of meconium. All virus-positive chicks were saved for DNA hybridization assays.

23.2 RESULTS

23.2.1 Introduction of p-BGH4 DNA into Unfertilized Ova

After surgery, which resulted in exposure of the chicken ovary (see Materials and Methods, section 23.1.1), pBGH-4 DNA was injected through a capillary pipette into five to 10 of the largest ova. Between 1 and 100 µg of linear pBGH-4 DNA was injected either in the presence of DEAE-Dextran or as calcium phosphate precipitate. Approximately 1,000 ova were injected in 100 artificially inseminated hens. Of the 400 offspring produced and analyzed, none were found to possess DNA specific for pBGH-4 in the genomic DNA in red blood cells (RBCs) as determined by blot hybridization analyses, and none expressed detectable bGH in serum as assayed by radioimmunoassay (RIA) (Kopchick et al. 1985).

23.2.2 Transfection of pBGH-4 DNA into Day-1 Blastoderms

pBGH-4 DNA was introduced into day-1 blastoderms as circular or linear DNA molecules in the presence or absence of DEAE-Dextran (200 μg/ml) or as calcium phosphate precipitates. Of 1,500 animals generated via this protocol, seven animals were found to have elevated levels of serum bGH (between 20–40 ng/ml). pBGH-4 DNA was not found in genomic DNA in RBCs of these seven animals as determined by hybridization blot analysis. These animals were grown to maturity and mated with control animals. Offspring (approximately 200) derived from these crosses were negative when analyzed for serum bGH and also negative for pBGH-4 DNA in their genomic DNA. The seven founder chickens were sacrificed and various tissues removed (liver, muscle, brain, kidney, intestine). Hybridization blot analysis of preparations of DNA from these tissues showed that three chickens possessed DNA that hybridized to the *Pvu*II fragment of bGH, namely, animal E1198, E1196, and C7161 (Figure 23-3). Animal E1198 possessed about 30 copies of bGH DNA per cell in the preparation from muscle and approximately one to two copies per cell in preparations of DNA from liver and kidney cells. Animal E1196 possessed more than 30 copies of bGH DNA per cell in DNA derived from brain tissue. Animal C7161 contained bGH DNA in genomic DNA derived from intestinal tissue (>30 copies per cell) and kidney tissue (approximately two copies per cell).

23.2.3 Infection of Precursors to Chicken Germ Cells with an Exogenous Avian Leukosis Retrovirus

Since germline transgenic chickens containing the *bGH* gene were not generated by the above "transfection" procedures, we attempted to use avian retroviruses as gene-transducing agents. However, before placing genes such

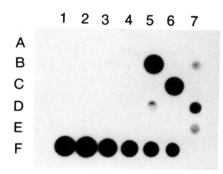

FIGURE 23-3 Dot-blot analysis. Hybridization dot-blot analysis of DNA derived from seven chickens that expressed bGH in serum (20–50 ng/ml). Lanes 1 to 7 were loaded with 20 μg of DNA derived from chickens H5117, F560, E1189, D9862, E1198, E1196, and C7161, respectively. Rows A to E represent tissues from which the DNA was derived: A, liver; B, muscle; C, brain; D, kidney; and E, intestine. Row F represents pBGH-4 DNA at copy numbers ranging from 80 (lanes 1 and 2) to 5 (lane 6). A *Pvu*II fragment of bGH-specific DNA was used as hybridization probe, as described in Materials and Methods.

as *bGH* into avian leukosis virus (ALV) vectors, we determined the ability of the ALV-subgroup A virus to infect chicken germ cells or precursor germ cells. Infection of chicken germ cells by exogenous virus rarely occurs. Recent evidence indicates that exogenous virus from subgroup A can indeed infect germ cells of a laboratory strain of chicken that is free of endogenous virus (Salter et al. 1987). The chicken line used in the present studies possesses numerous sequences of endogenous retroviral DNA (data not shown).

The general protocol involves the generation of a group of founder female chickens (G_0) that "shed" exogenous virus via the oviduct (Figure 23-4). Embryonic cells of the developing blastoderm of G_1 individuals are exposed to the exogenous virus as the blastoderm passes through the oviduct. Male offspring (G_1) produced from such viremic females were assayed for their ability to pass the exogenous viral DNA sequences to offspring (G_2). Since viremic males do not (or very rarely) congenitally pass virus to offspring, viremic G_2 offspring produced from viremic G_1 males would indicate germline transmission of the exogenous virus to the G_2 individuals.

Exogenous virus (10^5–10^6 particles per 50 μl) was introduced into either day-1 chicken blastoderms or day-1 chickens. G_0 female chickens were selected that shed exogenous virus via the oviduct. The frequency of pro-

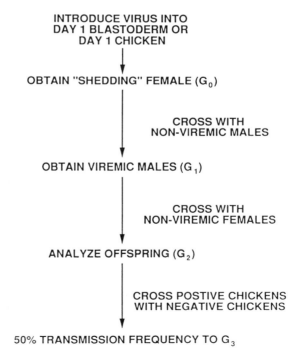

FIGURE 23-4 Protocol for generation of G_2 transgenic chickens.

duction of these G_0 "shedding" females is shown in Tables 23–1 and 23–2.

G_0 virus-positive females were crossed with nonviremic males and G_1 offspring were screened for exogenous virus. G_1 viremic males were selected by assaying serum for infectious virus and RBC-DNA by hybridization blots. The frequency of production of the G_1 males is shown in Tables 23–1 and 23–2. Dot-blot hybridization analyses of these G_1 males revealed that less than one exogenous provirus was present per genome in the RBC DNA (data not shown). This result is consistent with the hypotheses that G_1 males are mosaic relative to exogenous retroviral DNA sequences.

Fourteen males positive for exogenous virus were mated individually with females negative for exogenous virus. Hybridization blot analysis of RBC-DNA derived from these animals is shown in Figure 23–5. Both exogenous and endogenous virus-specific DNA hybridization probes were used (as described in Materials and Methods). These crosses generated a total of 316 G_2 animals whose RBC-DNA was analyzed by DNA hybridization. A typical analysis of DNA samples from RBC of G_2 chickens is shown in Figure 23–6 in which G_2 individuals that were positive for exogenous virus can be identified. Generally, positive birds were found to possess one or two copies of the exogenous viral DNA per cell. Frequencies of transmission of sequences of exogenous retroviral DNA from G_1 males to G_2 individuals

TABLE 23–1 Efficiencies of Production of G_0 Viremic Spafas Females (Shedders) and G_1 Viremic Males

	Positive/Total Number	Efficiency of Production (%)
G_0 viremic females (shedders)	3/50	6.00
G_1 viremic males	28/240	11.66
G_1 viremic males that pass gene to progeny	7/28 (minimum)	25.00
	7/9 (to date)	77.77

TABLE 23–2 Efficiency of Production of G_0 Viremic Hubbard Line 139 Females (Shedders) and G_1 Viremic Males

	Positive/Total Number	Efficiency of Production (%)
G_0 viremic females (shedders)	6/132	4.54
G_1 viremic males	62/797	7.78
G_1 viremic males which pass gene to progeny	4/62 (minimum)	6.40
	4/5 (to date)	80.00

284 Introduction of Recombinant DNA into Chicken Embryos

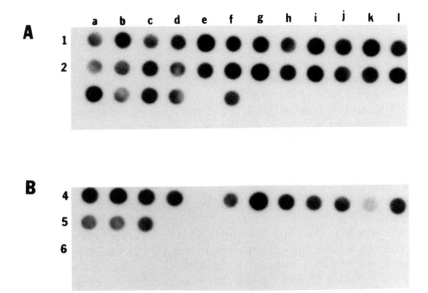

FIGURE 23-5 Probes specific for exogenous and endogenous virus. DNA (20 μg) from 15 viremic males, rows 1 and 4 (lanes a to l) and rows 2 and 5 (lanes a to c), and 14 nonviremic females, rows 2 and 5 (lanes d to l) and row 6 (lanes a, b, c, d, and f), was applied to hybridization membranes and exposed to hybridization probes specific for (A) endogenous avian retroviruses or (B) exogenous avian retroviruses.

ranged from 0–40% (Table 23-3). Frequencies of transmission from G_2 individuals to subsequent generations (G_3, G_4, and G_5) were approximately 50% (data not shown).

23.2.4 Southern Blot Analysis of DNA from Transgenic Chickens

For the insertion of foreign DNA into a chicken germline, viremic female chickens (G_0) were produced that shed exogenous virus via the oviduct. As the fertilized egg passes through the oviduct, embryonic cells are exposed to the virus. If precursors to the germ cell are infected by the virus, offspring (G_1) should be generated that are capable of passing the viral DNA to the next generation (G_2). A pedigree study by Southern blot analysis would verify this prediction and also reveal valuable information about the timing of viral infection during early embryonic development of the G_1 chickens and the frequency of transmission of viral DNA to the progeny (G_2). For this purpose, DNA was isolated from a G_1 viremic male that was mated to a nonviremic female. DNA from 10 G_2 viremic offspring that resulted from this cross was isolated and subjected to Southern blot analysis. Results are

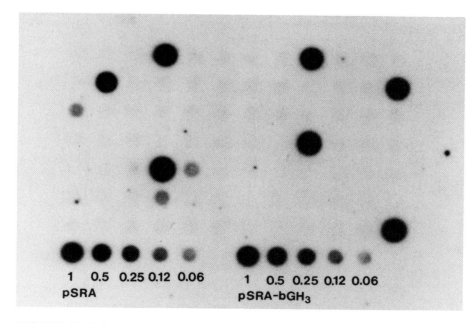

FIGURE 23-6 Dot-blot analysis of G_2 offspring. DNA (20 μg) from 96 G_2 offspring was examined for DNA specific for exogenous virus by hybridization analysis. From this figure, seven individuals were determined to be positive. Lighter dots (equivalent to about 0.12 copies of DNA) were seen in three cases. DNA standards of pSRA (ranging from 1 copy/cell to 0.06 copies/cell) were spotted on the hybridization membrane.

shown in Figure 23-7. No discrete bands or smears of DNA can be seen in the case of the G_0 chicken (lane A), indicating that the circulating RBC were not infected by the virus, even though the female chickens were shedding the virus through the oviduct. A smear of DNA can be detected in the case of a G_1 chicken (lane B), indicating heterogeneous sites of integration of viral DNA in RBC DNA. All 10 viremic G_2 offspring gave discrete bands of junction fragments, indicating that all the RBCs of a given chicken had the same site of viral integration, thereby confirming the predicted transmission through the germline. While most of the G_2 viremic chickens had one site of integration, one chicken (lane J) had two sites of integration, as confirmed by analysis of RBC DNA with another restriction enzyme. Lane J in Figure 23-8 clearly reveals four junction fragments indicative of two viral integration events. Furthermore, none of the 10 viremic G_2 chickens gave the same pattern of integration, indicating that each chicken was derived from a different germ-cell precursor. These results showed that although a significant proportion of the precursors to germ cells were infected by the virus, the infection took place at a relatively late stage in the differ-

TABLE 23-3 Frequency of Transmission of Viral DNA by Viremic G_1 Males to G_2 Progeny[a]

G_1 Male (Band no. 1)	Nonviremic female (Band no.)	G_2 Progeny Positive/Total	Frequency of transmission (%)
95-289	39	2/20	10.0
95-286	43	1/14	7.1
95-294	50	3/28	10.7
96-259	36	4/17	23.5
96-260	38	3/13	23.0
96-246	42	7/28	25.0
96-273	44	0/23	0
96-268	49	14/39	35.9
97-211	46	0/29	0
90-627	47	0/25	0
91-258	48	3/22	13.6
91-218	40	2/5	40.0
91-181	41	1/28	3.4
98-232	37	1/21	4.8

[a]The numbers reflect results of progeny tested to date. Testing of progeny from these birds, as well as of other G_1 viremic males, is ongoing.

entiation of germ cells. Evidence for this statement is provided by the fact that no identical patterns of integration of DNA can be detected in the offspring. A pedigree study involving another viremic G_1 and four viremic G_2 offspring gave similar results.

To obtain further information about viral DNA distribution and patterns of integration of viral DNA in various tissues, a G_1 viremic bird and its G_2 viremic offspring were sacrificed. DNA isolated from various tissues was subjected to Southern blot analysis (Figure 23-9). Results showed that all tissues from G_2 viremic chicken gave identical patterns of DNA, indicating that all tissues were derived from a single embryonic cell that contained integrated provirus, as would be expected from transmission from G_1 to G_2 birds through the germline. However, no discrete bands were detected in the DNA from various tissues of the G_1 viremic chicken (data not shown).

Comparison of Southern blot patterns of DNA from G_2 and G_3 transgenic chickens in three different lines showed identical patterns for parent and offspring (Figure 23-10), indicating stable transmission of the integrated genes from parent to offspring.

23.2.5 Resistance of Transgenic Chickens to a Challenge by Subgroup A Rous Sarcoma Virus

We inserted viral DNA (from a transformation-defective, Schmidt-Ruppin A strain of RSV) into the germline of chickens. Transgenic chickens were expected to be refractory to subsequent infection by ALV since they produce

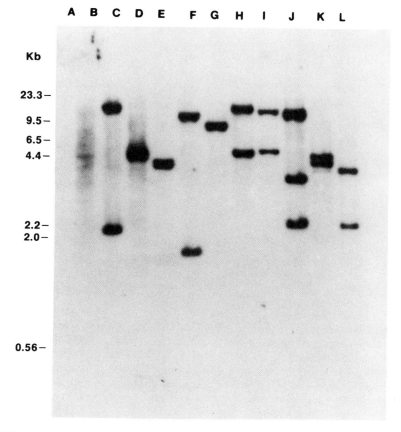

FIGURE 23-7 Southern hybridization analysis of DNA from family members from the G_0 and G_1 generations. Chicken genomic DNA was prepared as described in Materials and Methods. After digestion with a 10-fold excess of *Hind*III, 10 μg of DNA was resolved by electrophoresis on a 1% agarose gel. After transfer to a membrane filter, hybridization was carried out as described in Materials and Methods using a probe specific for exogenous virus. Lanes A through L represent the following individuals: A, 559 G_0 female; B, 96-268 G_1 male; C, 13-279 G_2; D, 13-285 G_2; E, 13-297 G_2; F, 13-291 G_2; H, 13-306 G_2; I, 13-314 G_2; J, 13-311 G_2; K, 13-289 G_2; and L, 13-296 G_2. Sizes of marker proteins are indicated in kilobase pairs (kb).

the envelope protein that interferes with the ability of the virus to infect chicken cells. To test this possibility, we challenged chickens with Subgroup A strain of RSV and compared the development of tumors in transgenic and nontransgenic chickens.

Our results indicated that such transgenic chickens are resistant to Subgroup A virus (Table 23-4). We challenged chickens by injection of Subgroup A virus in one wing and Subgroup B virus in the other wing.

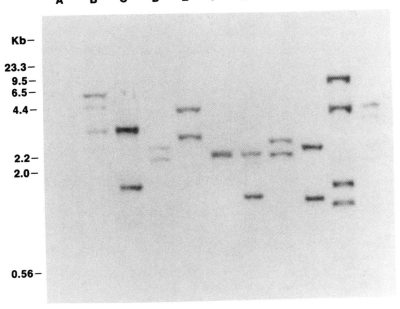

FIGURE 23-8 Southern hybridization analysis of DNA from family members from the G_1 and G_2 generations. Chicken genomic DNA was prepared as described in Materials and Methods. After digestion with a 10-fold excess of *Pvu*II, 10 μg of DNA was resolved by electrophoresis on a 1% agarose gel. After transfer to a filter membrane, hybridization was performed as described in Materials and Methods using a probe specific for exogenous virus. Lanes A to K represent the following individuals: A, 96-268 G_1 male; B, 13-279 G_2; C, 13-285 G_2; D, 13-297 G_2; E, 13-289 G_2; F, 13-291 G_2; G, 13-295 G_2; H, 13-306 G_2; I, 13-314 G_2; J, 13-311 G_2; and K, 13-296. Sizes of marker proteins are indicated in kilobase pairs (kb).

TABLE 23-4 Comparison of Development of Tumors in Transgenic and Control Chickens Challenged with Subgroup A and B Viruses

			Incidence of Tumors			
			Subgroup A		Subgroup B	
	Number of Birds	Viremic	2 wks	3 wks	2 wks	3 wks
Transgenic	24	+	0/24	1/23	19/24	18/23
Control	11	−	9/11	9/11	9/11	8/11

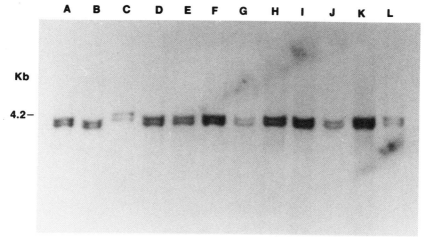

FIGURE 23-9 Southern hybridization analysis of genomic DNA from different tissues of a G_2 offspring, 13-297 G_2. Chicken genomic DNA from various tissues of chicken 13-285 G_2 was prepared as described in Materials and Methods. After digestion with a 10-fold excess of *Hind*III, 10 μg of DNA was resolved by electrophoresis on a 1% agarose gel. After transfer of fractionated DNA to a filter membrane, hybridization was carried out as described in Materials and Methods using a probe specific for exogenous virus. DNA was derived from the following tissue: A, ovary; B, muscle; C, lung; D, heart; E, intestine; F, spleen; G, brain; H, kidney; I, liver; J, stomach; K, oviduct; and L, bursa. A molecular weight marker of 4,200 base pairs is indicated (4.2 kb).

Transgenic chickens were resistant to Subgroup A virus, but susceptible to Subgroup B virus (see Table 23-4).

23.3 DISCUSSION

We have attempted to insert DNA into the chicken genome using three different approaches. Our first attempt involved insertion of recombinant DNA into unfertilized follicles of female chickens. Although the surgical procedures and manipulation of follicles did not affect ovulation and the capacity of the egg for fertilization, no offspring positive for recombinant DNA were obtained by this protocol. One possible explanation of these negative results is that the inserted DNA was degraded in the unfertilized egg. When DNA is exposed to this type of tissue and incubated for 1 hour at 37°C in vitro, the DNA is cleaved (data not shown). We have not attempted to introduce viruses using this protocol.

The second protocol for the production of transgenic chickens involved insertion or "transfection" of recombinant DNA (pBGH-4) into day-1 fer-

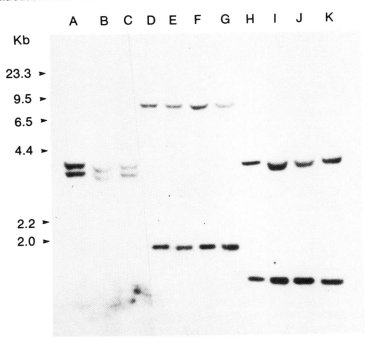

FIGURE 23-10 Southern hybridization analysis of DNA from G_2 parents and G_3 offspring. Preparation of genomic DNA, digestions with *Hind*III, electrophoresis on 1% agarose gel, transfer, and hybridization protocols are described in Materials and Methods. DNA from G_2 parents 1, 2, and 3 is represented in lanes A, D, and H, respectively. DNA from G_3 offspring of G_2 parent 1 are shown in lanes B and C; G_3 offspring from G_2 parent 2 are shown in lanes E, F, and G; and G_3 offspring from G_2 parent 3 are shown in lanes I, J, and K. Molecular sizes of marker proteins are indicated in kilobase pairs (kb).

tilized embryos. At this stage of development, the embryo is composed of approximately 60,000 cells (Spratt and Haas 1961). Furthermore, primordial germ cells are indistinguishable from somatic cells (Ginsberg and Eyal-Giladi 1986 and 1987). Therefore, in all probability, transgenic animals generated by the procedure would be mosaic relative to the cells that take up the DNA and also mosaic relative to the site(s) of integration of the DNA. Recombinant DNA, pBGH-4 (1–100 µg), was introduced into the fertilized blastoderm as either circular or linear molecules in the presence or absence of DEAE-Dextran or as a calcium phosphate precipitate. Seven animals expressed detectable levels of bGH in serum; however, none of the birds passed the bGH-specific DNA to their offspring. Presumably, the seven birds possessed the DNA in somatic tissue. To test this possibility, the birds were sacrificed and DNA from various tissue was analyzed. In three of the birds, DNA was detected. Bird E1196 had approximately 30 copies of bGH-specific DNA in cells from brain tissue. Bird E1198 had approximately 30 copies of bGH-specific DNA per cell in muscle tissue and one to two copies of bGH-specific DNA per cell in kidney tissue. Animal C7161 had bGH-specific DNA in kidney tissue (approximately five copies per cell), intestinal tissue (approximately one copy per cell), and muscle tissue (approximately one copy per cell). Thus, the ability to "transfect" cells of chicken blastoderm appears to be relatively inefficient. However, a few of the cells of certain blastoderms apparently took up and integrated the DNA and the foreign DNA was expressed in the adult birds. It appears that different types of cell in the blastoderm can take up DNA, since a variety of tissues in the positive birds contained the DNA. It must be pointed out that, in our analyses, only birds that secreted bGH into serum at detectable levels (>10 ng/ml) or birds with bGH-specific DNA in blood cells were classified as positive. Cells of the blastoderm that integrated DNA but did not differentiate to blood cells or cells that could not express the bGH gene would result in the adult birds being classified as negative. In order to establish whether precursors to germ cells in the blastoderm could be "transfected" by this procedure, all progeny of the 1,500 birds generated in this experiment should have been tested. However, facilities were not available for such an ambitious experiment. Nonetheless, we have established that cells in the blastoderm can be "transfected" with recombinant DNA, even though the frequency of transfection precursors to germ cells was very low.

Our third protocol for our attempts at generating germline transgenic animals was developed using exogenous retroviruses. The ability of retrovirus to infect somatic embryonic cells of the chicken (Hippenmeyer et al. 1988) or precursors to germ cells and the passage of viral DNA to offspring of a laboratory strain of chicken free of endogenous virus have been documented (Salter et al. 1987). In addition, the ability of a replication-defective reticuloendotheliosis virus vector to infect germ cells has been demonstrated (Bosselman et al. 1989). In our study we attempted to determine whether a replication-competent subgroup A retrovirus could infect precursors to germ cells of a bird that carried many copies of endogenous viral DNA sequences (data not shown).

A Schmidt-Ruppin A viral vector was used in this study. The general protocol of the study is presented in Figure 23–4. Viruses (10^4–10^5 infectious units) were brought in contact with day-1 chickens or day-1 blastoderms and adult female birds were assayed for the ability to pass the exogenous virus via the oviduct. Our ability to generate the G_0 "shedding" females was approximately 6% (see Table 23–1) for Spafas leghorn chickens and approximately 5% for Hubbard 139 leghorn birds (see Table 23–2). The ability of the "shedding" females to infect fertilized embryos (G_1) as they passed through the oviduct was approximately 6% for the Spafas birds and 4% for the Hubbard birds. When these results were tabulated for production of G_1 viremic males, the production efficiencies were 11.6% (see Table 23–1) and 7.8% (see Table 23–2), respectively. G_1 males positive for exogenous virus were mated to virus-negative females and G_2 offspring were analyzed. Eleven of 14 viremic G_1 males were able to pass exogenous virus-specific DNA to offspring with transmission frequencies ranging from 3.4 to 40% (see Table 23-3). These frequencies are higher than those reported previously (Salter et al. 1987) and may be attributed to the route of production of G_1 males; that is, through the oviduct of G_0 females.

Transmission of the exogenous viral genes from G_1 males to G_2 individuals was confirmed by Southern hybridization analyses. DNA derived from G_1 viremic males showed a heterogenous mixture of molecules that hybridized to the probe specific for the exogenous virus. The probe used in this study was derived from the U_3 region of the viral LTR (see Materials and Methods). This observation is consistent with the occurrence of many different infectious events in the embryonic cells of the developing G_1 individual. DNA derived from G_2 offspring generated two strongly hybridizing, junction fragments in the case of each of the virus-positive individuals (see Figures 23-7 and 23-8). These results demonstrate that the viral DNA sequences were passed via the germline from G_1 sires to G_2 progeny. These results also confirm the mosaic nature of viral integration into the germ cells of G_1 males. Presumably, many different precursors to germ cells were infected with the virus. We have not found identical patterns of hybridizing bands for any G_2 progeny derived from any single G_1 male (data not shown). Furthermore, in one case, it seems that one precursor to germ cells was infected by two viruses (see Figure 23-7, lane J; Figure 23-8, lane J) since four junction bands were seen on Southern analysis.

If the G_2 birds that carried exogenous virus were produced by way of a sire's germ cell that possessed an integrated copy of an exogenous retrovirus, all cells derived from a G_2 positive bird should possess the same pattern of integration of viral DNA. This lineage was confirmed by analysis of DNA isolated from a variety of tissues of a G_2 bird. The Southern hybridization pattern was identical for all tissues examined (see Figure 23-9). Furthermore, G_2 positive birds produced in this manner should pass the exogenous viral DNA to approximately 50% of progeny, and the Southern hybridization pattern in the positive offspring should be identical to that of the sire. This identity was confirmed by analyses of families where G_2 sires were mated with nonviremic females and the G_3 individuals were analyzed. Approximately 50% of offspring possessed exogenous viral DNA (data not shown). In addition, the Southern hybridization patterns from the three G_2 sires and their respective G_3 positive offspring were identical (see Figure 23-10).

Thus, exogenous replication-competent subgroup A retrovirus can infect germ cells or precursors to germ cells in a developing embryo. Apparently, the virus infects the precursors to germ cells at a relatively late stage in development. The efficiency of production of G_1 males that pass genetic information to offspring via the germline is relatively high. Therefore, the necessary techniques are now available for genetic manipulation of the germline of chickens.

REFERENCES

Bosselman, R.A., Hsu, R.Y., Boggs, T., et al. (1989) *Science* 243, 533–535.
Eyal-Giladi, H., and Kochav, S. (1976) *Develop. Biol.* 49, 321–337.
Ginsburg, M., and Eyal-Giladi, H. (1986) *J. Embryol. Exp. Morph.* 95, 53–71.
Ginsburg, M., and Eyal-Giladi, H. (1987) *Development* 101, 209–219.
Hammer, R.E., Pursel, V.G., Rexroad, C.E., et al. (1985) *Nature* 315, 680–683.
Hippenmeyer, P.J., Given, G.K., and Highkin, M.K. (1988) *Nucl. Acid Res.* 16, 7619–7632.
Kochav, S., Ginsburg, M., and Eyal-Giladi, H. (1980) *Develop. Biol.* 79, 296–308.
Kopchick, J.J., Malavarca, R.H., Livelli, T.J., and Leung, F.C. (1985) *DNA* 4, 23–31.
Lannett, E.H., and Schmidt, N.J. (1980) in *Diagnostic Procedures for Viral and Rickettsial Infections (4th ed.)*, pp. 2–65, Aneum Public Health Association, New York.
Mindur, C., Krawczyk, E., and Wezyk, S. (1985) *Bri. Poult. Sci.* 26, 527–529.
Palmiter, R.D., and Brinster, R.L. (1986) *Annu. Rev. Genet.* 20, 465–499.
Salter, D.W., Smith, E.J., Hughes, S.H., Wright, S.E., and Crittendon, L.B. (1987) *Virology.* 157, 236–240.
Spratt, N.T., and Haas, H. (1961) *J. Exp. Zool.* 147, 57–93.

CHAPTER 24

Gene Transfer in Fish

Kevin S. Guise
Anne A.R. Kapuscinski
Perry B. Hackett, Jr.
Anthony J. Faras

Classical systems of genetic selection have led slowly to the development of strains of several commercially important species of fish that have improved growth characteristics. While heavy emphasis has been placed on the rainbow trout (*Salmo gairdneri*), many species have been virtually ignored in terms of such selection processes. Progress in the manipulation of growth characteristics via selection is a slow process, with rapid enhancement in early generations followed by a slowing of the rate of improvement as the natural variation is "exhausted" by optimization of the choice of possible alleles at each relevant locus. Effects of single genes on growth are difficult to follow during such selection because of the selective pressure placed on all the genes that participate in determining the selected parameter, be it feeding efficiency, growth rate, or age at maturity. The effect of single genes may be assayed by the injection, usually intramuscular, of the product of a growth-related gene, for example, growth hormone (GH). Attempts have been made for years to improve the growth rate of fish by treatment with pituitary extracts or purified bovine, ovine, porcine, or piscine GH, and all have produced increased rates of growth, some of which have been truly dramatic. Beginning with the studies of Pickford and

Thompson (1948) and Pickford (1954), it has been shown that hypophysectomized fish (killifish, *Fundulus heteroclitus*) responded to injections of piscine or bovine GH. Higgs et al. (1975) showed a 30–66% increase in weight over control values at eight weeks of age in coho salmon (*Oncorhynchus kisutch*) after injections of bovine GH or implantation of pellets of GH. Komourdjian et al. (1978) demonstrated the effects of porcine GH on rainbow trout, and Adelman (1977 and 1982) reproduced the results using carp GH and bovine GH in carp (*Cyprinus carpio*). The work of Clarke et al. (1977) showed similar effects on *Tilapia mossambica* and *Oncorhynchus nerka* (sockeye salmon) with injection of either Tilapia and bovine GH. Clarke's group also showed that increased levels of GH facilitated the adaptation of salmon to salt water. Komourdjian et al. (1976) confirmed this observation in young Atlantic salmon (*Salmo salar*) treated with porcine GH. In addition to the substantially increased rates of growth, improved feeding efficiency was also noted by Markert et al. (1977) in coho salmon, where injections of bovine GH significantly increased both food intake and protein conversion. Recombinant rainbow trout GH has been shown to produce the same growth effect in rainbow trout (Agellon et al. 1988), as have recombinant chicken and bovine GH in Pacific salmon (*Oncorhynchus kisutch*) (Gill et al. 1985). Purified bovine GH has produced not only significant growth, but record-breaking growth, in grass pickerel (*Esox americanus vermicalatus*) in a study by Weatherly and Gill (1987). Grass pickerel in excess of the North American record length of 38.1 cm were produced through administration of bovine GH. In this latter study, the potential for rapid growth appeared to be correlated with the ability to recruit new muscle fibers into the growing axial muscle mass. All of the above studies involved injection of GH, a process that is both laborious and risky because of the need for repeated injections, and one that introduces a certain amount of stress which may cloud assessment of the effects of the injected hormone. Feeding studies utilizing GH as a feed supplement have shown little success, and only one report (Degani and Gallagher 1985) has indicated increased growth and food conversion in American elvers (*Anguilla rostrata*). No method for the mass administration of growth factors has yet proven to be economically feasible, nor have any methods shown great promise as techniques for the analysis of the effects on growth of products of various growth-promoting genes. Indeed, to date, little work has been devoted to the examination of other growth-promoting peptide hormones or releasing factors by injection or feed supplementation; only GH has been subjected to extensive testing.

Molecular genetics and gene transfer have potential for the introduction of significant new sources of variation within a species through the introduction of extra copies of cloned genes into the genome and germline of that species. Since extra copies of a gene generally result in increased levels of the encoded enzyme or protein, the effect on the organism into which these genes have been introduced can be dramatic. The classic experiment in mice by Palmiter et al. (1982) showed a doubling of the mature size in transgenic mice that contained extra copies of genes for either rat or human GH. Since the genes were incorporated into the germline, this trait was passed in true Mendelian fashion to the offspring, producing a strain of larger mice. Faster growing mice have little enhanced economic value, but increased growth rates in fish are of significant economic interest.

Gene transfer in fish by microinjection has produced results similar to those in mice. Zhu et al. (1985) first reported the transfer of a human GH construct into goldfish (*Carassius auratus*) and subsequently the expression, transmission, and significantly enhanced growth rates in transgenic fish have been reported (Maclean et al. 1987). Similar transfers of GH genes have been accomplished by Chourrout et al. (1986) in rainbow trout, Zhu et al. (1986) in loach, Dunham et al. (1987) in channel catfish (*Ictalurus punctatus*), McEvoy et al. (1987) in Atlantic salmon, Maclean et al. (1987) and D.A. Powers (see Chapter 25, in this book) in rainbow trout, and by our group in northern pike (*Esox lucius*). Expression of genes has been a problem, with few fish expressing the transferred genes when mammalian promoter/enhancer elements are used. We have found that a viral promoter/enhancer element derived from the avian Rous sarcoma virus (RSV) provides enhanced expression as compared to such elements from mammalian sources. This RSV construct has been associated with expression, in our hands, in approximately 50% of the transgenic fish when linked to the neomycin resistance gene *neo* and transferred into goldfish (Yoon et al. 1990). Very few promoter/enhancer elements have been tested for their utility in fish; namely, RSV (Yoon et al. 1990; G.W. Stuart, personal communication; D.A. Powers, Chapter 25 of this book), mouse metallothionein (Zhu et al. 1985; Chourrout et al. 1986; McEvoy 1988), and SV40 (Simian Virus 40) (Stuart et al. 1988) with only RSV showing significant promise.

Since gene transfer by microinjection in fish has been proven effective, the search for fish growth-promoting genes has intensified. The effort to clone and express piscine genes that promote growth in fish is a worldwide effort. GH genes have been cloned from rainbow trout (Agellon and Chen 1986; Agellon et al. 1988), gilthead seabream (*Sparus aurata*; Cavari et al. 1988), chum salmon (*Oncorhynchus keta*; Sekine et al. 1985), coho salmon (Nicoll et al. 1987; Gonzalez-Villasenor et al. 1988), chinook salmon (C.L. Hew, personal communication), yellowtail (*Seriola quinqueradiata*; Watahiki et al. 1988), tuna (*Thunnus thynnus*; Sato et al. 1988), and red seabream (*Pagus major*; Momota et al. 1988). Chum salmon prolactin has been cloned by Song et al.(1988) and represents the start of efforts to clone other genes for piscine peptide hormones. One gene that is not related to growth, namely, the gene for the antifreeze protein of winter flounder (*Pseudopleuronectes americanus*), has been cloned and transferred by Fletcher et al. (1988) into Atlantic salmon. In winter flounder, this gene in tandem arrays produces the major antifreeze protein that depresses the freezing temperature of the fish, allowing them to survive at temperatures at which fish freeze before their salt-water environment freezes. This gene for the antifreeze protein has economic potential if it can be expressed at sufficiently high levels to protect transgenic salmonid species, which normally lack the gene, from similar freezing temperatures. Such a change would extend the range of aquaculture of salmonids in sea-cages.

24.1 GENE TRANSFER BY MICROINJECTION

To date, microinjection has been the only successful technique for transfer of genes in fish. All of the reported transgenic fish have been produced by variants of the same techniques as those used in mice (see Table 24-1 for a list of successful efforts at transfer of genes in fish). Microinjection in fish is accomplished by one of two procedures, depending upon the ease of removal or softening of the egg chorion. Many fish eggs have an opaque chorion, which limits the ability to visualize the target for the injection of the DNA. Additionally, most fish eggs have a chorion that hardens upon fertilization and exposure to water. The extent of this hardening in water

TABLE 24-1 Successful Gene Transfer in Fish

Species	Construct	Reference
Xiphophorus	Pigment gene?	Vielkind et al. 1982
Goldfish	mMT/hGH	Zhu et al. 1985
Loach	mMT/hGH	Zhu et al. 1986
Medaka	chick crystallin	Ozato et al. 1986
Rainbow trout	SV40/hGH	Chourrout et al.1986
Channel catfish	mMT/hGH	Dunham et al. 1987
Atlantic salmon	mMT/rGH	McEvoy et al. 1987
Rainbow trout	mMT/rGH	Maclean et al. 1987
Atlantic salmon	mMT/hGH	Rokkones et al. 1988
Rainbow trout	mMT/hGH	Rokkones et al. 1988
Atlantic salmon	Winter flounder antifreeze protein, genomic	Fletcher et al. 1988
Zebrafish	SV40/hygromycin	Stuart et al. 1988
Atlantic salmon	mMT/β-galactosidase	McEvoy et al. 1988
Tilapia	mMT/hGH	Brem et al. 1988
Zebrafish	SV40/CAT	Stuart et al. 1988
Northern pike	RSV/bGH	J. Schneider et al. personal comm.
Goldfish	RSV/neo	Yoon et al. 1990
Rainbow trout	RSV/tGH cDNA	Powers, Chapter 25, this book

[a]mMT, mouse metallothionein; hGH, human growth hormone; rGH, rat growth hormone; bGH, bovine growth hormone; tGH, rainbow trout growth hormone; RSV, Rous sarcoma virus; and SV40, Simian virus.

varies between species, with those species that spawn in turbulent water (rapids, etc.) tending to have tougher chorions than those fish that spawn in still water (Hallerman et al. 1989).

Trypsin is effective in dechorionating goldfish and carp eggs (Zhu et al. 1985). We have not yet found an effective dechorionation treatment for walleye (*Stizostediun vitrium*). Although protease type XXV (Sigma) will dechorionate eggs of the northern pike, mortality rates frequently exceed 50% in dechorionated eggs (Hallerman et al. 1989). Chorions of eggs of salmonids (trout and salmon) and northern pike also are resistant to all tested enzymes. Microinjection of northern pike and salmonid eggs has largely been via "blind" injection, that is, with little ability to visualize the position of the needle tip. Several techniques are used to overcome the problem of water-hardening of the chorion. In our laboratory, short strong needles are used to pierce the water-hardened chorion or the partially softened chorion produced by enzymatic treatment (Hallerman et al. 1989). In other laboratories, a hole is bored through the chorion with a broken glass needle and then a second needle is inserted to introduce the solution of DNA (Chourrout et al. 1986). Other groups have injected through the micropyle, the entry tunnel used by the sperm during fertilization (Fletcher et al. 1988).

Microinjection of prepared vectors into goldfish, our model system, is currently being performed in the following manner in our laboratory:

1. Fish are induced to ovulate spontaneously by manipulation of photoperiod and temperature, with injections of carp pituitary extract if necessary (Stacy et al. 1979).
2. Fish are artificially spawned, with eggs and milt collected separately.
3. Eggs are fertilized by mixing with milt in well water.
4. Ten minutes after fertilization, eggs are dechorionated by trypsinization (Zhu et al. 1985; Yoon et al. 1990).
5. Dechorionated eggs are washed in Holtfreter's (Grand et al. 1941) solution that contains fetal bovine serum and are transferred to a black background for microinjection.
6. Plasmid vectors are linearized and dissolved in standard TE buffer.
7. Borosilicate glass needles (inner-tip diameter, 2 µm) are filled with the solution of plasmid and microinjection is performed with a constant flow rate of solution through the needle. The injection volume is controlled by timing of the interval between insertion and withdrawal. A routine injection consists of approximately 1×10^6 copies of plasmid in 2 nl.
8. DNA is released into the center of the germinal disc prior to the first cleavage in dechorionated eggs.
9. Microinjected eggs are allowed to develop in Holtfreter's solution until the blastula stage, and in well water thereafter.

In case of northern pike, eggs and milt are stripped from fish caught in the wild, without induction. Chorionated pike eggs are injected with linearized plasmid using a short, strong glass needle and a constant flow rate. Microinjection is an efficient method for transfer of DNA into fish, especially if the target nucleus can be visualized by dechorionation of the egg or oriented by low-speed centrifugation. However, relatively few eggs can be conveniently injected per day; in our laboratory we can process only 1,000 dechorionated goldfish eggs or a few hundred pike eggs per day. The labor needed to produce the large number of transgenic fish from which to select the most optimal transfer event is, thus, significant. We are trying to develop a mass transfer technology for use in fish such that a large number of eggs (10^4–10^6 per treatment) can be treated at one time, with a subsequent selection system to detect transgenic fish. Three main systems of mass transfer are currently being investigated: sperm binding, electroporation, and lipofection.

24.1.1 Sperm Binding

Binding or adsorption of plasmid DNA to fish sperm, with the utilization of the DNA/sperm complex to fertilize the egg, is a possible mode of production of transgenic fish. Infectious hematopoietic necrosis virus can be transmitted vertically via this technique of sperm binding (Mulcahy and Pascho 1984). In preliminary experiments in our laboratory, radiolabelled plasmid DNA has survived three cycles of washing and centrifugation by association with fish sperm in a nonactivating semen-extender buffer. Up to 10% of the labeled DNA remained associated with the sperm, supporting the hypothesis that the adsorbed DNA might survive transit through the micropyle by the sperm during fertilization and be carried by the sperm into the egg. This procedure has the potential for allowing transfer of genes into a large mass of eggs with little manipulatory effort. Transfer efficiencies are expected to be low and, thus, it is preferable that a selectable marker, such as *neo* (resistance to the neomycin analog G418; Southern and Berg 1982), or a gene that shows early phenotypic expression be used. Hormones would fall into the latter class of genes, with a phenotypic increase in growth rate being the criterion for selection.

24.1.2 Electroporation

The technique of electroporation for gene transfer has been very successful in bacteria, mammalian tissue culture, and plant protoplasts. Such use of electric current to effect the transfer of single genes into eukaryotic cells may be modifiable for use in fish. We have defined electroporation conditions that do not kill the eggs. It remains to be seen whether these conditions still generate sufficient changes in membranes to allow uptake of DNA. Dechorionation appears to be conceptually necessary if electroporation is to have a significant chance of success. Again, as with the sperm binding, masses of eggs should be treatable in a short period of time, with little manipulative effort, and selection could be applied by consideration of phenotype.

24.1.3 Lipofection
Lipofection for gene transfer depends on the encapsulation of DNA within a phospholipid bilayer. The encapsulated DNA is delivered by membrane fusion of the liposome with the cell (fish egg) membrane. Efficiencies in cell culture do not exceed those of other methods of mass transfer, such as transfer mediated by calcium phosphate (Fraley et al. 1981), but the technique has intriguing potential. Liposomes encapsulating a recombinant DNA construct have been injected intravenously into adult rats. The transgene was expressed transiently in the liver and spleen of the recipient animals (Nicolau et al. 1983 and 1987). There are indications that insertion of glycolipids into the liposome may allow the DNA to be targeted to a specific type of cell, such as the nonphagocytizing cells of the liver or spleen. With proper consideration of the lectin components of germline cells, specific liposomes might be targeted towards eggs, sperm or their progenitor cells, producing transgenic germ cells. Detection of transgenic offspring from such treated fish would depend on selection of the growth-enhancement phenotype or on the use of a selectable marker. While several groups, including ours, are attempting this mode of transfer, no success has yet been reported.

24.2 SUCCESS OF TRANSFER AND EXPRESSION OF TRANSGENES

While more than a dozen groups worldwide (see Table 24–1) have reported successful gene transfer in fish, few groups have reported high levels of gene expression. Transfer efficiencies reported for fish range from 3% (Brem et al. 1988) to over 70% (Z. Zhu, personal communication) with 10–50% being the typical range of efficiency in our group.

Most groups working on transfer of fish genes are using mammalian GH constructs with promoter/enhancer regions of mammalian origin. These constructs, which are identical or virtually identical to those used in studies of growth in transgenic mice, are not expressed at significant levels in many species of fish. Zhu et al. (1986) have reported expression of a mouse metallothionein/human GH fusion gene in the loach, which produced significant enhancement of growth. The transgene appears to be stably incorporated and has been transmitted to the F_1 generation in standard Mendelian fashion. Zhu has also reported that the same construct, in carp, has produced carp that grow at seven times the normal rate during at least some stage of their development. Expression does not appear to be constant during development, with growth rates falling as the fish mature. It is unknown whether the cause of this phenotypic change is related to developmental control of expression of the transgene or to a declining response to GH as the fish mature.

Viral promoters (SV40 and RSV) have been used by Stuart et al. (1988), D.A. Powers (see Chapter 25), and by our group. They permit more efficient expression than does the mammalian metallothionein promoter. RSV was chosen because of its ability to promote expression in both avian and mammalian cells, which indicated a good probability of effective promotion of expression in fish. Furthermore, RSV appears, in avian systems, to promote specifically the expression of genes in muscle and connective tissue, which are the desired targets for tissue-specific expression of growth genes. In our studies, about 50% of transgenic fish (northern pike and goldfish) show evidence of expression when RSV is used as the promoter.

In goldfish, we have used the RSV promoter to express neomycin resistance (neo) and have produced transgenic fish that survived one week of challenge with G418 (Yoon et al. 1990), thereby demonstrating the potential utility of *neo* as a selectable marker in production of transgenic fish. In northern pike, an RSV/bovine GH construct is producing transgenic fish that are growing significantly faster than their full sibs. At eight months after hatching, the transgenic fish are almost 60% larger on the average than the control full sibs, with some individuals being twice the weight of their full sib controls. Individual fish are marked (by freeze-branding, fin-clipping, or use of microtags) so that transgenic fish and control fish can be co-cultured to reduce variations in growth due to environment.

24.3 PRODUCTION OF STERILE TRANSGENIC FISH

The key to the proprietary protection of a valuable transgenic fish broodstock is the capacity to produce sterile offspring selectively for sale to growout facilities. Three primary techniques are used to produce functional sterility in fish: polyploidization, interspecies hybridization, and hormonal sterilization. Protocols for the combination of two or more of these techniques must be developed to ensure the sterility of the treated offspring. Such protocols are necessitated by the 99% sterilization rate produced by any single, currently available method, which leaves a small window for the escape of fertile offspring. Combination of two or more different techniques should reduce the fertile fish in a sterilized population to acceptable levels. Production of sterile fish is critical not only for proprietary protection of a transgenic broodstock, but also for the biological limitation of any possible environmental impact of transgenic fish that might escape from an aquaculture facility.

The three sterilization procedures noted above can be briefly described as follows:

1. Polyploidization: Temperature shock or pressure treatment of newly fertilized eggs prevents the expulsion of the secondary polar body, creating a 3N genome (Thorgaard 1983; Purdom 1984). Temperature shock is easily applied by dipping the eggs in a warm bath for 10–20 seconds. Pressure treatment requires more specialized equipment for application of high pressure (1000 psi or more) to the eggs.
2. Interspecies hybridization: Many fish species, including most salmonid species, will produce viable interspecific hybrids, with some crosses being quite sterile. Sterile transgenic × nontransgenic or transgenic × transgenic (of different salmonid species) hybrids may be produced for protection of broodstock.
3. Hormonal sterilization: Treatment of alevin or early fry stages of salmonids with androgens produces sterility in the adults. This technique has been successfully applied to coho, chum, and kokanee salmon (Donaldson and Hunter 1983), and it is simple and effective.

These three methods of sterilization should be refined for use with all commercially important species of fish and tested singly and in combination with each other to assess their synergistic effects.

24.4 FUTURE OF GENE TRANSFER TECHNOLOGY IN FISH

Our goal is to produce transgenic fish with enhanced growth characteristics. The ideal fusion genes for this project are composed of fish-derived promoter/enhancer elements linked to a piscine growth gene. We hope to create, for aquaculture, a transgenic fish that contains only extra DNA sequences that were originally derived from the same species, i.e., fish without any heterologous genes and no viral or mammalian promoters. While such an "all-fish" transgenic fish may grow no better or faster than a transgenic fish with, for instance, a bovine-derived growth gene, the concept of foreign genes in an organism destined for human consumption is repugnant to some groups and agencies. An all-fish system should ease some of those concerns.

Applications of gene transfer technology will have major effects on aquaculture and fishery-stocking programs in the next decade. The growing importance of aquaculture (for example, the U.S. production of fish was valued at $525 million in 1985) will exert a major impact on the economics of food production in the USA and elsewhere. For aquaculture, a faster growing, more feed-efficient fish would be of obvious utility, and indeed may provide the difference between loss and profit for many currently, financially marginal firms, especially in the colder climates found in the North-Central region of the USA. Aquaculture operations are world-wide in distribution, with concentrations in Europe, Southeast Asia, China, Japan, South America, and Canada, as well as in the USA. Thus, the successful production of growth-enhanced recombinant strains of fish would have wide application

TABLE 24-2 Laboratories in the USA Involved in Fish Gene Transfer

Institution	Species
Johns Hopkins/University of Maryland	Trout, carp
University of Minnesota	Goldfish, zebrafish, salmon, trout, northern pike, walleye
University of Southern Illinois	Walleye
Auburn	Catfish
University of Oregon	Zebrafish
University of Montana	Trout
SIBIA	Trout
Battelle Institute	Trout/salmon

TABLE 24-3 Laboratories Involved in Fish Gene Transfer Worldwide

Country	Species
In Operation	
Canada	Trout, salmon
England	Trout
France	Trout, salmon
Germany	Tilapia, Xiphophorus
Ireland	Salmon
Israel	Tilapia
Japan	Medaka
Norway	Salmon
People's Republic of China	Carp, loach
United States	Various
In Building Phase or in Progress, Extent Unknown	
India,	
Indonesia,	
Hungary,	
Malaysia,	
Thailand,	
USSR	

and major economic potential. Transfer of growth genes into fish is being viewed by developing countries as a major means of increasing available protein supply (FAO Workshop, 1988) and is a major goal of the biotechnology program of the People's Republic of China. Interest in genetic engineering of fish is accelerating, with groups in Ontario, Newfoundland, and British Columbia, Canada; Japan; Norway; France; China; England; Israel; Ireland; India; the Soviet Union; several countries in Southeast Asia; and the United States proceeding along similar lines (Tables 24–2 and 24–3).

Because of the relative ease of gene transfer in fish as compared to mammals and birds, we envision that the first practical use of transgenic technology will indeed be in fish rather than mammalian or avian farm species. We are already well on the way to the fulfillment of this prediction.

REFERENCES

Adelman, I.R. (1977) *J. Fish Res. Board Can.* 34, 509–515.
Adelman, I.R. (1982) *Prog. Fish-Cult.* 44, 94–97.
Agellon, L.B., and Chen, T.T. (1986) *DNA* 5, 463–67.
Agellon, L.B., Emery, C.J., Jones, J.M., et al. (1988) *Can. J. Fish. Aquat. Sci.* 45, 146–151.
Brem, G., B. Brenig, Horstgen-Schwark, G., and Winnacker, E-L. (1988) *Aquaculture* 68, 209–219.
Chourrout, D., Guyomard, R., and Houdebine, L-M. (1986) *Aquaculture* 51, 143–150.
Clarke, W.C., Farmer, S.W., and Hartwell, K.M. (1977) *Gen. Comp. Endocrinol.* 33, 174–178.
Degani, G., and Gallagher, M.L. (1985) *Can. J. Fish. Aquat. Sci.* 41, 185–189.
Donaldson, E.M., and Hunter, G.A. (1983) in *Fish Physiology, vol. 9B*, (Hoar, W.S., Randall, D.J., and Donaldson, E.M., eds.), pp. 405–435, Academic Press, New York.
Dunham, R.A., Eash, J., Askins, J., and Townes, T.M. (1987) *Trans. Am. Fish. Soc.* 116, 87–91.
FAO Workshop (1988) *Proceedings of the Regional Workshop on Biotechnology and Animal Production and Health in Asia*, Bangkok, Thailand, Food and Agriculture Organization (FAO), United Nations and Kasetsart.
Fletcher, G.L., Shears, M.A., King, M.J., Davies, P.L., and Hew, C.L. (1988) *Can. J. Fish. Aquat. Sci.* 45, 352–357.
Fraley, R., Straubinger, R.M., Rule, G., Springer, E.L., and Papahadjopoulos, D. (1981) *Biochemistry* 20, 6978–6987.
Gill, J.A., Sumpter, J.P., Donaldson, E.M., et al. (1985) *Bio/Technology* 3, 643–646.
Gonzalez-Villasenor, L.I., Zhang, P., Chen, T.T., and Powers, D.A. (1988) *Gene* 65, 239–245.
Grand, C.G., Gordon, M., and Cameron, G. (1941) *Cancer Res.* 1, 660–666.
Hallerman, E., Schneider, J.F., Gross, M.L., et al. (1989) *Trans. Am. Fish. Soc.* 117, 456–460.
Higgs, D.A., Donaldson, E.M., Dye, H.M., and McBride, J.R. (1975) *Gen. Comp. Endocrinol.* 27, 240–253.
Komourdjian, M.P., Burton, M.P., and Idler, D.R. (1978) *Gen. Comp. Endocrinol.* 34, 158–162.
Komourdjian, M.P., Saunders, R.L., and Fenwick, J.C. (1976) *Can. J. Zool.* 54, 531–535.
Maclean, N., Penman, D., and Zhu, Z. (1987) *Bio/Technology* 5, 257–261.
Markert, J.R., Higgs, D.A., Dye, H.M., and MacQuarrie, D.W. (1977) *Can. J. Zool.* 55, 74–83.
McEvoy, T., Stack, M., Barry, T., et al. (1987) *Theriogenology* 27, 258 (Abstract).

McEvoy, T., Stack, M., Keane, B., et al. (1988) *Aquaculture* 68, 27–37.
Momota, H., Kosugi, R., Hiramatsu, H., et al. (1988) *Nuc. Acids Res.* 16, 3107.
Mulcahy, D., and Pascho, R.J. (1984) *Science* 225, 333–335.
Nicolau, C., Legrand, A., and Grosse, E. (1987) *Meth. Enzymol.* 149, 157–176.
Nicolau, C., LePape, A., Soriano, P., Fargette F., and Juhel, M.F. (1983) *Proc. Natl. Acad. Sci. USA* 80, 1068–1072.
Nicoll, C.S., Steiny, S.S., King, D.S., et al. (1987) *Gen. Comp. Endocrinol.* 68: 387–399.
Ozato, K., Kondoh, H., Inohara, H., et al. (1986) *Cell Differentiation* 19, 237–244.
Palmiter, R.D., Brinster, R.L., and Hammer, R.E. (1982) *Nature* 300, 611–615.
Pickford, G.E. (1954) *Endocrinology* 55, 273–287.
Pickford, G.E., and Thompson, E.F. (1948) *J. Exp. Zool.* 109, 367.
Purdom, C.E. (1984) in *Oxford Reviews of Reproductive Biology, vol. 6* (Clarke, J.R., ed.), pp. 303–340, Oxford University Press, Oxford, England.
Rokkones, E., Alestrom, P., Skjervold, H., and Gautvik, K.M. (1989) *J. Comp. Physiol.* B 158, 751–758.
Sato, N., Watanabe, K., and Murata, K., (1988) *Biochim. Biophys. Acta* 949, 35–42.
Sekine, S., Mizukumi, T., Nishi, T., et al. (1985) *Proc. Natl. Acad. Sci. USA* 82, 4306–4310.
Song, S., Trinh, K.-Y., Hew, C.L., et al. (1988) *Eur. J. Biochem.* 172, 279–285.
Southern, P.J., and Berg, P. (1982) *J. Mol. Appl. Genet.* 1, 327.
Stacy, N.E., Cook, A.F., and Peter, R.E. (1979) *J. Fish. Biol.* 15, 349–361.
Stuart, G.W., McMurray, J.V., and Westerfield, M. (1988) *Development* 103, 403–412.
Thorgaard, G.H. (1983) in *Fish Physiology, vol. 9B* (Hoar, W.S., Randall, D.J., and Donaldson, E.M., eds.), pp. 405–435, Academic Press, New York.
Vielkind, J., Haas-Andela, H., Vielkind, U., and Anders, F. (1982) *Mol. Gen. Genet.* 1985, 379–389.
Watahiki, M., Tanaka, M., Masuda, N., et al. (1988) *Gen. Comp. Endocrinol.* 70, 401–406.
Weatherly, A.H., and Gill, H.S. (1987) *Aquaculture* 65, 55–66.
Yoon, S.J., Hallerman, E.M., Gross, M.L., et al. (1990) *Aquaculture* 85, 21–33.
Zhu, Z., Li, G., He, L., and Chen, S. (1985) *Z. Agnew. Ichthyol.* 1, 31–34.
Zhu, Z., Xu, K., Li, G., Xie, Y., and He, L. (1986) *Kexue Tongbao Acad. Sin.* 31, 988–990.

CHAPTER 25

Studies on Transgenic Fish: Gene Transfer, Expression, and Inheritance

Dennis A. Powers
Lucia Irene Gonzalez-Villasenor
Peijun Zhang
Thomas T. Chen
Rex A. Dunham

Gene transfer is currently employed as a standard procedure for generating transgenic animals. In addition to addressing fundamental research questions, this technology offers the possibility of generating economically important animal species. Furthermore, transgenic animals can be used as bioreactors to produce large quantities of important proteins (Patton et al. 1984; Bishop et al. 1986; Gordon et al. 1987; Pittius et al. 1988; Simons et al. 1988; Van Brunt et al., 1985; Cundiff 1983; Wagner and Murray 1985).

The application of gene-transfer technology to fish offers the opportunity to bring about a radical change in aquaculture and in efforts to restock fisheries. It also provides unparalleled opportunities for the use of fish as

---

This research was supported by grants from the National Science Foundation (DCB–86–42247) and the MSG program (R/F 47).

experimental models in attempts to address fundamental questions in the areas of neurobiology, developmental biology, reproductive biology, toxicology and cancer research (reviewed by Powers, 1989). The goal of our laboratory is to extend these new frontiers by studying the transfer and expression of foreign genes in fish. We are addressing a number of basic research objectives, some of which have the potential to accelerate growth rate, to enhance the efficient utilization of food, and to increase resistance of fin- and shell-fish to disease.

The administration of natural growth hormone (GH) by either intraperitoneal or muscular injection has been shown to enhance the growth of fish (Wagner et al. 1985; Gill et al. 1985; Weatherly and Gill 1987). More recently, some of us have demonstrated that the injection of biosynthetic GH from the rainbow trout (rt) into yearling trout resulted in marked increases in length and weight (Agellon et al. 1988a). While such experiments clearly demonstrate that exogenous GH enhances somatic growth of cultured fish, the experiments may not be cost-effective because they are laborious and require multiple injections, and the continuous handling of fish increases the risk of infection. Alternatively, the generation of fast-growing strains could be achieved by constructing transgenic fish that carry exogenous genes for GH.

Previously, we determined the molecular structure of coho salmon (cs) GHcDNA (Gonzalez-Villasenor et al. 1988), and we determined the nucleotide sequence of both the rt GH gene and rtGH cDNA (Agellon et al. 1988b and 1988c). Recently we succeeded in transferring genes for rt, cs, and human GH into carp and catfish. Herein, we describe the integration, expression, and inheritance of a chimeric plasmid that contains the rainbow trout GHcDNA (construct 1) fused to the Rous sarcoma virus (RSV) long terminal repeat (LTR) promoter in the carp, *Cyprinus carpio* (L.). In addition, we present data on the introduction and integration of chimeric plasmids that contain the structural genes of a rtGHcDNA (construct 2), csGHcDNA, MThGH, and luciferase, respectively, into carp (*Cyprinus carpio*) and/or catfish (*Ictalurus punctatus*).

25.1 MATERIALS AND METHODS

25.1.1 Construction of Chimeric Plasmids
Several plasmids based on pRSV-2 (Gorman et al. 1982; McKnight and Tjian 1986), containing the rt GHcDNA (construct 1), rt GHcDNA (construct 2), cs GHcDNA, and luciferase sequences, respectively, were constructed as follows: pRSVrtGHcDNA (construct 1) is a 5.23-kbp construct containing a *Bam*HI-*Hind*III fragment isolated from the plasmid pGHFl (L.B. Agellon and T.T. Chen, unpublished data) and inserted into the *Sal*I site of pRSV-2. It harbors 580 bp of the LTR sequence, 950 bp of the rtGHcDNA sequence, and 3.7 kbp of the pRSV flanking sequence (Figure

25–1A). pRSVrtGHcDNA (construct 2) is about six nucleotides shorter than construct 1 (see Figure 25–1B). pRSVcsGHcDNA is a 5.5-kbp recombinant plasmid containing the 1.3-kbp csGHcDNA sequence derived from the clone pCSGH36 (Gonzalez-Villasenor et al. 1988) (see Figure 25-1C). pRSV2-Luc is a 6.0-kbp recombinant plasmid (see Figure 25–1D) containing the 1.89-kbp structural gene for luciferase (Ow et al. 1986) inserted at the *Hind*III site of the pRSV2 plasmid. A pBR322 derived construct containing the metallothionein-human GH fusion gene, pMThGH (Dunham et al. 1987), was also used for microinjection (see Figure 25-1E). Prior to microinjection, the recombinant clones were linearized by digestion with the appropriate endonuclease.

25.1.2 Microinjection of Eggs

Fish were spawned artificially by induction with carp pituitary extract. Eggs were stripped from a mature female into a Petri dish. Milt and water were added to the eggs with gentle stirring to enhance fertilization and dispersal.

Fish embryos were microinjected by use of a dissecting microscope and a micromanipulator. About 20 nl, containing approximately 10^6 copies of the chimeric plasmid, were delivered into the germinal disc with a 2- to 10-μm glass pipette. The flow of the DNA solution was controlled with hydraulic pressure from a microinjector. Embryos were incubated in static water at about 20°C. All fish were heat-branded, weighed, and fin-clipped.

25.1.3 Extraction of Genomic DNA

Genomic DNA was extracted from pectoral fin clips of presumptive transgenic fish and their offspring by the methods of Maniatis (1982) with some modifications (Zhang et al., 1990). Pectoral fin tissue was lysed in 4 ml of 10 mM Tris buffer (pH 8.0), which contained 0.1 M EDTA (pH 7.0), 0.5% SDS and 200 μg/ml proteinase K, and digested at 50°C for 15 hours. Samples were phenol-extracted and digested with RNase (100 μg/ml) for 3 hours at 37°C. DNA was reextracted with phenol and chloroform, and it was then dialysed against several changes of TE buffer (10 mM Tris-1 mM EDTA, pH 8.0). After precipitation in ethanol, the DNA pellet was dissolved in TE buffer and the concentration of DNA was determined.

25.1.4 Dot-Blot and Southern Blot Analyses

Genomic DNA (18 μg) was denatured in 0.5 M NaOH at 37°C for 10 minutes and spotted onto a nylon membrane using a commercial dot-blot apparatus. The membrane was soaked in 1.5 M NaCl, 0.5 M Tris-HCl (pH 7.2), 1 mM EDTA for 5 minutes. After air-drying, the membrane was ultraviolet-irradiated for 5 minutes.

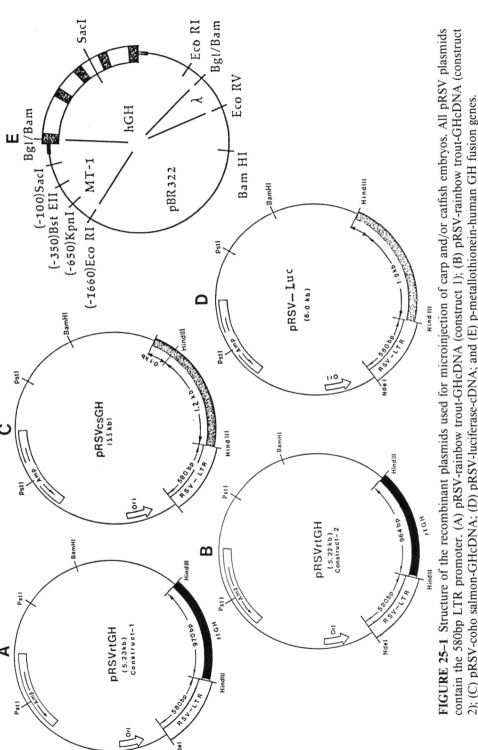

FIGURE 25-1 Structure of the recombinant plasmids used for microinjection of carp and/or catfish embryos. All pRSV plasmids contain the 580bp LTR promoter. (A) pRSV-rainbow trout-GHcDNA (construct 1); (B) pRSV-rainbow trout-GHcDNA (construct 2); (C) pRSV-coho salmon-GHcDNA; (D) pRSV-luciferase-cDNA; and (E) p-metallothionein-human GH fusion genes.

For Southern blot analyses, DNA (18 μg) was digested with HindIII or BamHI under conditions indicated by the supplier (Boehringer Manheim) and subjected to electrophoresis on a 0.8% agarose gel. DNA samples in the gel were transferred to nylon membrane by diffusion and treated as above.

Estimates of copy number were based on a haploid DNA content of 1.7 pg for *Cyprinus carpio* and 1.2 pg for *Ictalurus* (Hinegardner and Rosen 1972). The molecular weights of catfish and carp genomic DNA and the corresponding chimeric plasmids were determined and used to estimate the amount of exogenous DNA, equivalent to one DNA molecule, present in known amounts of fish genomic DNA. The total numbers of DNA copies were estimated from the optical density profiles of dot-blot autoradiographs scanned under visible light.

The RSV-LTR and/or the insert of interest [rtGHcDNA (1), rtGHcDNA (2), csGHcDNA, luciferase, and MThGH] were isolated from the corresponding construct and labelled with [α-^{32}] dCTP either by nick-translation (Maniatis et al. 1982) or by random priming (Feinberg and Vogelstein 1984). Each insert was then purified through a Sephadex G-50 spun-dialysis column and used as probes. Prehybridization was carried out in 20 mM Tris-HCl (pH 7.5), 0.1% SDS, 5× SSC, 10× Denhardt's, and 100 μg/ml of denatured calf thymus DNA at 42°C with constant shaking for 3 hours. Hybridization was carried out in the same buffer supplemented with denatured probe at 42°C with constant shaking overnight. Membranes were washed twice in a solution of 2× SSC and 0.1% SDS for 15 minutes at room temperature, and twice in a solution of 0.2× SSC, 0.1% SDS for 15 minutes at 65°C.

25.1.5 Immunobinding Assay

Quantitative determination of rt GH polypeptide in the red blood cells of transgenic carp that harbored the pRSVrtGHc-DNA (construct 1) was carried out by a radioimmunobinding assay on Terasaki plates, using an antiserum raised against chum salmon GH as a probe. This trout-specific antiserum (probe) was prepared by extensive absorption of an antibody specific for chum salmon GH to carp pituitary extract. One milliliter of the antiserum was added to 200 μl (1 mg/ml) of pituitary extract, mixed, incubated on ice for one hour, and centrifuged for 30 minutes. This procedure was repeated several times until the chum salmon GH antibody no longer reacted with the carp pituitary extract.

Red blood cell extracts of transgenic and control fish were prepared by subjecting the red cells to several cycles of thawing and freezing in 50 mM Tris-HCl buffer (pH 9.0) that contained 0.2% Triton X-100. Ten microliters of each cell extract were placed in the wells of the Terasaki plate precoated with poly-L-lysine (10 μg/ml). Following fixation with 0.5% glutaraldehyde, each sample was reacted with the antiserum against chum salmon GH, prepared as described above, and [^{125}I] goat antirabbit IgG. X-ray film was directly exposed to the air-dried plate. In each assay, a standard curve was constructed using known amounts of purified chum salmon GH as an antigen. The amount of immunoreaction in each sample was determined by scanning the autoradiograms.

25.2 CONSTRUCTION OF TRANSGENIC FISH

25.2.1 Fast-Growing Transgenic Carp

25.2.1.1 Integration of pRSVrtGHcDNA (Construct 1) in Carp. A total of 1,746 carp embryos at different developmental stages were microinjected with *Bam*HI linearized pRSVrtGHcDNA (1) (see Figure 25-1A). Approximately 37% of the injected embryos hatched after 4 days of incubation, of which 57% of fry survived at least 90 days (Table 25-1). No differences in rates of hatching or survival were observed.

Genomic DNA of the survivors was analyzed for integration of pRSVrtGHcDNA by dot-blot and Southern blot hybridization. Of 365 fish, 20 were found to be transgenic by dot-blot analysis. This value represents an overall rate of integration of about 5.5%, which is similar to that observed in zebrafish (Stuart et al. 1988). The maximum frequency of integration of the foreign gene (15.6%) was observed in embryos microinjected at the two-cell stage. The amount of pRSVrtGHcDNA present in the fin tissue of positive fish, ranged from one to five DNA copies per haploid genome (Table 25-2).

The results of dot-blot analysis were confirmed by digesting 10 randomly selected, positive samples of DNA with *Bam*HI or *Hind*III, with subsequent hybridization to both ^{32}-P-labeled RSV-LTR and/or rtGHcDNA. Representative results of this experiment are presented in Figure 25-2. Discrete bands from genomic DNA of transgenic fish, 24L, 27L, and 131L, hybridized to RSV-LTR, rtGHcDNA, or both. An approximately 5.2-kbp fragment, co-migrated with the *Bam*HI linearized pRSVRTGHcDNA, was detected with both probes in the *Bam*HI digestions of genomic DNA of fish 24L and 131L (see Figure 25-2). Since these transgenic fish contain at least two copies

TABLE 25-1 Hatching, Survival, and Frequency of Integration of Foreign DNA By Carp Embryos Microinjected with pRSV-rtGHlcDNA (Construct 1)

Eggs and Embryos	*Number*
Total eggs microinjected	1746
Total eggs hatched	642
Embryos survival: 21 days	524
Embryos survival: 90 days	365
Samples of genomic DNA analyzed	365
Embryos carrying the foreign DNA	20

TABLE 25-2 Representative Data for Gene Copy Number, Levels of rtGH, and Body Weight of 10 Transgenic Carp That Contain pRSVrtGHcDN (Construct 1)

Fish Number	Developmental Cell-stage[a]	pRSVrtGHcDNA Molecules[b]	ng GH/mg P[f] in RBC[c]	Wet Weight (g)[d]
20L	Early-I	2	8.0	129
27L	Early-I	5	ND[e]	171
34L	Early-I	4	47.7	147
36L[g]	Early-I	1	47.5	163
94R[g]	II	1	48.8	71
OR	II	3	64.2	75
O4R[g]	II	2	28.6	44
O7R[g]	II	1	8.9	127
O8R	II	3	73.8	94
131L[g]	IV	2	89.1	88

[a]Cell-stage at which embryos were microinjected.
[b]Copy number per haploid genome.
[c]Nine transgenic fish analyzed by immunoblotting expressed the foreign gene.
[d]Mean weight (g) of transgenic fish (N = 7) at the early one-cell stage, 143.4 ± 31 g; mean weight (g) of nontransgenic siblings (N = 64) at the early one-cell stage, 116.1 ± 40 g; weight difference between transgenic and nontransgenic fish is 22% at $p < 0.05$.
[e]ND, not determined (fish died).
[f]P, protein.
[g]Male transgenic fish generating offspring.

of the exogenous DNA, the additional 4.3-kbp *Bam*HI fragment in fish 131L may be a second copy of the foreign gene integrated at a different site in the genome. In the case of fish 27L, where five copies of the foreign gene were found, additional bands were observed after the *Bam*HI digestion. These bands may correspond to a single copy of the foreign gene integrated at different sites in the chromosome. The differences observed among the bands suggest that they carry genomic DNA junction fragments of variable sizes. Further analysis of the DNA restriction patterns of the four transgenic fish shows that a *Hind*III fragment of about 3.0-kbp hybridized to both RSV-LTR and rtGHcDNA sequences, suggesting that the *Hind*III site between those sequences has been modified. Since we also observed a 900-bp fragment of DNA that co-migrated with the rtGHcDNA sequence in the *Hind*III digests, we assume that at least one unmodified copy of the exogenous DNA was also integrated into the fish chromosome. Thus, genomic DNA of transgenic fish may carry both modified and unmodified copies of donor DNA.

25.2.1.2 Expression of pRSVrtGHcDNA (construct 1) in Red Blood Cells of Transgenic Carp.

Rainbow trout GH was not detected in samples of serum from transgenic fish, because the rtGHcDNA used in the gene-transfer

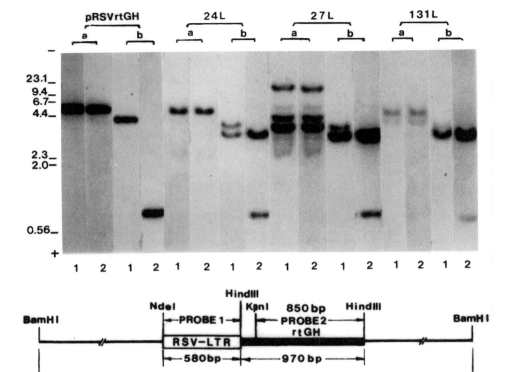

FIGURE 25-2 Representative Southern blot analysis of genomic DNA isolated from presumptive transgenic carp. The diagram at the bottom shows the linearized 5.2-kb pRSVrtGHcDNA (construct 1) used for microinjection. A 580-bp NdeI-HindIII fragment containing the RSV-LTR sequence was used as probe 1. An 850-bp KpnI-HindIII fragment containing the rtGHcDNA sequence was used as probe 2. BamHI-(a) or HindIII-(b) digested genomic DNA (18 μg) was subjected to electrophoresis on a 0.8% agarose gel, transferred to nylon membrane, and hybridized with either probe 1 or 2, as indicated at the bottom of each lane by 1 or 2, respectively. BamHI- and HindIII-digested recombinant plasmid was used as positive control. (From: Zhang et al. 1990.)

studies did not contain the signal peptide sequence. Thus, instead, we examined the expression of pRSVrtGHcDNA (1) in the red blood cells (RBC) of nine individual transgenic carp. Although all nine samples tested gave a positive immunoreaction, there was a particularly high level of expression in three samples (see Table 25-2). The levels of rtGH vary from 8.0 ng/mg protein in fish 20L to 89.1 ng/mg protein in fish 131L. There was no correlation between the number of foreign DNA molecules integrated and the levels of rtGH expressed in erythrocytes. These results are in agreement with those observed in mouse (Palmiter et al. 1982).

The body weight of each fish was measured at 90 days of age. The mean body weight of all transgenic fish was 125 ± 45 g, which is not significantly different from that of their nontransgenic siblings (129 ± 49 g). However, the mean body weight of transgenic individuals derived from embryos microinjected at the early one-cell stage was 143 ± 31 g. This figure represents a 22% difference and is significantly different ($p < 0.05$) from the mean body weight (116 ± 40 g) of nontransgenic siblings.

25.2.1.3 Inheritance of pRSVrtGHc DNA (Construct 1) in Carp.

Four male transgenic fish (P_1) carrying the pRSVrtGHcDNA sequences were crossed to one nontransgenic female and the resulting progeny (F_1) were allowed to develop for three months. A fraction of the total offspring derived from each male (04R, 94R, 36L, and 131L) was randomly sampled and assayed for the presence of pRSVrtGHcDNA by dot-blot analysis. About 32.3 and 42.3% of the progeny (F_1) analyzed from fish 131L and 94R, respectively, were found to carry the pRSVrtGHcDNA sequence. While most of the F_1 progeny from fish 36L died, the four survivors inherited the foreign gene sequence. None of the F_1 progeny of fish 04R received the foreign DNA from their father, suggesting that the pRSVrtGHcDNA sequence was not integrated into the germ line of fish 04R. Transgenic progeny carried about one to three copies of the foreign DNA per haploid genome.

Genomic DNA samples of progeny 9F, 16F, and 17F derived from male parent 131L were digested with *Hind*III and hybridized to rtGHcDNA. The hybridization patterns are shown in Figure 25-3. While offspring 9F and 17F show identical *Hind*III restriction patterns, progeny 16F show fragments that are slightly larger. These restriction patterns are different from those observed with the genomic DNA of parent 131L digested with *Hind*III (see Figure 25-2), suggesting that some may have occurred during the transmission of pRSVrtGHcDNA from the F_0 to F_1 generation.

The mean body weight of progeny (F_1) derived from two transgenic males (F_0) was also measured at three months of age. The positive F_1 progeny of transgenic fish 131L and 94R were 20.8% ($p < 0.05$) and 40.1% ($p < 0.001$) larger than their controls (nontransgenic siblings), respectively. About 32 and 46% of transgenic offspring derived from male parent 131L and 94T, respectively, were larger than the largest control (Table 25-3).

25.2.2 Transgenic Catfish

25.2.2.1 Integration of pRSV-Coho Salmon-GHcDNS in Channel Catfish, *Ictalurus punctatus*.

A total of 4,984 catfish embryos at five developmental stages were microinjected with pRSVcsGHcDNA linearized with *Bam*HI (see Figure 25-1C). Approximately 9.5% of the injected embryos hatched, of which 43.8% of fry survived 8 to 10 weeks (Table 25-4). While the rates

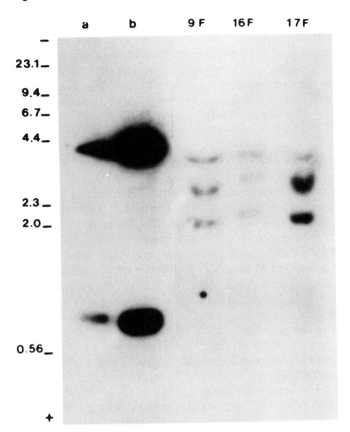

FIGURE 25-3 Southern blot analysis of DNA samples isolated from the F_1 generation of transgenic fish (F_o) derived from crosses between transgenic males and nontransgenic females. Genomic DNA (18 μg) from three progeny derived from transgenic male parent 131L was digested with *Hind*III, subjected to electrophoresis on a 0.8% agarose gel, transferred to nylon membrane, and hybridized with probe 2 (see bottom panel of Figure 25-2). pRSVrtGH [30 pg (a) and 300 pg (b)] was used as positive control. (From: Zhang et al. 1990.)

of hatching and survival of fish embryos microinjected at the early one-cell stage were higher than those observed at later stages of development, the highest percentage of integration of foreign DNA (33.3%) was observed in embryos microinjected at the two-cell stage. These results are in agreement with those obtained in common carp.

Genomic DNA of the survivors was analyzed for integration of pRSVcsGHcDNA by dot-blot and Southern blot hybridization. Of 134 samples analyzed, 14 were found to contain the foreign sequence by dot-blot analysis (Figure 25-4, top). This value represents an overall rate of inte-

25.2 Construction of Transgenic Fish

TABLE 25-3 Mean Weight and Range of Weights at 90 Days of Progeny Derived from Transgenic Fish 131L and 94R, with Percentage of Fish That Inherited the Foreign DNA

	Progeny of parent 131L		Progeny of parent 94R	
	Transgenic	Non-transgenic	Transgenic	Non-transgenic
N	31	65	11	15
% inheritance	32.3	—	42.3	—
Mean weight (g)	120.6[a]	99.3[b]	206.0	147
Standard Deviation	17.4	14.7	45.2	48
Weight range (g)	95–173[c]	65–129	115–283[d]	67–228
% difference	20.8[e]	—	40.1[f]	—

[a]N = 28; [b]N = 38; [c]32%, and [d]46% of transgenic progeny were larger than largest control. Transgenic progeny were larger than nontransgenic progeny at [e]$p < 0.05$ and [f]$p < 0.001$, respectively. (From: Zhang et al. 1990.)

TABLE 25-4 Percentage of Fish Hatching and Surviving and Integration Frequency for Catfish Embryos Microinjected with pRSV-coho Salmon-GHcDNA at Five Developmental Stages

Developmental Stage	Number Injected	Percent Hatched	8–10 Weeks Survival	Number Analyzed	Integration Frequency
One-cell					
Early	991	16.35%	53.7%	35	2.85%
Intermediate	1203	9.89%	45.4%	36	0.0%
Late	1147	8.11%	30.1%	28	14.28%
Two-cell	990	4.54%	44.4%	18	33.33%
Four-cell	653	8.57%	33.9%	17	17.65%
Control (nonmicroinjected)	325	19.69%	—	—	

gration of about 10.4%. The amount of foreign DNA present in the tissue of positive fish ranged from one to nine copies of the DNA per haploid genome, and the body weight ranged from 42 to 101 g. The Southern blot analysis of three positive samples is shown in the lower part of Figure 25-4. DNA samples were digested with *Hind*III, *Apa*I, and *Sca*I, with subsequent hybridization to ^{32}P-labeled csGHcDNA released from the construct by digestion with *Hind*III. Discrete bands from genomic DNA of transgenic catfish 31E, 38E, and 141E hybridized to csGHcDNA.

A fragment of approximately 1.2 kbp that co-migrated with the *Hind*III-linearized pRSVcsGHcDNA was detected in the *Hind*III-digested genomic DNA of fish 31E and 38E (see Figure 25-4). While this result suggests that

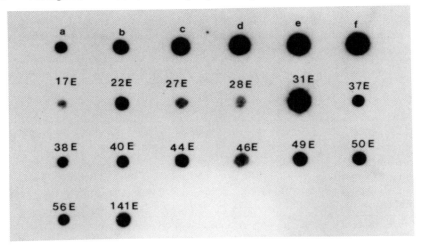
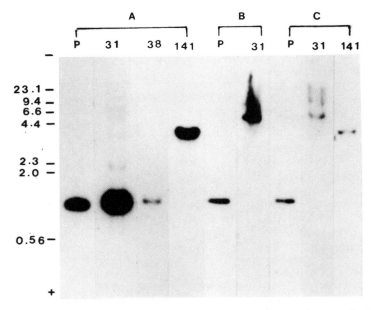

FIGURE 25-4 Dot-blot and Southern blot analysis of genomic DNA isolated from presumptive transgenic catfish. A 1.2-kb *Hind*III fragment containing the cs GHcDNA sequence was used as probe. Top: Genomic DNA (18 μg) isolated from fin tissue was denatured and spotted onto nylon membrane (17E to 141E) together with 30 pg (a), 60 pg (b), 120 pg (c), 180 pg (d), 240 pg (e), and 300 pg (f) of pRSVcsGHcDNA as standards. Bottom: *Hind*III- (A), *Apa*I- (B), and *Sca*I (C) digested genomic DNA (18 μg) was subjected to electrophoresis on a 0.8% agarose gel, transferred to nylon membrane, and hybridized with ^{32}P-labeled cs GHcDNA sequence. *Hind*III-digested DNA from bacteriophage lambda was used as a standard. The csGHcDNA insert was used as a positive control (p). Numbers (31, 38, 141) indicate fish samples.

these fish have integrated intact copies of the foreign DNA in their genomes, fish 141E seems to have a modified HindIII site in the pRSVcsGHcDNA sequence because of the 1.2-kbp fragment that corresponds to the csGHcDNA sequence was not released by digestion with HindIII. A 3.4-kbp fragment was observed instead.

Digestion of genomic DNA with ScaI generated a fragment of about 4.0 kbp. Since this fish contains at least three copies of the foreign sequence, the absence of the HindIII 1.2-kbp fragment and the differences between the hybridization signals of the HindIII-generated 1.2-kbp fragment and the ScaI-generated 4.0-kbp fragment suggests that the foreign DNA has been modified at the HindIII site and integrated as a single molecule at different sites into the fish genome. A fragment of genomic fragment joined to the ScaI and HindIII digested pRSVcsGHcDNA sequence, respectively, may account for the size of the 4.0- and 3.4-kbp fragments.

In the case of fish 31E, where nine copies of the foreign gene were found, seven bands were observed after the digestion with HindIII, including the 1.2-kbp fragment, and three bands were obtained after digestion with ScaI and ApaI, respectively. This pattern of hybridization was consistent with a head-to-tail, tandem array of foreign DNA integrated into the fish genome. All the predicted fragments that would result from digestion of the genomic DNA with HindIII (1.2 and 4.2 kbp), ScaI (5.5 kbp), and ApaI (5.5 kbp) were observed (see Figure 25–4B). Since fish 31E integrated nine foreign DNA molecules into the genome, the additional fragments observed in the HindIII (2.2, 3.4, 4.6, and 6.8 kbp), ApaI (9 and 11 kbp), and ScaI (9 and 11 kbp) digestions could represent genomic DNA junction fragments of variable size, generated by digesting with the corresponding enzyme at a site of the integrated DNA and the next site in the fish genomic DNA. Restriction enzyme analysis using several endonucleases is under way to confirm this pattern of integration of the foreign DNA.

25.2.3 Microinjection of Carp and/or Catfish with Chimeric Plasmids that Contain the pRSVrtGHcDNA (Construct 2) and MThGH

Data from microinjection of fish embryos with recombinant plasmids that contain different GHs are presented in Table 25–5. While 1,500 common carp microinjected with the pRSVrtGHcDNA (construct 2) (see Figure 25–1B) had a survival rate of 25.54% at 90 days, and no integration of exogenous DNA was observed, microinjection of channel catfish with the same construct resulted in seven transgenic fish out of a total of 120 fish analyzed by dot-blot hybridization. This value represents a rate of integration of foreign DNA of about 5.8%, a value similar to that observed in zebrafish (Stuart et al. 1988).

A total of 566 carp embryos were microinjected with pMThGH (see Figure 25–1E). About 41% of the injected embryos hatched, of which 49.13%

TABLE 25-5 Microinjection of Common Carp and Channel Catfish with Recombinant Plasmids That Contain the pRSVrtGHcDNA (Construct 2) and pMThGH

Recombinant Plasmid	Host Fish	Number Injected[a]	Percent Hatched	90-day Survival	Number Analyzed	Integration Frequency
pRSVrtGHcDNA[b]	Carp	1500	9.1%	25.84%	35	0.0%
pRSVrtGHcDNA[c]	Catfish	5849	4.7%	58.9%	120	5.7%
pMThGH	Carp	566	40.98%	49.13%[c]	123	4.87%

[a]Fish embryos were injected at 5 developmental stages.
[b]Construct 2.
[c]Percentage of fish surviving at 21 days.

of fry survived at least 21 days. Of 123 samples of DNA analyzed for integration of pMThGH, six were found to be transgenic by dot-blot hybridization. This value represents a rate of integration of foreign DNA of about 4.7% (Table 25-5), which is again close to that observed in zebrafish (Stuart et al. 1988).

The recombinant plasmid pRSV-Luc (see Figure 25-1D) was also microinjected into carp embryos. We are interested in using this sequence as a reporter gene (Ow et al. 1986) to identify transgenic animals that carry foreign DNA. Currently, we are analyzing about 400 samples of DNA for integration of the luciferase sequence in order to determine whether this exogenous gene has been integrated into the carp genome. Further experiments will include the determination of expression of this gene in transgenic carp.

25.3 TRANSGENIC FISH CONTAINING GENES FOR GROWTH HORMONES

25.3.1 Fast-Growing Carp Containing pRSVrtGHcDNA (Construct 1)

Transgenic animals usually integrate foreign DNA into one of the host chromosomes at an early stage of embryonic development (Gordon and Ruddle 1981; Palmiter et al. 1982a). As a result. exogenous DNA is generally transmitted through the germline (Gordon and Ruddle 1985). In some cases, expression of foreign genes has been detected (Brinster et al. 1981; Wagner et al. 1981).

While integration of exogenous DNA usually results in multimers in tandem, in a head-to-tail array at a single chromosomal site (Palmiter et al. 1982b; Dunham et al. 1987), examples of single-copy integration at multiple sites have also been observed (Wagner et al. 1983). In our present study, analysis of the genomic DNA revealed different phenotypes with respect to restriction pattern which were consistent with single-copy integration at multiple sites (Wagner et al. 1983).

Since pRSVrtGHcDNA was introduced into carp embryos at different developmental stages, germline mosaicism is expected in the resulting transgenic fish. The degree of mosaicism usually determines whether a foreign DNA sequence will be present in the germline, such that the organism is able to pass it on to its progeny (Flytzanis et al. 1985; Etkin and Pearman 1987; Stuart et al. 1988). In this study, F_1 progeny of transgenic males were mosaic for the foreign gene.

While the offspring of parents 131L and 94R were associated with a 32.3 and 43% frequency of germline transmission, respectively, the transgenic male 04R did not carry the foreign DNA. The original transformed progenitor cells of the male parent 94R may be primordial to the whole germline, because the ratio of transgenic to nontransgenic progeny is about 1 to 1 since approximately 50% of the progeny received the foreign gene. Alternatively, more than one of the original cells of male parent 131L were primordial to the germline, because less than half of the progeny were transgenic. The ratio of progeny ratio was 1 transgenic to 3 normal.

If we assume that transgenic fish have multiple sites of integration of pRSVGHcDNA within their germlines, we might expect variation in the restriction patterns of foreign DNA between siblings derived from the same parent. Since different *Hind*III restriction patterns were observed among F_1 progeny of male 131L, the pRSVrtGHcDNA sequence must be integrated at different regions of the chromosome of fish 131L. This mode of inheritance of foreign DNA has also been observed in mice (Etkin and Pearman 1987).

While there was expression of the rtGH gene in the RBC of nine fish tested, there was no correlation between the number of copies of the DNA in the tissue examined and the level of expression. Expression of GH was reflected by increased growth in transgenic fish derived from one-celled embryos. The gain in body weight (above that of controls) seems to be dependent on the developmental stage at which the foreign DNA was transferred, and it appears to be correlated to the amount of rtGH polypeptide present.

We have shown that low levels of biosynthetic GH, rather than high doses, injected into rt induce rapid gains in weight and length in GH-treated fish (Agellon et al. 1988a). Such results are in agreement with this study. Transgenic fish with lower levels of rtGH grew more rapidly than those with higher levels of rtGH. By contrast, the absence of any increase in body weight in the transgenic fish with high levels of rtGH may also indicate that GH is not produced in target tissues. Such possibilities are under investigation.

An increase, relative to controls, in the body weight of offspring was also observed. A mean gain in body weight of about 30%, with respect to controls, was estimated in transgenic progeny derived from parents 131L and 94R.

25.3.2 Transgenic Catfish Containing pRSV-Coho Salmon-GHcDNA

While studies in transgenic carp that carried pRSVrtGHcDNA (1) revealed no differences in rates of hatching or survival at different developmental stages, studies of transgenic catfish that carry pRSVcsGHcDNA revealed higher rates of hatching and survival for embryos microinjected at the early one-cell stage. Furthermore, the highest rate of integration of foreign DNA was obtained from both carp and catfish embryos microinjected at the two-cell stage.

Data from Southern blot analysis of transgenic catfish DNA revealed two modes of integration of the foreign gene: 1) single-copy integration at multiple sites, as observed in transgenic carp, and 2) integration in head-to-tail, tandem array at different sites in the chromosome. The latter integration pattern has been observed in other animal systems (Palmiter et al. 1982b; Dunham et al. 1987).

25.3.3 Luciferase Gene as a Reporter Gene

Detection of foreign DNA sequences in genomic DNA from presumptive transgenic animals is usually carried out by Southern blot hybridization. This procedure, which requires purified DNA, is extremely tedious, expensive, and time-consuming when a large number of samples has to be analyzed. Use of the luciferase gene as a reporter gene may provide an alternative and simpler method for detection of the integration of foreign genes.

Recently, the luciferase gene has been successfully used as a powerful reporter gene for assessing expression in transgenic plants (Ow et al. 1986) and promoter activity in mammalian cells (deWet et al. 1987). Assays of the expression of luciferase provide a rapid, sensitive, and inexpensive method for monitoring promoter activity. Luciferase can be detected immunochemically, with a luminometer, or by simple exposure of reactions to X-ray film when production of light is to be detected. We are exploring the possibility of using luciferase as a reporter gene for detection of transgenic animals that harbor foreign sequences. We are currently analyzing the genomic DNA of several hundred samples from carp microinjected with a chimeric plasmid that contains the structural gene for luciferase fused to the RSV-LTR promoter.

25.3.4 Applications of Genetically Engineered Fish

Aquaculture and the fishing industry are multibillion dollar businesses. Although production of fish in the USA has increased dramatically in recent years, consumption has accelerated more rapidly and, as a result, the USA imports more fish each year than it can produce, a factor that contributes to our growing trade deficit. The use of transgenic fish to enhance the U.S. aquaculture industry could significantly improve our international competitiveness and, in the process, help reduce the trade deficit. Several countries in Asia and Europe are committing hundreds of millions of dollars to bring transgenic-fish technology to the market place. Japan alone has committed $200 million to expand their involvement in marine biotechnology, part of which involves transgenic fish research and implementation. Although the USA has been a major innovator in the development of this exciting field, this country may soon be buying transgenic fish from Japan, Norway, and other countries unless there is a significant increase in funding and research focused on transgenic fish.

REFERENCES

Agellon, L.B., Emery, C.J., Jones, J.M., et al. (1988a) *Can. J. Fish. Aquat. Sci.* 45, 146–151.
Agellon, L.B., Davies, S.L., Chen, T.T., and Powers, D.A. (1988b) *Proc. Natl. Acad. Sci. USA* 85, 5136–5140.
Agellon, L.B., Davies, S.L., Lin, C.M., Chen, T.T., and Powers, D.A. (1988c) *Mol. Reprod. Develop.* 1, 11–17.
Bishop, J.O., Archibald, A.L., Clark, A.J., et al. (1986) in *Animal Breeding Research Organization Report* (Slee, J., ed.), pp. 22–26, Haddington, England.
Brinster, R.L., Chen, H.Y., Trumbauer, M.E., Senear, A.W., Warren, R., and Palmiter, R.D. (1981) *Cell* 27, 223–231.
Cundiff, L. (1983) *Proc. Annu. Reciprocal Meat Conf. Am. Meat Sci. Assn.* 36, 11–17.
DeWet, J.R., Wood, K.V., DeLuca, M., Helinski, D.R., and Subramani, S. (1987) *Mol. Cell. Biol.* 7 (2), 725–737.
Dunham, R.A., Cash, J., Askins, J., and Townes, T.M. (1987) *Trans. Am. Fish. Soc.* 116, 87–91.
Etkin, L.D. and Pearman, B. (1987) *Development* 99, 15–23.
Feinberg, A.P., and Vogelstein, B. (1984) *Anal. Biochem.* 137, 266–267.
Flytzanis, C.N., McMahon, A.P., Hough-Evans, B.R. et al. (1985) *Dev. Biol.* 108, 431–442.
Gill, J.A., Sumpter, J.P., Donaldson, E.M., et al. (1985) *Biotechnology* 3, 643–646.
Gonzalez-Villasenor, L.I., Zhang, P., Chen, T.T., and Powers, D.A. (1988) *Gene* 65, 239–246.
Gordon, J.W., and Ruddle, F.H. (1985) *Gene* 33, 121–136.
Gordon, J.W., and Ruddle, F.H. (1981) *Science* 214, 1244–1246.
Gordon, K., Lee, E., Vitale, J.A., et al. (1987) *Biotechnology* 5, 1183–1187.
Gorman, C.M., Merlino, G.D., Willingham, M.C., Pastan, I., and Howard, B.H. (1982) *Proc. Natl. Acad. Sci. USA* 79, 6777–6781.
Hammer, R.E., Pursel, V.G., Rexroad, C.E. Jr., et al. (1985) *Nature* 315, 680–683.
Hinegardner, R., and Rosen, D.E. (1972) *Am. Natur.* 106, 621–644.
Inouye, S., Vlasuk, G.P., Hsiung, H., and Inouye, M. (1984) *J. Biol. Chem.* 259, 3729–3733.

Maniatis, T., Fritsch, E.F., and Sambrook, J. (1982) *Molecular Cloning: A Laboratory Manual*, Cold Spring Harbor Laboratory, Cold Spring Harbor, New York.
McKnight, S., and Tjian, R. (1986) *Cell* 46, 795–805.
Ow, D.W., Wood, K.V., DeLuca, M., et al. (1986) *Science* 234, 856–859.
Palmiter, R.D., Brinster, R.L., Manner, R.E., et al. (1982a) *Nature* 300, 611–615.
Palmiter, R.D., Chen, H.J., and Brinster, R.L. (1982b) *Cell* 29, 701–710.
Patton, S., Welsch, U., and Singh, S. (1984) *J. Dairy Sci.* 67(6), 1323–1326.
Pittius, C.W., Hennighausen, L., Lee, E., et al. (1988) *Proc. Natl. Acad. Sci. USA* 85, 5874–5878.
Powers, D.A. (1989) *Science* 246, 352–358.
Simons, J.P., Wilmut, I., Clark, A.J., et al. (1988) *Biotechnology* 6, 179–183.
Stuart, G.W., McMurray, J.V., and Westerfield, M. (1988) *Development* 103, 403–412.
Van Brunt, J., Fargher, R.C., Brown, J.C., and McKeown, B.A. (1985) *Gen. Comp. Endocrinol.* 60, 27–34.
Wagner, G.F., Covarrubias, L., Stewart, T.A., and Mintz, B. (1983) *Cell* 35, 647–655.
Wagner, T.E., Hoppe, P.C., Jollick, J.D., et al. (1981) *Proc. Natl. Acad. Sci. USA* 78, 6376–6380.
Wagner, T.E., and Murray, F.E. (1985) *J. Anim. Sci.* 61(3), 25–37.
Weatherly, A.H., and Gill, H.S. (1987) *Aquaculture* 65, 55–66.
Zhang, P., Hayat, M., Joyce, C., et al. (1990) *Molec. Reproduc. Develop.* 25, 3–13.
Zhu, A., Xu, K., Li, G., Xie, Y., and He, L. (1986) *Xexue Tongbao*, Academia Sinica, Wukan, People's Republic of China 31, 988–990.

CHAPTER 26

Transgenic Rats: A Discussion

Jan Heideman

Approximately one year prior to the writing of this chapter, after many months of work, Dr. Amy Moser and I finally began to achieve a workable level of efficiency in the fledgling transgenic mouse facility that we had developed. It seemed that we encountered and had to solve every possible problem associated with making transgenic mice. We had finally defined all the doses and schedules of hormones, the culture media, the peculiarities of mouse strains, and the collection of embryos, and the techniques of injection and transfer to the extent that we felt we were in a position to provide a useful service. We eagerly anticipated making a wide variety of transgenic mice. We were then introduced to Dr. Michael Gould of the Department of Human Oncology at the University of Wisconsin who informed us that a number of investigators there and elsewhere had a very great need for transgenic rats. We were not initially eager to work with rats. We had worked with mice for many years, but neither of us had any prior experience with rats. As dedicated mouse geneticists, we regarded rats as unfamiliar and as animals whose embryos were too delicate to work with. We were not eager to test and modify all the techniques that had taken us so long to develop for mice. In the course of the past year, however, we

---

We thank Dr. Michael Gould of the Department of Human Oncology at the University of Wisconsin for many stimulating discussions and Dr. Henry Pitot who, with Dr. Gould, is financing this research.

have worked extensively with rats, and we expect to have some transgenic rats in the very near future. This chapter includes a description of the development of the various techniques we have perfected. Throughout the following discussion, unless otherwise specified, "we" refers to Amy Moser and myself.

26.1 NEED FOR TRANSGENIC RATS

The production of transgenic animals of any kind is not trivial, and the adaptation of transgenic technology for use in a new species is an extremely time-consuming and difficult undertaking. Therefore, an important preface to our discussion of the techniques used to make transgenic rats is a discussion of why the availability of transgenic rats is important.

The rat is a useful alternative species to the mouse, and there is the possibility that a given result in the mouse might not hold true for any other species. In fact, there are many results that are only true for a particular strain of mouse. Furthermore, as convenient as the small size of mice is for some purposes, some physiological manipulations are simply impossible in mice because of their very small size. For instance, the cannulation of certain blood vessels may be easy in rats but nearly impossible in mice. It is also much easier to collect blood and tissue samples from rats in amounts sufficient for biochemical studies.

A somewhat more specific reason for using rats is that models for certain diseases are available in rats but not in mice. For example, the rat provides a much better model for human mammary cancer than does the mouse. Most mammary tumors in the mouse are associated with mouse mammary tumor viruses, whereas human and rat mammary tumors are generally not due to viruses and are frequently similar in origin.

Genes that affect susceptibility to mammary cancer have been identified only in certain strains of rats and no similar genes are known to exist in mice. In Wistar-Furth (WF) rats there is a gene that enhances susceptibility to mammary tumors and Copenhagen rats have a gene that suppresses susceptibility. Both genes are autosomal and dominant. The enhancer is expressed exclusively in mammary epithelium, and the suppressor also appears to be expressed primarily in mammary epithelium. When they are present together, the suppressor is dominant over the enhancer. Both genes act at some point distal to the early effects of carcinogens, but the enhancer enhances the development of chemically or hormonally induced tumors only and has no effect on spontaneous tumors (Gould 1986).

A number of expression promoter-oncogene constructions are available that cause tumors in specific tissues in transgenic mice. Our initial goal in making transgenic rats is to produce rats with a transgene construction that has been shown to produce mammary tumors in mice. It will then be possible to study the effects of the tumor enhancer and suppressor genes of the rat on these tumors of well-characterized origin. It will also be possible to study the ways in which the enhancer and suppressor genes affect the tumorogenicity of various types of oncogene construction. Because the tumor enhancer genes have been characterized in WF rats, this is the strain with which we are currently working. Unfortunately, these WF rats are inbred and have relatively poor reproductive performance. Some of the techniques described below were developed for work with low-ovulating rats.

26.2 RATS VERSUS MICE

We were introduced to rats primarily by Jane Barnes of the Department of Radiation Biology at the University of Wisconsin, from whom we received extensively valuable assistance and advice. Our early experiences with rats involved unlearning much of what we had assumed to be true from our considerable experience with mice. Because many of the chapters in this book deal with transgenic mice, we shall emphasize herein those differences between rats and mice that affect the production of transgenic animals. Investigators who work with rats are already familiar with this information, but those who have not worked with rats and might be interested in doing so may find it valuable.

In many strains of mice, the female reproductive cycle at almost any age can be completely reprogrammed by injections of pregnant mare serum gonadotropin (PMSG) and human chorionic gonadotropin (hCG). The time of day at which ovulation and formation of the pronucleus occurs can be precisely controlled by the timing of the hormone injections. Adult animals superovulate almost as well as prepuberal ones, and an average of 20 to 30 normal embryos per donor is a reasonable expectation; often the yield is even better. After mating, female mice usually retain their vaginal plugs for 12 or more hours, so it is easy to tell which potential donors will yield fertilized eggs and which potential recipients are really psuedopregnant. Surgery on mice is routinely performed using injected anesthetics on the stage of a dissecting microscope, and incisions can be closed using only wound clips. Rats are less convenient to work with. Only prepuberal rats superovulate well, and the timing of ovulation and formation of pronucleus is closely tied to the lighting schedule. If it is necessary to use adult rats for mating on a certain day, one must either check vaginal smears until sufficient females in estrus are found, or administer an agonist of luteinizing hormone-releasing hormone (LHRH) for one entire estrous cycle before the intended date of mating. Unfortunately, lengths of cycles can vary. Rats usually retain their vaginal plugs for only a few hours after mating, so it is necessary either to check plugs in the early hours of the morning or to recheck smears for the presence of sperm on the day after mating. Obtaining psuedopregnant rats is more complicated than obtaining pseudopregnant mice because one must either find a plug left by a vasectomized male, or trust that the presence of leukocytes in a smear indicates sufficient stimulation to induce psuedopregnancy. Mechanical or electrical stimulation can also be used, but consistent success with these methods requires experience.

Some researchers have reported good success with pentabarbital anesthesia in rats, but in our hands, a lethal dose in one rat might be insufficient to induce anesthesia in another of the same age, sex, and weight. Use of inhalant anesthesia is more controllable, but requires both frequent monitoring and experience. Furthermore, a rat does not fit on the stage of a standard dissecting microscope. Finally, if, as is often the case, animal housing is located some distance from surgical and microinjection equipment, the transport of rats can become a real logistical problem. We had to adapt to all these previously well-characterized differences between rats and mice before we could tackle the more interesting and novel problems associated with making transgenic rats.

26.3 COLLECTION AND TRANSFER OF EMBRYOS

The first step in developing transgenic rats is learning to collect and handle rat embryos and transfer them to the oviducts of psuedopregnant recipients. Many of the standard techniques used in the mouse (Brinster et al. 1985; Hogan et al. 1986) must be modified for use in the rat. Aspects of mouse techniques that do work in the rat will be mentioned only briefly here.

A rat's skin cannot be peeled off as can the skin of a mouse. The incision must be made close to the organ to which access is needed. The volume of intestines and fat is much larger in a rat than in a mouse, while the reproductive tract is only slightly larger. It easiest to collect embryos by removing the oviducts through an incision in the flank. It is not as easy to visualize rat embryos through the oviductal wall as is the case in the mouse, but rat oviducts are very easy to flush. Rat embryos look similar to mouse embryos, although rat embryos appear paler than mouse embryos under a dissecting microscope and, at higher magnification, they have a slightly more granular cytoplasm and less regular and uniform polar bodies and pronuclei.

26.4 TRANSFER OF EMBRYOS TO RAT OVIDUCTS

To prepare recipient rats for oviductal transfer, we have induced anesthesia with Halothane U.S.P. (2-bromo–2-chloro–1,1,1-trifluoroethane) and maintained it with Metofane (methoxylfluorane). The incision in the skin can be made in the middle of the lower back, and the body wall can then be penetrated directly over the ovarian fat pad, as in the mouse. However, the body wall of the rat is much too thick to permit visualization of any underlying structures and some care and anatomical knowledge is needed to place the opening correctly. The ovarian fat pad is very large in an adult rat. The best place at which to grasp it for oviductal transfer is on the exact opposite side of the ovary from the oviduct. This approach minimizes tension on the ovarian bursa, thereby facilitating access to the infundibulum. Those who have performed oviductal transfers in the mouse will know that the infundibulum in the mouse usually sticks up almost vertically from behind the coils of the oviduct and can be brought into view through the bursa by gentle probing and manipulation. It is then usually possible to find a nonvascular area of the bursa that can be maneuvered close to the opening of the infundibulum to permit a bloodless incision in the bursa. In the rat, the bursa is much more vascular, and there is a thick capillary bed overlying the oviduct. In addition, the infundibulum almost always lies horizontally in a deep groove between the oviduct and the ovary.

To perform the transfers, we suggest the following procedure. Inspect the bursa and determine the location of the largest nonvascular area close to the oviduct. Make an incision close to the oviduct. Gently and gradually stretch the opening and move it around until the infundibulum is visible. Some bleeding is almost unavoidable, but a small amount of blood can be allowed to clot and blotted away without affecting the success of the transfer. It is much easier to insert the transfer pipet into the infundibulum if about 4 mm of the tip has been bent at an angle of approximately 30°. The pipet can then be moved down through the incision and the tip inserted into the horizontally oriented infundibulum. (This type of bent transfer pipet was suggested to me by Sally Camper for use in mice.) When the embryos have been placed in the oviduct, gently return the tract to the body, close the opening in the body wall with a single stitch, and close the skin with wound clips. About 80% of the nonmicroinjected rat embryos that have been transferred in this way in our laboratory have developed into normal baby rats.

26.5 TECHNIQUE FOR MICROINJECTION OF RAT EMBRYOS

Although rat embryos look similar to mouse embryos, microinjecting them is quite different in a number of ways. The largest difference is that the vitelline and pronuclear membranes in the rat are tremendously flexible compared to those in the mouse. (One consequence of this flexibility is that rat embryos, with their membranes intact, sometimes ooze out of the zona pellucida through the sperm-penetration slit.) In mouse embryos, if the injection pipet is new and sharp and is correctly aimed at a pronucleus, only a gentle poke is needed to insert it. Not so in the rat! A pipet that would enter a mouse embryo with barely a touch must be pushed all the way through a rat embryo and out the other side into the center of the holding pipet before the membranes are pierced. The tip must then be quickly pulled back so that the DNA solution is expelled into a pronucleus rather than into the cytoplasm. In the mouse, it is quite important that the center of the holder, the equatorial plane of the egg, the pronucleus to be injected, and the injection pipet all be in one line. In the rat this orientation is absolutely essential. A slight misalignment is permissible in the mouse, but not, apparently, in the rat.

A relatively small opening in the top of the injection pipet makes it slightly easier to insert, and the resultant slower flow of DNA solution provides more time to pull the tip back and position it correctly inside a pronucleus. A faster flow of DNA means that more will be spilled into the cytoplasm. To prepare injection tips, thin-walled, microfilament capillary tubing is heated and pulled so that the opening is invisibly small (probably less than 0.01 μm). After the tip has been filled and is installed in the micromanipulator, the end should be brushed against the outside of the holding pipet until the opening is large enough to permit a flow that can be seen with phase-contrast optics. This manipulation results in a tip that is very sharp because it is freshly broken glass. This method is faster than grinding, and the tips produced are sharper than those that are pulled with a larger opening. Entering a rat embryo with a ground tip is almost impossible (M. Hoshi, personal communication).

A successful injection in the rat is easier to see than in the mouse. The rat pronucleus stays "blown up" for a longer time, and it undergoes a greater increase in brightness when viewed with phase-contrast optics. One must be aware, however, that it is possible to create DNA-filled vacuoles in the cytoplasm of rat embryos if the pipet tip is not placed precisely inside a pronucleus. It is useful to check in each case that the structure injected has nucleoli. Fortunately, nucleoli in rat embryos seem to have somewhat less of the aggravating tendency of mouse-embryo nucleoli to bind irreversibly to the injection pipet.

When microinjecting mouse embryos, there is often a lysis rate of about 10%. With rats the rate is usually about 15% in our hands, but it is apparent that more damage is done in microinjecting the rats. It is possible in rat embryos that the pronuclear contents stream out into the perivitelline space but the plasma membrane still heals, so that the embryo does not actually lyse. It is also possible that a fistula forms between the two pronuclei so that an exchange of pronuclear contents occurs. These oddities do not present a problem in mouse embryos since they have never been recognized in our laboratory.

We do not know whether we have produced any transgenic rats thus far. We have transferred several hundred embryos in which the volume of one pronucleus was conspicuously increased by the injected DNA solution, and we have had 16 pups born, 10 of which survived. It is probable that the percentage of microinjected rat embryos that develops into transgenic animals will be much lower than the percentage possible with mouse embryos, largely because rat embryos are so much more difficult to inject.

26.6 INCREASING THE YIELD OF EMBRYOS

More surviving pups from microinjected embryos might be obtainable with Sprague-Dawley (SD) rats, which are hardy, outbred, large-litter rats for which superovulation conditions are well-established and which mate very readily. When initially learning how to adapt mouse techniques for use in the rat, we did use SD rats, from which we found it relatively easy to collect as many embryos as necessary. For the reasons mentioned above, however, we concentrated on Wistar-Furth (WF) rats, which are small, inbred, low ovulators, and do not always mate readily. There is little in the literature relevant to superovulation conditions for WF rats. In our hands, 28- to 35-day-old WF females undergo vaginal opening and have proestrus smears 48 hours after injection of PMSG, but their vaginal opening is very small and they do not mate. Older prepubertal females respond unevenly to PMSG, but they do mate if they have proestrus smears and they yield about 2 to 3 times as many embryos as do naturally mated adults. In our hands, 10 IU of PMSG is the optimal dose, and hCG does not appear to increase the yield of usable embryos. To reduce the time spent checking smears, we have begun to use the LHRH agonist (Des-Gly10, D-Ala6, L-ProNHEt9) LHRH (Walton and Armstrong 1983) to synchronize the cycles of adult females and we have found that in addition to synchronizing them, this treatment increased the percentage of females that mated and doubled the yield of good-quality embryos. All our WF females have five-day cycles, so our current preparation of WF donors involves injection of 40 μg of LHRH agonist on day −5 (where mating occurs on the night of day 0). This procedure gives an average of 10 good-quality embryos per donor.

26.7 INCREASING THE SIZE OF LITTERS

The reproductive performance of our WF rats remains somewhat inconsistent and, because microinjection is very damaging to embryos, only a small proportion of microinjected embryos develops into pups. Our initial attempts to obtain pups from microinjected WF embryos failed to result in pregnancies because insufficient numbers were transferred. The first pregnancy produced only two pups, which were born one day late and did not survive. The second litter, of four, was also late and only one pup survived. To increase the size of litters, we now mate a few WF females with Copenhagen (Cop) males. Copenhagen rats have dominant hooded-coat markings, and Cop males mate and fertilize eggs at a consistently higher rate than do WF males. We transfer embryos from a Cop × WF mating (non-injected), together with the injected WF × WF embryos to produce litters of reasonable size. Injected pups can be distinguished from noninjected pups by coat color. About 10% of the injected embryos should develop into pups, and the optimal litter size in the SD recipients is about eight so that the appropriate number of noninjected embryos is easily calculated. The mixed strain litters are working out well, but quite a few of them will be required to satisfy the need for transgenic rats.

Note Added in Proof In June 1990 about 8% of rats we produced in collaboration with Dr. Henry Pitot (Department of Oncology, University of Wisconsin, Madison) were shown to be transgenic and to have an expected altered phenotype due to transgene expression.

REFERENCES

Brinster, R.L., Chen, H.Y., Trumbauer, M.E., Yagle, M.K., and Palmiter, R.D. (1985) *Proc. Natl. Acad. Sci. USA* 82, 4438–4442.
Gould, M.N. (1986) *Cancer Res.* 46, 1199–1202.
Hogan, B., Constantini, F., and Lacy, E. (1986) *Manipulating the Mouse Embryo: A Laboratory Manual*, Cold Spring Harbor Laboratory, Cold Spring Harbor, New York.
Walton, E.A., and Armstrong, D.T. (1983) *J. Reprod. Fertil.* 67, 309–314.

PART VI

Transgenics: Society, Commerce, and the Environment

CHAPTER 27

The Mouse that Roared
Lisa J. Raines

In April 1988, the U.S. Patent and Trademark Office awarded Harvard University the world's first patent for an animal, a mouse whose cells had been genetically engineered to carry a cancer-promoting gene. Although the man-made rodent may prove invaluable in cancer research, its place in history will likely stem more from its role as the symbol of a heated national debate.

Simmering for several years, the controversy began to boil in the spring of 1987, when the patent office announced its intention to award patents for "transgenic" animals, that is, animals created with genes from nonparent animals. Animal rights advocates claimed the new policy would increase animal suffering. Farm groups charged that it would hurt the family farm. Environmental activists said that it would upset the laws of nature. Religious leaders maintained that such patents violated divine law. And Jeremy Rifkin of the Foundation on Economic Trends, the perennial foe of biotechnology, expressed concern about all the above.

The critics turned up pressure on Congress to declare a moratorium on animal patenting. The Senate approved without debate a four-month moratorium, which was added as a rider to an appropriations bill, but the House

---

This chapter is reprinted, with permission, from *Issues in Science and Technology*, Volume IV, Number 4, © 1988 National Academy of Sciences, Washington, DC.

rejected the legislation. Rep. Charlie Rose (D-NC) then introduced a House bill calling for a two-year halt on animal patents. The bill, which has now gathered 59 co-sponsors, has been the subject of four hearings, but it is still not known if the House will vote on the measure. In the Senate, Mark Hatfield (R-OR), author of the proposal rejected last year, has introduced new legislation that would ban animal patenting indefinitely. So far, no Senate action has been scheduled.

27.1 AN ABUNDANCE OF USES

The Harvard mouse illustrates several of the potential applications of animal biotechnology. Transgenic animals can serve as models for studying disease—cancer, in this case. "Here is a strain of mice that have taken the first step toward cancer," Philip Leder, head of the team of geneticists that won the patent, said in *Science* magazine. "With them, you can learn some rather powerful and telling lessons about the genesis of cancer in vivo." For numerous other serious disorders, such as Alzheimer's disease, scientists have few or no animal models to use in their studies, and genetic engineering may offer a way to make models to order.

Bioengineered animals will also prove valuable in toxicology testing. Animals tailored to be hypersensitive to a variety of carcinogens, mutagens, and poisons will make efforts to screen for environmental hazards both faster and more accurate. As the Harvard patent claims for its product,

> [the mouse's] sensitivity will permit suspect materials to be tested in much smaller amounts than the amounts used in current animal carcinogenicity studies and thus will minimize one source of criticism of current methods, that their validity is questionable because the amounts of the tested material used are greatly in excess of amounts to which humans are likely to be exposed.

"Molecular farming," creating animals that produce useful human or animal protein products, is another wide-open area. For example, scientists have added a particular human gene to a mouse so that it manufactures a protein called tissue plasminogen activator in its breast milk. The protein, which can be easily extracted from the milk, dissolves blood clots effectively and was recently approved by the U.S. Food and Drug Administration for treatment of heart attack patients.

Conventional farming will also see many applications of animal biotechnology. Production will be made more efficient and foods healthier. For instance, sheep produce milk that is both more nutritious and better suited chemically to the manufacture of butter and cheese than is cow's milk, while cows produce milk in greater quantity. Researchers have now taken an important first step toward combining these traits, by inserting the appropriate sheep gene into mice. Studying how the gene functions in mice—

which are cheaper, easier to work with, and better understood genetically—will help provide the knowledge needed to eventually add the gene to cows. Scientists are also working to help farmers literally bring home more bacon, as well as other pork products. They have inserted a gene into pigs that boosts production of growth hormone. Farmers may thus be able to bring their animals to market in 17 weeks instead of the current 22-25 weeks, which is expected to increase profits substantially.

At least 21 applications for animal patents are now pending at the U.S. Patent Office, and more are expected as researchers work their way up the learning curve. But if Congress indeed approves a moratorium on such patenting, temporary or permanent, the pace of development in animal biotechnology will be dramatically slowed. Companies will be discouraged from investing their research dollars in this field. And since patentability is the quid pro quo for public disclosure of how an invention works, a halt will likely lead to secrecy by the few companies who choose to continue commercial work. A moratorium will ultimately undercut competitiveness of U.S. agriculture and unnecessarily hobble progress in disease treatment and prevention.

27.2 THE RULES CHANGE, SLOWLY

Federal law requires that, to qualify for a patent, an invention must be *novel, useful,* and *non-obvious.* That is, the invention must be the first of its kind, it must be functional, and it must constitute something more than a trivial extension of what was previously known.

These guidelines have generally excluded "products of nature"; microbes, plants, and animals that occur in the wild are unpatentable because they are not made by humans. Cultured strains of microbes and cross-bred varieties of plants and animals are *not* considered to be products of nature—human intervention was a necessary component of their creation—but such cultivated organisms produced by traditional techniques are often considered "obvious" and hence unpatentable. Actually, lower-order life forms have been patented since the advent of microbiology; in fact, Louis Pasteur obtained the first U.S. patent on a living invention, a cultured strain of yeast, roughly a century ago. But with the advent of the scientific revolution in practical genetics and the potential for a vast array of new products, the patent office became reluctant to apply its own precedents to novel products created by the new biotechnology.

However, the door opened in 1980. The U.S. Supreme Court, in *Diamond* versus *Chakrabarty*, ruled in favor of a microbiologist who had been denied patent protection for a genetically engineered strain of bacteria that could break down crude oil. After examining the legislative history of the patent statute and the record of judicial interpretation, the court determined that the microbes developed by Ananda Chakrabarty, then with General

Electric, met the criteria for patentability. His bacteria were unique, they were potentially useful in managing oil spills, and their invention would not have been obvious to a skilled scientist of the day. The fact that the invention was living was deemed irrelevant; "anything under the sun that is made by man," declared the Court, is patentable.

Still, the Patent Office maintained that genetically engineered plants and animals were unpatentable, arguing that the Supreme Court had acted only on the specific matter of microorganisms. Moreover, Patent Office officials were actually uncertain about how to interpret laws enacted by Congress to protect the rights of plant breeders. The Plant Patent Act of 1930 and the Plant Variety Protection Act of 1970 granted certain patent-like rights to the creators of new types of plants: Did this imply that Congress did *not* intend for the general patent statute to apply to plants, or did it mean that Congress simply wanted plant breeders to have some form of protection even when their invention failed to meet all the criteria to qualify for a patent? The Patent Office did not want to decide by itself. This question did not arise with respect to patenting animals because there were no related laws. But as long as animal biotechnology was still in its infancy, resolution of the overall issue was not deemed urgent.

Legal commentators frequently criticized the government's hard-line stand. In 1985, the American Bar Association's (ABA) Section on Patent, Trademark, and Copyright Law voted to oppose the Patent Office's decision not to patent novel plants and animals. Shortly thereafter, an inventor appealed the rejection of his patent application for a new plant to the Patent Office's Board of Patent Appeals and Interferences, and the board unanimously ruled that man-made plants were patentable.

But the government still wouldn't budge on animals, and in 1986 the ABA again voted against the Patent Office. Later that year the inventor of a novel oyster protested his patent denial to the appeals board. The board upheld the decision, but not because the invention was an animal. On the basis of this ruling, the Commissioner of Patents and Trademarks issued his landmark 1987 announcement that animals would henceforth be patentable (human beings were excepted, because the 14th Amendment to the Constitution prohibits the ownership of one person by another). The bar association adopted another resolution that opposes legislation calling for a moratorium on animal patents in August 1988.

27.3 THE NATURE OF PATENTS

Opponents of animal biotechnology are using the patent issue as a way to focus public attention on matters unrelated to the Patent Office's mandate. They are, in effect, demanding that patent law be revised to address concerns about product regulation, welfare, and morality. Many of the critics fail to understand what a patent is, and what it is not.

The government's authority to issue patents is derived from a provision in the Constitution that states

> The Congress shall have the power . . . to promote the progress of Science and the Useful Arts by securing for limited times to Authors and Inventors the exclusive right to their respective writings and discoveries.

The first Congress ruled that patents should be granted for a nonrenewable term of 14 years. This law has been revised periodically and today patents are issued for a 17-year term.

Congress and the Supreme Court have repeatedly underscored the purpose of the patent system as promoting progress. An inventor, in return for the right to exclude others from profiting from his creativity for a limited time, must disclose to the world how to make and use the invention. Exclusivity is offered as an incentive to risk the often enormous cost of research and development. And by making details of the invention public, other people benefit intellectually from the achievement, thereby advancing the state of the technology. Conversely, the inability to obtain patent protection is a significant disincentive for both investment and publication. Trade secrecy is the only avenue of protection for the inventor of an unpatented product, and this is often insufficient to prevent a competitor from pirating the technology. The competitor, who has no research and development investment to recoup, can then market generic versions of the product at a lower price.

However, a patent does not automatically grant an inventor the right to commercialize his or her product. Rather, it only prevents others from commercializing the invention without the inventor's permission. Thus, a patentee may not make, use, or sell his invention if legal restrictions exist or until any required premarket regulatory approval is obtained. For instance, a patented drug may not be sold until it is approved by the Food and Drug Administration, and a patented gun may under some circumstances be prohibited from sale.

But the patent system was not created to be a substitute for regulatory authority, which falls clearly within the domains of the various federal and state regulatory agencies. If existing regulations are inadequate to protect public health, the environment, and animal welfare, let them be changed. Indeed, proponents of animal biotechnology acknowledge that some applications should be regulated. For example, transgenic animals designed to carry pathogenic material, such as a mouse that has viral genes added to its genetic code, must be very carefully controlled if there is a risk of the infection spreading.

Yet critics persist in offering a number of philosophical arguments against patenting animals. For example, Jeremy Rifkin originated the claim that transgenic animals violate "species integrity." This he defines as the "right" of an animal "to exist as a separate identifiable creature." The concept has been echoed by the National Council of Churches, which objected to violating "creation's inherent structures and boundaries."

However, the concept of species integrity, like "creation science," runs counter to what is known about biology. As stated by the congressional Office of Technology Assessment, which has studied animal bioengineering: "There is no consistent or absolute rule that species are discretely bounded in any generally applicable manner." In the wild, species repeatedly recombine their genes through sexual reproduction. Individual genes are constantly being altered by mutation and natural selection. Viruses insert foreign genes into animal genomes. Occasionally, new species arise from old, or two related species interbreed to form a third species. And for domestic animals, the notion of species integrity is even less meaningful. Traditional crossbreeding has transformed species enormously over many centuries. With each new cross, thousands of genes are mixed.

Even a study prompted by the critics found no basis in the species-integrity argument. The President's Commission for the Study of Ethical Problems in Medicine and Biomedical and Behavioral Research was created in 1980 at the request of the National Council of Churches and others who wrote to President Carter calling for an examination of the moral issues raised by the Supreme Court's Chakrabarty decision. According to the commission's 1982 report, there are no specific religious prohibitions about "breaching species barriers." In fact, the report noted, "the very notion that there are barriers that must be breached prejudges the issues. The question is simply whether there is something intrinsically wrong with intentionally crossing species lines. Once the question is posed in this way, the answer must be negative, unless one is willing to condemn the production of tangelos by hybridizing tangerines and grapefruits or the production of mules by the mating of asses with horses."

Current techniques of genetic engineering are capable of inserting only a handful of genes into an animal with 50,000 to 100,00 or more genes. Such manipulations would not disrupt anything fundamental in the animal's architecture. A mouse with a human growth hormone gene is still fundamentally a mouse, though its growth characteristics may be enhanced.

27.4 ANIMALS AS TENNIS BALLS?

Critics also argue that animal patents devalue animal life. "The patenting of life reflects a dominionistic and materialistic attitude toward living beings that precludes a proper regard for their inherent nature," says John Hoyt, president of the Humane Society of the United States. And according to Rifkin, patenting animals "reduces the entire animal kingdom of this planet to the lowly status of commercial commodity, a technological product indistinguishable from electric toasters, automobiles, tennis balls, or any other patented product."

Such statements are no more or less true for patenting animals than they are for domesticating animals. Owning animals is legitimate and traditional in our culture, and human dominion over the animal kingdom is even a common biblical theme. Animals are bought and sold daily, valued in the marketplace on the basis of their rarity and utility. We eat them, wear them, perform biomedical research on them, put them on leashes. As bioethicist LeRoy Walters of Georgetown University's Kennedy Institute for Bioethics has suggested, patenting them "seems relatively benign." Of course, patented animals should be treated with care and compassion. Nevertheless, nobody has yet explained how owning a patented animal meaningfully differs from owning an unpatented animal.

The Humane Society argues, too, that transgenic animal research will result in what it calls a "dramatic increase in the suffering of animals." Certainly, it is possible for genetic engineering to produce an animal with characteristics that interfere with its normal functioning. But it is equally possible to create dysfunctional animals through traditional cross-breeding activities. For example, modern turkeys have been bred with breasts so large that it is impossible for them to mate, and numerous breeds of dogs now possess inbred health problems ranging from arthritis to kidney failure.

In fact, genetic engineering may ultimately lead to less animal suffering. One of the most promising areas of research is the development of farm animals that resist disease. For example, scientists may be able to create cattle that are resistant to a disease called shipping fever. This would reduce the chances of their suffering while being transported to feedlots or to market, as well as diminish a major source of economic loss for ranchers.

Biotechnology is also likely to provide some welcome alternatives to the ways that animals are used in laboratory testing. Because animals can be engineered to be extremely sensitive to various materials, just as the Harvard mouse was made sensitive to carcinogens, scientists will be able to obtain meaningful data using fewer animals. And lower-order animals such as mice might be substituted in some research for higher-order animals such as primates. For example, the National Institutes of Health recently developed a transgenic strain of mice whose genes contain a portion of the human immunodeficiency virus (HIV) implicated in AIDS. The virus sits quietly in the animals' cells, just as HIV can remain latent for years before flaring up and taking control of cellular operations. For certain kinds of AIDS research, these mice are the only substitutes for chimpanzees. And as Anthony Fauci, director of federal AIDS research, points out, the limited supply of laboratory animals is a serious impediment to progress in this important field.

But perhaps the most common argument offered by critics of patenting animals is that producing new life forms for profit "strikes many as morally offensive," in the words of the National Council of Churches. However, 68% of the American public and 81% of college graduates believe that it is *not* morally wrong to genetically engineer plants and animals, according to a recent Louis Harris survey. Most people, in fact, want to see the technology advance.

Of course, the government should not turn a deaf ear to ethical concerns simply because the views are held by a minority. But neither should the government deny protection to an entire class of intellectual property simply because some people find certain potential applications morally objectionable. Should the government forbid patenting of contraceptives because some people believe artificial birth control is immoral, or deny copyrights for books that some readers judge to be pornographic?

Even if the government were to consider opposition to animal patenting on moral grounds, it would have to consider as well the needs of those who would benefit from animal biotechnology. Patients with life-threatening disease, for example, need the medical advances that would certainly occur. "The lives of millions of cancer patients and their families depend on the continued development of research tools, including animals, that can help us develop safer and more effective treatments for this terrible disease," John E. Ultmann, chairman of the National Coalition for Cancer Research, wrote in a letter to the House of Representatives opposing the proposed moratorium.

Millions of people also stand to benefit from the improved nutritional value of products derived from genetically engineered animals. For instance, the National Research Council's (NRC) Board on Agriculture recently recommended the development of animals that have leaner meat. Lowering the fat content in the typical American diet, declared the NRC committee report, *Designing Foods*, will significantly reduce the number of deaths from cardiovascular disease, now the nation's leading killer.

27.5 DOWN ON THE FARM

The patentability of animals is being challenged on economic as well as moral grounds. Some agricultural activists claim that the new policy will boost costs, citing in particular the possibility that farmers and ranchers will have to pay royalties for progeny produced by patented animals. But a wise farmer is not going to buy a patented animal, or any other item, unless it will increase profits. That is just what transgenic breeds are being engineered to do. The new animals will provide meat and milk at a lower cost of production. They will be healthier. And they will yield premium products, such as meat that is low in calories and cholesterol, which will fetch higher prices.

Moreover, early signs suggest that companies are not likely to want a royalties-for-offspring arrangement. Enforcing the royalty policy would require periodically sending inspectors, armed with genetic screening kits, to every farm where transgenic animals have been purchased in order to determine which young animals possess the patented trait. The costs would probably exceed the value of the royalties collected. Instead, the free market will likely yield the solution. For example, farmers may be able to make "ordinary farm use" of the animals, including normal breeding activities, but they would be forbidden from commerical breeding that would compete with the patent holder.

Nor will biotechnology hurt the small farmer in particular. Unlike some other agricultural technologies, patented animals will not be likely to require an unduly large startup investment or offer economies of scale. Small farmers should be able to integrate the animals into their operations as quickly and easily as large farmers, and with similar effects on the bottom line. According to a report on patenting animals, published by the American Farm Bureau Federation, which represents 80% of U.S. farmers, "improved breeds that produce meat and milk with a lower cost of production or that resist common diseases should help the small farmer stay competitive by reducing farm costs and increasing the value of the products."

In fact, U.S. farmers in general are hard pressed by low commodity prices and foreign competition, not by innovation. They need animal biotechnology to help them compete in the world marketplace, which is influenced by Japanese beef policies and Australian lamb policies as well as by conditions at home. And most farmers seem to agree. "We believe that the majority of farmers in the United States are supportive of animal patents and that a moratorium would be detrimental to the long-term interest of agriculture," the Farm Bureau Federation stated in a recent letter to members of Congress.

Why is there now such a pressing need for animal patents, when traditional breeders have made great strides without their protection? Historically, patents were less necessary because the government, rather than the private sector, funded most basic and applied agricultural research. Virtually all research on improving breeds was conducted either by the U.S. Department of Agriculture or academic scientists working with public funds at land-grant colleges.

However, federal support for agricultural research has declined steadily since World War II. "This lack of commitment to financial support for basic research in agriculture has had cumulative and far-reaching impacts," noted a 1987 report by the National Research Council's Board on Agriculture. The private sector now plays a larger role in funding agricultural research than does the government, according to the report, accounting for $2.1 billion of the $4 billion spent in 1986. But companies have generally stayed away from animal breed-improvement research, largely because they had little incentive to invest in a technology they could not protect. And federal funding is no longer sufficient to keep U.S. researchers apace with foreign efforts.

27.6 TIME TO DECIDE

Although the United States is the first country to patent a transgenic animal, it is not the only nation that recognizes the patentability of transgenic animals. A 1987 survey by the World Intellectual Property Organization shows that 53 countries have not expressly excluded animals from patent protection, and patents are already permitted in 10 of these countries: Japan, Canada, Australia, Argentina, New Zealand, Turkey, Brazil, Greece, Hungary, and the Netherlands.

And the United States is not the world's scientific leader in this field. Although the U.S. biotechnology industry is superior overall, the United Kingdom and Ireland appear to lead in animal biotechnology. For example, British researchers have genetically engineered cattle, hogs, poultry, salmon, and caterpillars; Scottish researchers have developed sheep and mice; and Irish researchers have created new salmon. The Irish project, funded by the European Economic Community's Biotechnology Action Programme, has been pronounced "very successful from a commerical point of view" by the Irish Development Authority, which hopes to begin exporting fish within the next few years. Although these countries do not allow animals to be patented, their governments provide major financing for animal biotechnology research.

Although the United States is being left behind, some members of Congress are apparently still tempted to support a moratorium in the belief that more time is needed to study the matter, in spite of a year's worth of congressional fact-finding, during which more than 30 witnesses testified before the House Subcommittee on Courts, Civil Liberties, and the Administration of Justice. Indeed, the staff of subcommittee chairman Robert W. Kastenmeier (D–WI) has prepared a draft report, which concludes that a moratorium would be "unwise and unnecessary." Such a step, the report argues, "would stifle—if not extinguish—important innovations in medical research and agriculture without justification."

Even without a moratorium, Congress will always be free to place restrictions on animal biotechnology at any time and by a variety of methods. Congress may place a heavy tax on the sales of patented animals, revise regulatory laws to impede or prohibit sales of transgenic animals, or even modify patent law to reduce the rights of certain patent holders. But unless Congress can find a good reason to ban all forms of transgenic animals, a patent moratorium will be a mistake, discouraging U.S. companies from investing in animal biotechnology while foreign competitors push forward.

Banning animal patents would also undercut U.S. credibility in current international negotiations. U.S. negotiators are now trying to persuade a number of foreign countries, such as Brazil, to revise intellectual property policies that discriminate between classes of inventions. For instance, some countries deny protection to pharmaceuticals, computer software, microbes, and other inventions protected by the United States. This means U.S. inventors lose their protection when it comes to doing business in these markets. The United States now has the moral high ground by arguing that all such restrictions are inappropriate. But if the government begins adopting similar restrictions on patentable subject matter, how can our negotiators persuade other nations to abandon restrictions that protect other technologies?

The arguments against patentability are for the most part thinly veiled objections from a small number of groups opposed to biotechnology and large-scale commercial farming. On the other hand, the Patent Office's decision to patent novel animals is consistent with Supreme Court rulings regarding the scope of patentable subject matter and reflects the general opinion of the scientific and legal communities. And given the need to enhance U.S. industrial competitiveness, it constitutes sound public policy.

RECOMMENDED READING

American Farm Bureau Federation, National and Environmental Resources Division (1987) *Animal Patents: Agriculture's Perspective*, American Farm Bureau Federation, Park Ridge, IL.
Bent, S.A., Schwaab, D.G., Conlin, D.G., and Jeffrey, D.D. (1987) *Intellectual Property Rights in Biotechnology Worldwide*, Stockton Press, New York.
Patenting Higher Life Forms: Hearings Before the Subcommittee on Courts, Civil Liberties and the Administration of Justice of the Committee on the Judiciary (1988) 100th Congress, 2nd Session, U.S. Government Printing Office, Washington, DC.
President's Commission for the Study of Ethical Problems in Medicine and Biomedical and Behavioral Research (1982) *Splicing Life: A Report on the Social and Ethical Issues of Genetic Engineering with Human Beings*, U.S. Government Printing Office, Washington, DC.

INDEX

Acquired immunodeficiency syndrome. *See* AIDS
Actin gene, 56, 59–61
  cis-acting DNA sequences in, 56–58
β-Actin promoter, 103–123, 145
  activity in tissue culture cells, 120–121
  with deleted sequences, 121
β-Actin promoter-multidrug-resistance transgene, 104–108
  expression of drug resistance, 121–122
  production of mice carrying, 108–110
  transfection to human cells, 107
  transfection to mouse cells, 107–108
ADA. *See* Adenosine deaminase
Adenosine deaminase (ADA) deficiency, 193–194
Adenosine deaminase (ADA) gene, 196–201

Adrenal gland, in hGH transgenic mice, 243
AFP. *See* α-Fetoprotein
Agininosuccinic aciduria, 194
Agricultural economics, 342–343
AIDS, animal model of, 213–226
Albumin enhancer, 83, 85–86
Albumin gene
  activation and silencing in liver, 81–86
  regulatory elements of, 84–85
Albumin promoter, 84–86
Allelic exclusion, 11
ALV. *See* Avian leukosis virus
American elvers, 296
Angioneurotic edema, hereditary, 194
Anguilla rostrata. See American elvers
Animal patent, 335–345
Animal rights activists, 338–342
Antennapedia complex, 136. *See also* Homeobox gene
Antifreeze protein gene, 297–298

347

Antigen. *See specific antigens*
Antisense gene, 15
α1-Antitrypsin deficiency, 194–195
Apo A-1. *See* Apolipoprotein A-1
Apolipoprotein A-1 (apo A-1)
    quantification of, 228
    turnover in vivo, 229
Apolipoprotein A-1 (apo A-1) gene
    5' flanking sequences, 234
    in mice, 227–234
Aquaculture, 303, 307, 322–323
Atlantic salmon, 296–298
Avian leukosis virus (ALV)
    assay in blood, meconium, and cloacal swabs, 279–280
    infection of chicken germ cell precursors, 281–288, 291–293
    injection into blastoderm, 280
    propagation of, 277–278
    resistance to, 125–130
    titering of, 279
    tolerance to, 129–130
Avian leukosis virus (ALV) envelope gene, 125–130
5-Azacytidine, 225–226

B cells, of HIV transgenic mice, 221–224
B-cell receptor, 175–176
Base-pair substitution, in retrovirus, 26–27, 30–31
bGH. *See* Bovine growth hormone
Bithorax complex, 136. *See also* Homeobox gene
Blastoderm, chicken
    injection of ALV into, 280
    transfection of, 276–277, 281, 289–291
Bone marrow
    in gene therapy in primates, 198–201
    in gene transfer in mice, 198
    *MDR*1 gene expression in, 103–123
    as target for gene therapy, 191–192
    transplantation of, 191–192
Bovine growth hormone (bGH), 237
    injection into fish, 296
Bovine growth hormone (bGH) gene, 241
    in chicken, 278–293
    in fish, 298, 302
    hybridization probe for, 279
    in sheep, 260–262
    in swine, 252

c-*myc* gene, 12, 16
Carbamyl phosphate synthetase deficiency, 195
Carp, 296, 299, 301, 304, 308
    transgenic
        fast-growing, 312–315
        rtGH gene in, 312–315, 319–322
β-Casein gene, 65–73
    expression of
        during mammary development, 66–68
        during pregnancy, 71–73
        tissue-specificity of, 68–71
β-Casein promoter, 73
CAT gene. *See* Chloramphenicol acetyl transferase gene
Catfish, 304, 308
Cattle
    collection and analysis of fetal tissue from, 268–269
    embryo collection in, 266, 269
    embryo culture and transfer in, 267–268
    exogenous administration of GH to, 254
    transgenic, production by pronuclear microinjection, 265–272
CD4 protein, 213–214, 224
Cell lineage, genes for ablation of, 14
Central nervous system, homeobox gene expression in, 140–143
Channel catfish, 297–298
    transgenic
        csGH gene in, 315–319, 322
        rtGH gene in, 319–320
Chemotherapy regimen, testing of, 122
Chicken
    ALV infection of germ cell precursors, 281–288, 291–293
    blastoderm, transfection of, 276–277, 281, 289–291
    introduction of bGH gene into unfertilized ova, 276, 280, 289
    introduction of recombinant DNA into embryo, 275–293
    preparation of genomic DNA from, 278–279
    shedding of virus in oviduct, 282–288

transgenic
  ALV envelope genes in, 125–130
  bGH gene in, 278–293
  resistance to RSV, 286–289
Chimeric embryo, 94–95
Chinook salmon, 297
Chloramphenicol acetyl transferase (CAT) gene, 17, 61–62, 146–149
Cholesterol, plasma, 227–234
Chum salmon, 297, 303
cis-acting element, 7–8, 24
  of actin gene from rat muscle, 56–58
  of homeobox genes, 143–144
  regulation of HIV proviral DNA, 225
Citrullinemia, 194
Cloning, 4
Coagulating gland, in hGH transgenic mice, 243
Coagulation factor deficiency, 194
Coho salmon, 296–297, 303, 308
Coho salmon growth hormone (csGH) gene, 308–310
  in channel catfish, 315–319, 322
Collagen I gene, 13
Collagen II gene, 60
Complement factor deficiency, 194
Concatamer, 6–7
Cord blood cells
  in gene therapy, 192
  human, gene transfer into, 206–209
  nonhuman primate, gene transfer into, 209
Creatine phosphokinase gene, 56
Crystallin gene, 94
csGH gene. *See* Coho salmon growth hormone gene
Cyprinus carpio. See Carp

D17 cells, 21–32
Daunomycin resistance, 103–123
Developmental regulation
  albumin and AFP genes in liver, 81–86
  homeobox genes in, 135–151
  studies in transgenic mice, 16–17
Differentiation antigen, 153, 155
Diphtheria toxin A gene, 14
Directed oncogenesis, 12
Disease resistance gene, insertion into chicken germline, 125–130

Disease susceptibility, in livestock, 255
DNA
  amplification by polymerase chain reaction, 23, 28–30
  binding to fish sperm, 300
  injection into mouse embryo, 163
  liposome-encapsulated, 301
  methylation of, 60–61, 122, 225
  sequencing of, 23
Dot-blotting, 281–288, 309–311
Drug resistance, 103–123

Economic issues, 342–343
Ectoderm, embryonic, 92–93
Ejaculation, in hGH transgenic mice, 246
Elastase I promoter, 14
Electroporation, in fish eggs, 300–301
Elliptocytosis, 194
Embryo
  chicken, introduction of recombinant DNA into, 275–293
  collection of
    in cattle, 266, 269
    in rats, 328
  culture and transfer in cattle, 267–268
  integration of microinjected sequences into, 5–7
  transfer in rat, 328–329
Embryoid body, 90–93
Embryology, postimplantation, of ES cells, 90–93
Embryonic stem (ES) cells, 4
  establishment and maintenance of lines, 90–93
  expression of muscle-specific genes in, 59–61
  gene targeting in, 89–100
  nongermline chimeric embryos from, 94–95
  postimplantation embryology of, 90–93
  transgenesis by means of, 94
  targeted mutagenesis in, 45–53
  in vitro, 90–93
  in vivo, 93–95
Endogenous gene, suppression by transgene, 57–59
Endothelial cells, as target for gene therapy, 192–193

Enhancer, 7–9
  AFP, 83–84, 86
  albumin, 83, 85–86
  albumin-AFP locus, 82
  homeobox gene, 144
  interaction with steroid hormone
    receptor, 37
  μ, 10
  tissue specificity of, 83–84
ES cells. See Embryonic stem cells
Esox americanus vermicalatus. See
    Grass pickerel
Esox lucius. See Northern pike
Estrogen receptor. See Steroid
    hormone receptor
Estrogen response element, 38
Evolution, of homeobox genes, 137–
    139, 150–151

Fabry disease, 195–196
Fanconi's anemia, 192
Feed efficiency
  in fish, 296
  in sheep, 259
  in swine, 253–254
α-Fetoprotein (AFP) enhancer, 83–84,
    86
α-Fetoprotein (AFP) gene, 8
  activation and silencing in liver,
    81–86
α-Fetoprotein (AFP) promoter-
    proximal region, 84
Fetus
  gene transfer into hematopoietic
    cells of
      human, 206–209
      nonhuman primate, 206–209
  gene transfer into lambs in utero,
    201–206
Fibroblasts
  chicken, culture of, 277
  chicken embryo, 125–130
  mouse, homologous recombination
    in, 46
  as target for gene therapy, 193
Fish
  DNA binding to sperm, 300
  extraction of genomic DNA from,
    309
  feed efficiency in, 296
  growth of, 296
  transgenic
    applications of, 322–323

bGH gene in, 298, 302
future prospects for, 303–305
GH genes in, 307–323
hGH gene in, 297–298, 301
luciferase gene in, 322
production of, 295–305
rGH gene in, 298
sterile, 302–303
Fish eggs
  dechorionation of, 299
  electroporation in, 300–301
  lipofection of, 301
  microinjection of, 297–301, 309
Fishery-stocking program, 303, 307
Fluorescence cytometry, identification
    of MHC gene products with,
    164–170
Follicle-stimulating hormone, in hGH
    transgenic mice, 245
Food animal. See specific animals
Fucosidosis, 195–196

Gain-of-function mutation, 136, 145–
    146, 149
Gaucher disease, 194–196
Gene. See specific genes
Gene duplication, 79, 84, 139
Gene expression. See also Transgene
    expression
  of homeobox genes, 139–143
  of milk protein genes, 65–73
Gene targeting, 45–53
  in ES cells, 89–100
  future animal modeling via, 100
  in somatic cells, 96–97
Gene therapy, 13–14
  bone marrow-mediated
    in mice, 198
    in nonhuman primate, 198–201
  with fetal hematopoietic cells
    of human, 206–209
    of nonhuman primate, 206–209
  human candidates for, 190–198
  principles of, 189–210
  using tumor-infiltrating
    lymphocytes, 209–210
  in utero, with fetal lambs, 201–206
Gene-regulatory protein family, 37
Genetic disease. See specific diseases
Genetic engineering, 14
Germ cell, chicken, infection with
    ALV, 281–288, 291–293
Gilthead seabream, 297

α-Globin gene, 80
β-Globin gene, 9–10, 13, 59–61, 100
Glucocorticoid receptor. *See* Steroid hormone receptor
Glucocorticoid/progesterone response element, 38
GnRH gene. *See* Gonadotropin-releasing hormone gene
Goldfish, 298–299, 302, 304
Gonadotropin-releasing hormone (GnRH) gene, 14
Granulocytic actin deficiency, 194
Grass pickerel, 296
Group 2 gene, 75–76
Growth
 of fish, 296
 of hGH transgenic mice, 239, 241–243
 of transgenic swine, 251–256
Growth hormone, exogenous administration to livestock, 253–254
Growth hormone gene, 13, 80
 in fish, 307–323
 in sheep, 259–263
Gut, homeobox gene expression in, 145

H-2D^d protein, 178–184
H-2D^d/Q10^b gene
 construction of, 178–179
 in mice, 179–180
 quantitation of protein product, 180–183
H-2K^k protein, 182
Hair, of hGH transgenic mice, 238–239
Harvard mouse, 335–345
HBsAg gene. *See* Hepatitis B surface antigen gene
HDL. *See* High-density lipoprotein
Heat-shock protein, 39
Helix-turn-helix motif, 137
Hematopoietic cells, fetal, gene transfer into, 201–209
Hemoglobinopathy, 194
Hepatitis B, 13
Hepatitis B surface antigen (HBsAg) gene, 13, 17
hGH gene. *See* Human growth hormone gene
hGRF gene. *See* Human growth hormone releasing factor gene

High-density lipoprotein (HDL), 227–234
HIV. *See* Human immunodeficiency virus
ho gene, 16
Homeobox gene
 discovery of, 136–137
 evolution of, 137–139, 150–151
 expression of, 139–143
 functions of, 137
 homeobox-to-homeobox distance between, 143–144
 multiplex regulation of, 145–149
 nucleotide sequence of, 137–139
 regulation of, 143–144
 in transgenic mice, 135–151
Homeotic gene, 136
Homologous recombination, 45, 95–98
 in mouse fibroblasts, 46
hox genes, 52, 136. *See also* Homeobox gene
Hprt gene. *See* Hypoxanthine phosphoribosyltransferase gene
Human disease. *See also specific diseases*
 candidates for gene therapy, 190–198
 transgenic mice as models for, 12–13
Human growth hormone (hGH) gene
 in fish, 297–298, 301
 in mice, 237–247, 253
 in sheep, 259–261
 in swine, 252
Human growth hormone releasing factor (hGRF) gene
 in sheep, 260–262
 in swine, 252
Human immunodeficiency virus (HIV)
 pathological findings and viral expression, 225
 proviral DNA in mice, 213–226
 transcriptional regulation of, 225–226
Hunter syndrome, 191
Hurler syndrome, 191, 195
Hydrocortisone, control of milk protein genes, 71–73
hygro gene, 21–32
Hypercortisolism, without Cushing's syndrome, 40
Hypermutation, somatic, 11

Hypophosphatasia, 195
Hypoxanthine
  phosphoribosyltransferase
  (*Hprt*) gene, 10, 96–100
  targeted mutagenesis in mouse ES
  cells, 47–49

Ia antigen. *See* Immune-response-associated antigen
ICP4 promoter, 146–147
Ictalurus punctatus. *See* Channel catfish
IGF-1. *See* Insulin-like growth factor-1
IL-2. *See* Interleukin 2
Immune status, of HIV transgenic mice, 220–224
Immune-response-associated (Ia) antigen, 161–173
Immunoglobulin gene, rearrangement of, 11
Immunoglobulin heavy-chain gene, 10–11
Immunoglobulin light-chain gene, 8, 11
Immunological education, 175–184
Immunological tolerance, 129–130, 175–184
Immunology, transgenic mice in, 10–11
Immunomodulator, 155–156
Insertional mutagenesis, 15–16
Insertional mutation, 4
Insertional mutation rate, 31–32
Insulin, control of milk protein genes, 71–73
Insulin-like growth factor-1 (IGF-1), 239, 245
Insulin-like growth factor-1 (IGF-1) gene, 255–256, 262
int-2 gene, 52–53, 98
Intellectual property policy, 335–345
Interference, 125–130
  in vitro, 127–128
  in vivo, 127–129
Interferon, 155–156, 162, 172
Interferon-response element (IRE), 156–157
Interleukin 2 (IL-2), 178
Interspecies hybridization, 302–303
IRE. *See* Interferon-response element
Isotypic exclusion, 11

Kokanee salmon, 303

Lactation, in hGH transgenic mice, 241
Lamb. *See* Sheep
LDL gene. *See* Low-density lipoprotein gene
Lentivirus, mutation rate in, 32
Lesch-Nyhan disease, 13, 195, 201
Leukoencephalopathy, progressive multifocal, 13
Lifespan, of hGH transgenic mice, 239–240
Lipid, plasma, 228–233
Lipofection, of fish eggs, 301
Lipopolysaccharide (LPS), bacterial, 162, 166, 172
Lipoprotein, plasma, 228–233
Litter size, in rats, 331–332
Liver
  activation and silencing of gene transcription in, 81–86
  apo A-1 gene expression in, 227–234
  of HIV transgenic mice, 218–219
Livestock. *See specific animals*
Loach, 298, 301, 304
Long terminal repeat (LTR), 122, 225–226
Low-density lipoprotein (LDL) gene, 14
LPS. *See* Lipopolysaccharide
LTR. *See* Long terminal repeat
Luciferase gene, in fish, 322
Lung, of HIV transgenic mice, 218–219, 224
Luteal failure, in hGH transgenic mice, 240
Luteinizing hormone, in hGH transgenic mice, 243–245
Luteinizing hormone receptor, 245
Lymph node, of HIV transgenic mice, 218–219
Lymphocytes
  of HIV transgenic mice, 221–223
  tumor-infiltrating, gene therapy using, 209–210
Lymphoid tissue
  class I MHC gene expression in, 153–158, 175–184
  class II MHC gene expression in, 161–173
  of HIV transgenic mice, 218–224

Lymphokine, 172, 176
Lysosomal storage disease, 191, 196

Major histocompatibility complex (MHC) gene, 13
  class I
    in mice, 153–158, 175–184
    of miniature swine, 154–168
  class II, 8
    identification of gene products, 164–170
    in mice, 171–173
    upstream promoter elements of, 171–172
Major histocompatibility complex (MHC) protein, class I
  domains of, 176
  soluble, 177–184
Major urinary protein (MUP) gene, silent, expression in mice, 75–80
Mammary cancer
  animal models for, 326–327
  in hGH transgenic mice, 239–240
  susceptibility to, 326
Mammary gland, milk protein gene expression in
  during development, 66–68
  during pregnancy, 71–73
Maternal inheritance, 17
MBP gene. See Myelin basic protein gene
Medaka, 298, 304
Messenger RNA (mRNA)
  MDR, 122
  for milk proteins, 65–73
  MUP, 76
  PD1, 155
Metachromatic leukodystrophy, 195
Metallothionein-growth hormone gene, 237–247, 319–320
Methylation, of DNA, 60–61, 122, 225
MGR system. See Multiplex gene regulatory system
MHC gene. See Major histocompatibility complex gene
Microinjection, pronuclear. See Pronuclear microinjection
Middle T antigen gene, 94–95
Milk protein gene
  mRNA from, 65–73
  regulation of expression of, 65–73
Mitomycin C, 226
Molecular farming, 336
Moloney murine leukemia virus (MoMuLV), 4, 196–198
MoMuLV. See Moloney murine leukemia virus
Monoclonal antibody, for identification of MHC gene products, 164–170
Mouse
  bone marrow-mediated gene transfer in, 198
  chimeric, 4, 94–95
  DNA microinjection into embryo, 163
  homeobox genes of, 136
  transgenic
    ablation of cell lineages, 14
    albumin and AFP genes in, 82–86
    antisense genes in, 15
    apo A-1 gene in, 227–234
    class I MHC genes in, 153–158, 175–184
    class II MHC genes in, 161–173
    in developmental genetics, 16–17
    gene expression in, 7–15
    Harvard, 335–345
    hGH gene in, 237–247, 253
    HIV proviral DNA in, 213–226
    homeobox genes in, 135–151
    identification of, 163
    in immunology, 10–11
    insertional mutagenesis in, 15–16
    MDR1 gene in, 103–123
    as models for gene therapy, 13–14
    as models for genetic engineering, 14
    as models of human disease, 12–13
    in oncology, 11–12
    patenting of, 335–345
    production by pronuclear microinjection, 5–7
    rat muscle-specific genes in, 56–59
    regulatory gene analysis using, 135–151
    relationship between exogenous and endogenous genes, 57–59
    rGH gene in, 253
    silent MUP gene in, 75–80

mRNA. *See* Messenger RNA
MT promoter, 261
MT-1 promoter, 10, 14, 16
Multidrug-resistance (*MDR*1) gene.
    See also β-Actin promoter-
        multidrug-resistance transgene
        expression in transgenic mice, 103–123
    integration and inheritance in mice, 120
Multiplex gene regulatory (MGR)
    system, 145–149
MUP gene. *See* Major urinary protein gene
Muscle cells, terminal differentiation of, 56
Muscle-specific gene
    expression in ES cells, 59–61
    from rat, in transgenic mice, 56–59
Muscle-specific promoter, 61–63
Mutagenesis
    insertional. *See* Insertional mutagenesis
    targeted. *See* Targeted mutagenesis
Mutation
    gain-of-function, 136, 145–146, 149
    insertional, 4
Mutation rate
    base-pair substitution, 26–27, 30–31
    insertion, 31–32
    in lentivirus, 32
    in reticuloendotheliosis virus, 31–32
    in retrovirus, 21–32
    in RNA virus, 30–31
myc gene, 59
Myelin basic protein (MBP) gene, 15
Myogenesis, germline transformation to study, 55–63
Myosin light-chain 2 gene, 59

neo gene, 21–32, 45–53, 98, 196, 198, 202–209
    revertants of, 28–30
Neomycin phosphotransferase gene.
    See neo gene
Neuroendocrine function, in hGH transgenic mice
    female, 240–241
    male, 241–246
Neurofibromatosis, 13
Neurofilament L promoter, 146
Neurophysin, 70

Nonselectable gene, targeted
    mutagenesis in mouse ES cells, 49–53
Northern pike, 297–300, 302, 304
Nucleosome, 39

Oncogene, 12, 37
Oncogenesis, directed, 12
Oncology, transgenic mice in, 11–12
Oncorhynchus keta. See Chum salmon
Oncorhynchus kisutch. See Coho salmon; Pacific salmon
Oncorhynchus nerka. See Sockeye salmon
Ornithine transcarbamylase deficiency, 195
Ornithine transcarbamylase gene, 10
Osteogenesis imperfecta, 13
Ovulation
    in cattle, 266, 269
    in rats, 327, 331
Ovum, chicken, introduction of DNA into, 276, 280, 289

P-glycoprotein, 104
    expression in cells with *MDR*1 gene, 115–119
    function of, 122
Pacific salmon, 296
Pancreatic elastase-1 gene, 8
Patent
    purpose of, 339
    qualifications for, 337–338
    on transgenic animals, 335–345
Paternal inheritance, 17
pcd gene, 16
PCR. *See* Polymerase chain reaction
PD1 gene, 154–158
Percutaneous umbilical vessel blood sampling (PUBS), 192
pGH. *See* Porcine growth hormone
Phenylketonuria, 192, 194
Pituitary gland, milk protein gene expression in, 70
Plant Patent Act, 338
Plant Variety Protection Act, 338
Plasmid pGMDR, 122
Plasmid pHG1, 104–108
Plasmid pHG2, 104–108
Plasmid pJD216Neo(Am)Hy, 21–32
Plasmid pJD216NeoHy, 21–32

Plasmid pMThGH, 319–320
Plasmid pNL4-3, 214
Plasmid pNL432, 214
Plasmid pRSV2-Luc, 309–310
Plasmid pRSVCAT, 267–270
Plasmid pRSVcsGHcDNA, 309–310, 315–316, 322
Plasmid pRSVrtGHcDNA, 308–310, 312–315, 319–322
PNS. *See* Positive-negative selection
Polyadenylation signal, 122
Polymerase chain reaction (PCR), 23, 28–30, 98–100
Polyploidization, 302–303
Porcine growth hormone (pGH), injection into fish, 296
Porcine growth hormone (pGH) gene, 252
Position effect, negative, 79
Positive regulation, 82
Positive-negative selection (PNS), 50–52
Postimplantation embryology, of ES cells, 90–93
Poultry. *See* Chicken
Pregnancy, milk protein gene expression during, 71–73
Primate, nonhuman
  gene therapy via bone marrow transplantation in, 198–201
  gene transfer into fetal hematopoietic cells of, 206–209
Progesterone receptor. *See* Steroid hormone receptor
Progressive multifocal leukoencephalopathy, 13
Prolactin
  control of milk protein genes, 71–73
  in hGH transgenic mice, 240–241, 243–244
Prolactin gene, 297
Promoter, 7–8, 10. *See also specific promoters*
Promoter-proximal region, of AFP gene, 84
Pronuclear microinjection, 3
  in cattle, 265–272
  in fish, 297–301, 309
  integration of injected sequences into embryo, 5–7
  in rat, 329–330
  in sheep, 259–260
  in swine, 254

Propionyl CoA carboxylase deficiency, 195
Proto-oncogene, 12
Pseudogene, 76
Pseudopleuronectes americanus. See Winter flounder
Public opinion, 338–342
PUBS. *See* Percutaneous umbilical vessel blood sampling
puc sequence, 173
Purine nucleoside phosphorylase deficiency, 194

Q10^b protein, 177–184

Rabbit, transgenic, 14
Rainbow trout, 295–298, 307
Rainbow trout growth hormone (rtGH), immunobinding assay for, 311–312
Rainbow trout growth hormone (rtGH) gene, 298, 308–310
  in carp, 312–315, 319–322
  in channel catfish, 319–320
Rat
  embryo collection in, 328
  embryo transfer in, 328–329
  exogenous administration of GH to, 254
  litter size in, 331–332
  ovulation in, 327, 331
  pronuclear microinjection in, 329–330
  transgenic
    need for, 326–327
    production of, 325–332
Rat growth hormone (rGH) gene, 239–240, 242–243
  in fish, 298
  in mice, 253
  in swine, 252
Receptor. *See specific receptors*
Recombination, homologous. *See* Homologous recombination
Red blood cells, carp, rtGH gene expression in, 313–315
Regulatory gene, analysis using transgenic mice, 135–151
Repressor, 82
Reproduction, in hGH transgenic mice
  female, 240–241

male, 241–246
Reticuloendotheliosis virus, mutation rate in, 31–32
Retroviral vector
  advantages of, 61
  ALV in chicken germ cells, 281–284, 291–293
  site of integration of, 285–288
  transfer of muscle-specific promoter, 61–63
  in treatment of human disease, 196–198
Retroviral vector N2, 197, 202–210
Retroviral vector SAX, 197–206
Retrovirus, 4
  JD216NeoHy expression of *neo* and *hygro* genes, 24–26
  mutation rate in, 21–32
    base-pair substitution, 26–27, 30–31
    insertion mutations, 31–32
  neo revertants in, analysis of genomic sequence, 28–30
  single cycle of replication of, 23–25
rGH gene. *See* Rat growth hormone gene
RNA virus, mutation rate in, 30–31
Rous sarcoma virus (RSV)
  mutation rate in, 22
  resistance in transgenic chickens, 286–289
Rous sarcoma virus (RSV) promoter, 302
Royalty policy, 342–343
RSV. *See* Rous sarcoma virus
rtGH. *See* Rainbow trout growth hormone

Salmo gairdneri. *See* Rainbow trout
Salmo salar. *See* Atlantic salmon
Salmon, 299, 304. *See also specific types of salmon*
Sandhoff disease, 195–196
Sanfilippo A syndrome, 191
Segmentation gene, 136
Self-regulating gene, 150
Seminal vesicle, in hGH transgenic mice, 243–244
Seriola quinqueradiata. *See* Yellowtail
Severe combined immunodeficiency disease, 193
Sex-dependent transgene expression, 122

Sheep
  exogenous administration of GH to, 254
  feed efficiency in, 259
  gene therapy for lambs in utero, 201–206
  transgenic, 14
    bGH gene in, 260–262
    GH genes in, 259–263
    hGH gene in, 259–261
    hGRF gene in, 260–262
    production by pronuclear microinjection, 259–260
    survival of injected embryos in, 260–261
    transgene transmission in, 262
Silent gene, 75–86
Skin, of HIV transgenic mice, 216–218, 224
Skin grafting, 165–170, 173
Sockeye salmon, 296
Somatic cells
  gene targeting in, 96–97
  gene transfer to, 189–210
Somatic hypermutation, 11
Southern blotting, 268–269, 284–288, 309–311
Sparus aurata. *See* Gilthead seabream
Species integrity, 339–340
Sperm
  fish, DNA binding to, 300
  in hGH transgenic mice, 241, 245–246
Spinal cord, homeobox gene expression in, 140–143
Spleen, in hGH transgenic mice, 243
Spleen necrosis virus, mutation rate in, 21–32
Spleen necrosis virus-based vector, 21–32
SRE. *See* Steroid response element
Sterility
  in hGH transgenic mice, 240, 245–246
  in transgenic fish, 302–303
Steroid hormone, gene regulation by, 35–36
Steroid hormone receptor
  activation by hormones, 39–40
  binding to DNA, 37
  dimers of, 39
  interaction with heat-shock protein, 39

interaction with steroid response
element, 39–40
mutations in, 41
as *trans*-acting element, 35–41
Steroid response element (SRE), 37
interaction with steroid hormone
receptor, 39–40
sequence of, 38
Stizostedium vitrium. See Walleye
Submaxillary gland, silent MUP gene
expression in, 75–80
SV40 promoter, 302
Swine
disease susceptibility in, 255
exogenous administration of GH
to, 253–254
feed efficiency of, 253–254
growth of, 254
miniature, class I MHC genes of,
153–158
as models for human disease, 255
transgenic, 14
bGH gene in, 252
enhanced growth performance in,
251–256
hGH gene in, 252
hGRF gene in, 252
production by pronuclear
injection, 254
rGH in, 252

T antigen gene, 12
T cells, of HIV transgenic mice, 221–
224
T-cell receptor, 175–177
T-cell receptor gene, 11
Targeted mutagenesis
in ES cells, 45–53
of *HPRT* gene in mouse ES cells,
47–49
of nonselectable gene in mouse ES
cells, 49–53
Tay Sachs disease, 195–196, 201
Testicular feminization, 40–41
Testis, in hGH transgenic mice, 243
Testosterone, in hGH transgenic
mice, 245
β-Thalassemia, 13, 193–194
Thunnus thynnus. See Tuna
Thy-1 gene, 8–9
Thymic education, 175–184
Thymic tolerance, 175–184
Thymus

in hGH transgenic mice, 243
in HIV transgenic mice, 218, 224
Thyroid hormone receptor, 37
Thyroid hormone response element,
38
TIL. *See* Tumor-infiltrating
lymphocytes
Tilapia, 296, 298, 304
Tissue plasminogen activator (TPA)
gene, 14, 336
Tissue-specific gene
for ablation of cell lineages, 14
AFP gene in liver, 83–84
albumin gene in liver, 84–85
apo A-1 gene in liver, 227–234
class I MHC genes in lymphoid
tissue, 153–158, 175–184
class II MHC genes in lymphoid
tissue, 161–173
hGH gene in neuroendocrine
organs, 240–247
hGH gene in reproductive organs,
240–247
homeobox genes in central nervous
system, 140–143
lesions in HIV transgenic mice,
216–219
*MDR*1 gene in bone marrow, 103–
123
milk protein genes, 66–71
MUP gene in submaxillary gland,
75–80
in muscle development, 55–63
tk gene, Herpes, 10, 16, 50–52
Toxicology testing, 336
TPA gene. *See* Tissue plasminogen
activator gene
trans-acting factor, 9, 24
for muscle-specific genes, 56
regulation of HIV proviral DNA,
225
steroid hormone receptor, 35–41
Transactivator line, 146–149
Transcription factor, 137, 148, 150
Transcriptional regulation
of HIV, 225–226
in liver, 81–86
Transfection, of chicken blastoderm,
276–277, 281, 289–291
Transgene
copy number of, 8, 109, 120, 165–
166, 239, 262, 281, 311, 319
integration of, 5–7
site of integration of, 8–10

structural changes after insertion of, 6–7
suppression of endogenous genes by, 57–59
Transgene expression
aberrant, 10
apo A-1 gene in mice, 227–234
class I MHC genes in mice, 153–158, 175–184
class II MHC genes in mice, 161–173
GH genes in fish, 301–302, 307–323
GH genes in sheep, 261–262
hGH gene in mice, 237–247
HIV provirus in mice, 225–226
homeobox genes in mice, 144–149
*MDR*1 gene in mice, 103–123
in mice, 7–15
rat muscle-specific genes in mice, 56–59
sex-dependent, 122
silent MUP gene in mice, 75–80
Transgenic animal
definition of, 3
patenting of, 335–345
Transplantation antigen, 153–155
Transponder line, 145–149
Transport protein, 104
Tropomyosin gene, 56
Troponin gene, 56
Trout, 299, 304
Tumor-infiltrating lymphocytes (TIL), gene therapy using, 209–210
Tuna, 297
Tyrosine aminotransferase gene, 13

Upstream promoter element, of MHC class II genes, 171–172

v-*erb* A gene, 37
v-*rel* gene, 31
Virus. *See specific viruses*
Vitamin D-resistant rickets, 40
VP16, 146–149

Walleye, 299, 304
WAP gene. *See* Whey acidic protein gene
Western blotting, 180–182
Whey acidic protein (WAP) gene, 65–73
expression of
during mammary development, 66–68
during pregnancy, 71–73
tissue-specificity of, 68–71
promoter/upstream region of, 70–73
Whey acidic protein (WAP)-tissue plasminogen activator (TPA) gene, 70–72
Winter flounder, 297–298

X chromosome
inactivation of, 17
transgenes in, 17
Xenopus, homeobox genes in, 145
Xiphophorus, 298, 304

Yellowtail, 297

Zebrafish, 298, 304
Zinc finger, 37, 40
Zona drilling, 16